Energy, Environment, and Sustainability

Series Editor

Avinash Kumar Agarwal, Department of Mechanical Engineering, Indian Institute of Technology Kanpur, Kanpur, Uttar Pradesh, India

AIMS AND SCOPE

This books series publishes cutting edge monographs and professional books focused on all aspects of energy and environmental sustainability, especially as it relates to energy concerns. The Series is published in partnership with the International Society for Energy, Environment, and Sustainability. The books in these series are edited or authored by top researchers and professional across the globe. The series aims at publishing state-of-the-art research and development in areas including, but not limited to:

- Renewable Energy
- Alternative Fuels
- Engines and Locomotives
- Combustion and Propulsion
- Fossil Fuels
- Carbon Capture
- Control and Automation for Energy
- Environmental Pollution
- Waste Management
- Transportation Sustainability

Review Process

The proposal for each volume is reviewed by the main editor and/or the advisory board. The chapters in each volume are individually reviewed single blind by expert reviewers (at least four reviews per chapter) and the main editor.

Ethics Statement for this series can be found in the Springer standard guidelines here https://www.springer.com/us/authors-editors/journal-author/journal-author-helpdesk/before-you-start/before-you-start/1330#c14214

More information about this series at http://www.springer.com/series/15901

Swatantra P. Singh · Karthik Rathinam ·
Tarun Gupta · Avinash Kumar Agarwal
Editors

Pollution Control Technologies

Current Status and Future Prospects

Springer

Editors
Swatantra P. Singh
Department of Environmental Science and Engineering
Indian Institute of Technology Bombay
Mumbai, Maharashtra, India

Karthik Rathinam
Mechanical Process Engineering/Water Technology
University of Duisburg-Essen
Duisburg, Germany

Tarun Gupta
Department of Civil Engineering
Indian Institute of Technology Kanpur
Kanpur, Uttar Pradesh, India

Avinash Kumar Agarwal
Department of Mechanical Engineering
Indian Institute of Technology Kanpur
Kanpur, Uttar Pradesh, India

ISSN 2522-8366 ISSN 2522-8374 (electronic)
Energy, Environment, and Sustainability
ISBN 978-981-16-0857-5 ISBN 978-981-16-0858-2 (eBook)
https://doi.org/10.1007/978-981-16-0858-2

This Springer imprint is published by the registered company Springer Nature Singapore Pte Ltd.
The registered company address is: 152 Beach Road, #21-01/04 Gateway East, Singapore 189721, Singapore

Preface

Water and air pollution has been receiving remarkable consideration due to increasing population, industrialization, and urbanization. Moreover, extensive usage of various compounds in modernized processes has also led to an increase in various toxic pollutants in water as well as in the atmosphere. Water and air pollution has shown a wide range of adverse effects on both human beings and the environment.

The International Society for Energy, Environment, and Sustainability (ISEES) was founded at Indian Institute of Technology Kanpur (IIT Kanpur), India, in January 2014 with an aim to spread knowledge/awareness and catalyze research activities in the fields of energy, environment, sustainability, and combustion. The society's goal is to contribute to the development of clean, affordable, and secure energy resources and a sustainable environment for the society and to spread knowledge in the above-mentioned areas and create awareness of the environmental challenges, which the world is facing today. The unique way adopted by the society was to break the conventional silos of specializations (engineering, science, environment, agriculture, biotechnology, materials, fuels, etc.) to tackle the problems related to energy, environment, and sustainability in a holistic manner. This is quite evident by the participation of experts from all the fields to resolve these issues. ISEES is involved in various activities such as conducting workshops, seminars, and conferences in the domains of its interests. The society also recognizes the outstanding works done by the young scientists and engineers for their contributions in these fields by conferring them awards under various categories.

The Fourth International Conference on 'Sustainable Energy and Environmental Challenges' (IV-SEEC) was organized under the auspices of ISEES from November 27 to 29, 2019, at NEERI, Nagpur. This conference provided a platform for discussions between eminent scientists and engineers from various countries including India, USA, China, Italy, Mexico, South Korea, Japan, Sweden, Greece, Czech Republic, Germany, the Netherlands, and Canada. In this conference, eminent speakers from all over the world presented their views related to different aspects of energy, combustion, emissions, and alternative energy resource for sustainable development and cleaner environment. The conference presented one high-voltage plenary talk by Mrs. Rashmi Urdhwareshe, Director, Automotive Research Association of India (ARAI), Pune.

The conference included 28 technical sessions on topics related to energy and environmental sustainability including 1 plenary talk, 25 keynote talks, and 54 invited talks from prominent scientists, in addition to 70+ contributed talks, and 80+ poster presentation by students and researchers. The technical sessions in the conference included fuels, engine technology and emissions, coal and biomass combustion/gasification, atomization and sprays, combustion and modeling, alternative energy resources, water and water and wastewater treatment, automobile and other environmental applications, such as sustainability, nuclear energy, sensors, and other environmental challenges, such as clean fuels and other environmental challenges, water pollution and control, biomass and biotechnology, waste to wealth, environmental microbiology, and environmental biotechnologies. One of the highlights of the conference was the rapid-fire poster sessions in (i) engine/fuels/emissions, (ii) environment, and (iii) biotechnology, where 50+ students participated with great enthusiasm and won many prizes in a fiercely competitive environment. 300+ participants and speakers attended this three-day conference, where 12 ISEES books published by Springer, Singapore, under a special dedicated series 'energy, environment, and sustainability,' were released. This was the third time in a row that such a significant and high-quality outcome has been achieved by any society in India. The conference concluded with a panel discussion on 'Balancing Energy Security, Environmental Impacts and Economic Considerations: Indian Perspective,' where the panelists were Dr. Anjan Ray, CSIR-IIP Dehradun; Dr. R. R. Sonde, Thermax Ltd.; Prof. Avinash Kumar Agarwal, IIT Kanpur; Dr. R. Srikanth, National Institute of Advanced Studies, Bengaluru; and Dr. Rakesh Kumar, NEERI, Nagpur. The panel discussion was moderated by Prof. Ashok Pandey, Chairman, ISEES. This conference laid out the roadmap for technology development, opportunities, and challenges in energy, environment, and sustainability domain. All these topics are very relevant to the country and the world in the present context. We acknowledge the support received from various funding agencies and organizations for the successful conduct of the Fourth ISEES conference, where these books germinated. We would therefore like to acknowledge SERB, Government of India (special thanks to Dr. Sandeep Verma, Secretary); NEERI Nagpur (special thanks to Dr. Rakesh Kumar, Director); CSIR; and our publishing partner Springer (special thanks to Swati Mehershi).

The editors would like to express their sincere gratitude to large number of authors from all over the world for submitting their high-quality work in a timely manner and revising it appropriately at a short notice. We would like to express our special thanks to Dr. Shiju Abraham, Dr. Ibrahim M. A. ElSherbiny, Dr. Hasmath Farzana, Dr. Subbaiah Muthu Prabhu, Dr. Amal Raj, and Dr. Elanchezhiyan, who reviewed various chapters of this monograph and provided their valuable suggestions to improve the manuscripts.

This book is based on pollution control technologies available today to deal with water and air pollution. This book includes the removal of a variety of pollutants including chromium, uranium, pesticides, and arsenic from the water using adsorption technique. In addition, this book deals with the sampling and removal of microplastics using various techniques. Moreover, this book includes few chapters on the role of membrane technology in water and wastewater treatment. Few chapters provide informations about particulate matter air pollution and its control techniques. Chapters include recent results and focused more on the current trends in water and air pollution sector. In this book, readers will get the idea about the availability of various pollution control techniques, which will help them in addressing various issues related to water and air pollution. We hope that the book would be of great interest to the professionals and postgraduate students involved in environmental science and engineering research.

Mumbai, India — Swatantra P. Singh
Duisburg, Germany — Karthik Rathinam
Kanpur, India — Tarun Gupta
Kanpur, India — Avinash Kumar Agarwal

Contents

Editors and Contributors

About the Editors

Dr. Swatantra P. Singh is an environmental engineer with experience in membrane fabrication, environmental nanotechnology, fate, and transport of pollutants and emerging contaminants in the environment. Currently, he is an assistant professor in the Environmental Science and Engineering Department at Indian Institute of Technology (IIT) Bombay, India. He has developed a key technology to fabricate the membranes for water purification and print graphene in-situ in a single step. He has four US patents (two granted and two provisional) on membrane and laser-based graphene fabrication techniques. He has authored 19 journal articles, one book, and three book chapters. He recently won the INAE Young Engineer Award (2020) and ISEES Young Scientists Award (2020).

Dr. Karthik Rathinam is currently leading adsorption technology group at the Chair for Process and Mechanical Engineering/Water Technology, University of Duisburg-Essen, Germany. His research interest is mainly focused on water and wastewater treatment including precipitation of inorganic minerals, development of antiscaling and antifouling membranes, development of porous eco-friendly based adsorbents, thermal reactivation of activated carbon, emerging contaminants removal, etc. After his Ph.D. from the Gandhigram Rural Institute, India, he worked as a postdoctoral researcher in Zuckerberg Institute for Water Research, Israel and BASF-SE, Germany. He is serving as a reviewer for several Scopus indexed journals. He is a recipient of many national and international awards and has published 18 research articles in a highly reputed peer-reviewed journals and one book chapter.

Prof. Tarun Gupta is doctorate from Harvard University, USA, and M.Tech. from Indian Institute of Technology (IIT) Bombay, India. He has authored more than 125 ISI indexed journal publications, 3 books, 10 book chapters, 4 patents, and has been reviewer of more than 36 journals. He has guided 6 Ph.D. and 34 M.Tech. theses. A submicron aerosol sampler designed, developed and evaluated by him at Indian Institute of Technology (IIT) Kanpur has been commercialized by Envirotech. He has developed a high volume fine PM sampler and transferred technology to BARC. He is currently P. K. Kelkar research fellow and selected member of INYAS and INAE Associate. He has recently won INAE Innovator and Entrepreneur Award (2018), VNMM award (2017), NASI-SCOPUS Award (2015), INSA Medal for Young Scientist (2011), INAE Young Engineer Award (2009) and IEI Young Engineer Award (2008). He is currently N. C. Nigam Chair Professor and serving as Associate Dean of Research and Development at IIT Kanpur.

Prof. Avinash Kumar Agarwal joined the Indian Institute of Technology (IIT) Kanpur, India in 2001 after working as a post-doctoral fellow at the Engine Research Center, University of Wisconsin at Madison, USA. His interests are IC engines, combustion, alternate and conventional fuels, lubricating oil tribology, optical diagnostics, laser ignition, HCCI, emissions and particulate control, and large bore engines. Prof. Agarwal has published 290+ peer reviewed international journal and conference papers, 42 edited books, 78 books chapters and has 10,000+ Scopus and 15,300+ Google scholar citations. He is a Fellow of SAE (2012), Fellow of ASME (2013), Fellow of ISEES (2015), Fellow of INAE (2015), Fellow of NASI (2018), Fellow of Royal Society of Chemistry (2018), and a Fellow of American Association of Advancement in Science (2020). He is recipient of several prestigious awards such as Clarivate Analystics India Citation Award-2017 in Engineering and Technology; NASI-Reliance Industries Platinum Jubilee Award-2012; INAE Silver Jubilee Young Engineer Award-2012; Dr. C. V. Raman Young Teachers Award-2011; SAE Ralph R. Teetor Educational Award-2008; INSA Young Scientist Award-2007; UICT Young Scientist Award-2007; INAE Young Engineer Award-2005. Prof. Agarwal received Prestigious Shanti Swarup Bhatnagar Award-2016 in Engineering Sciences. For his outstanding contributions, Prof. Agarwal is conferred upon Sir J. C. Bose National Fellowship (2019) by SERB.

Contributors

Avinash Kumar Agarwal Department of Mechanical Engineering, IIT Kanpur, Kanpur, UP, India

Gopika Ashokan School of Water Resources, Indian Institute of Technology Kharagpur, Kharagpur, India

Najmul Haque Barbhuiya Environmental Science and Engineering Department (ESED), Indian Institute of Technology Bombay, Mumbai, India

S. Gayathri Department of Civil and Environmental Engineering, Indian Institute of Technology Tirupati, Tirupati, India

Tarun Gupta Department of Civil Engineering, Indian Institute of Technology Kanpur, Kanpur, Uttar Pradesh, India

Subrata Hait Department of Civil and Environmental Engineering, Indian Institute of Technology Patna, Patna, Bihar, India

Kritika Jashrapuria Environmental Science and Engineering Department (ESED), Indian Institute of Technology Bombay, Mumbai, India

Perumal karthikeyan Department of Chemistry, The Gandhigram Rural Institute —Deemed to Be University, Gandhigram, Dindigul, Tamil Nadu, India

Shihabudheen M. Maliyekkal Department of Civil and Environmental Engineering, Indian Institute of Technology Tirupati, Tirupati, India

Vasudev Malyan Environmental Science and Engineering Department, Indian Institute of Technology, Bombay, India

Y. S. Mayya Department of Chemical Engineering, Indian Institute of Technology, Bombay, India

Sankaran Meenakshi Department of Chemistry, The Gandhigram Rural Institute —Deemed to Be University, Gandhigram, Dindigul, Tamil Nadu, India

Chettithodi Poovathumkuzhi Nabeena Pocker Sahib Memorial Orphanage College, Malappuram, Kerala, India

M. S. V. Naga Jyothi Department of Civil and Environmental Engineering, Indian Institute of Technology Tirupati, Tirupati, India

Tabish Nawaz Environmental Science and Engineering Department, Indian Institute of Technology Bombay, Mumbai, Maharashtra, India

Darakhshan Nayyar Environmental Science and Engineering Department, Indian Institute of Technology Bombay, Mumbai, Maharashtra, India

Saba Noore Department of Food Engineering, CSIR-Central Food Technological Research Institute, Mysore, India

Neha Parashar Department of Civil and Environmental Engineering, Indian Institute of Technology Patna, Patna, Bihar, India

T. Pushparaj Gandhi Department of Civil and Environmental Engineering, Indian Institute of Technology Tirupati, Tirupati, India

Karthik Rathinam Chair for Mechanical Process Engineering and Water Technology, University of Duisburg-Essen, Duisburg, Germany

Manoranjan Sahu Environmental Science and Engineering Department, Indian Institute of Technology, Bombay, India

Himanshu Sharma Department of Civil Engineering, Indian Institute of Technology Kanpur, Kanpur, Uttar Pradesh, India

Anubhav Pratap Singh Faculty of Land and Food Systems (LFS), University of British Columbia, Vancouver, BC, Canada

Swatantra P. Singh Environmental Science and Engineering Department (ESED), Centre for Research in Nanotechnology and Science (CRNTS), Indian Institute of Technology Bombay, Mumbai, India

Alok Sinha Department of Environmental Science and Engineering, Indian Institute of Technology (Indian School of Mines), Dhanbad, India

Palliyalil Sirajudheen Department of Chemistry, The Gandhigram Rural Institute—Deemed to Be University, Gandhigram, Dindigul, Tamil Nadu, India;
Pocker Sahib Memorial Orphanage College, Malappuram, Kerala, India

Vigneshwaran Sivakumar Department of Chemistry, The Gandhigram Rural Institute (deemed to be University), Dindigul, Tamil Nadu, India

Manoj Kumar Tiwari School of Water Resources, Indian Institute of Technology Kharagpur, Kharagpur, India

Shubhanshu Tripathi Department of Civil Engineering, Indian Institute of Technology Kanpur, Kanpur, Uttar Pradesh, India

Shivangi Upadhyay Department of Environmental Science and Engineering, Indian Institute of Technology (Indian School of Mines), Dhanbad, India

Sivakumar Vigneshwaran Department of Chemistry, The Gandhigram Rural Institute—Deemed to Be University, Gandhigram, Tamil Nadu, India

Dina Zaman School of Water Resources, Indian Institute of Technology Kharagpur, Kharagpur, India

Chapter 1
Pollution Control Technologies: Current Status and Future Prospects

Swatantra P. Singh, Karthik Rathinam, Tarun Gupta, and Avinash Kumar Agarwal

Abstract Water pollution and air pollution remediation is an important task to avoid their unnecessary impacts on environment and living species. Nowadays, there are several technologies employed to control the water and air pollution. Therefore, the main aim of this book is to provide an overview and future prospect of the pollution control strategies adopted in different sectors including water and air. This book contains 12 chapters and each chapter has its own purpose to discuss the current scenario of control technologies in remediation of various toxic metal ions, pesticides, micropollutants, radioactive pollutants, and dyes. Moreover, this book highlights the pollution control technologies adopted in food processing and desalination as well. Furthermore, it provides valuable informations about particulate matter air pollution and its control techniques. Overall, this book offers an updated literature for researchers and academicians working in field of water and air pollution and their control techniques.

Keywords Water pollution · Emerging contaminants · Air pollution · Particulate matter · Control technology

---

S. P. Singh (✉)
Environmental Science and Engineering Department, IIT Bombay, Mumbai 400076, India
e-mail: swatantra@iitb.ac.in

K. Rathinam
Chair for Mechanical Process Engineering and Water Technology, University of Duisburg-Essen, 47057 Duisburg, Germany

T. Gupta
Department of Civil Engineering, IIT Kanpur, Kanpur, UP 208016, India

A. K. Agarwal
Department of Mechanical Engineering, IIT Kanpur, Kanpur, UP 208016, India

S. P. Singh et al. (eds.), *Pollution Control Technologies*, Energy, Environment, and Sustainability, https://doi.org/10.1007/978-981-16-0858-2_1

1.1 Introduction

Water and air pollution is the most responsible factors related with several health diseases, and thus, it is imperative to find suitable remediation technology to clean the air and water. Although several precautionary measures have been taken to control the water and air pollution nowadays, no comprehensive solution is made available. This chapter ties with the chapters dealing with various water and air pollution remediation technologies in this book.

Pesticides are widely used nowadays to increase crop productivity and yields. Nevertheless, they are potentially toxic to the environment and the cause of both acute and chronic adverse effects on humans. Additionally, some of the old banned pesticides remain in the soil and water for many years. There are several strategies suggested for the pesticide control worldwide. Chapter 2 delivers a general overview of the types of pesticides, the usage of pesticides across the world, and their several adverse effects on human beings. In parallel, this chapter also provides information about the different physical and chemical techniques explored for pesticide degradation technologies. Eventually, this chapter highlights the potential application of nanomaterials as photocatalysts for the degradation of various persistent organic pollutants in water.

Due to the extensive industrial developments and population growth, the energy requirement is also steadily increased. Apart from coal, oil, and other fossil fuel sources, nuclear power is considered as one of the main alternative energy sources today. Nuclear energy has been produced through the splitting of the uranium atoms. Uranium (U), a radionuclide, is often used in nuclear fuels, nuclear reactors, and nuclear weapons across the world. Nevertheless, the continuous usage of U leading to cumulative environmental pollution. Moreover, U ore processing and mining, natural cycles, and usage of U containing fertilizers are also adding up to the U pollution. Chapter 3 summarizes the development of various pollution control techniques including photocatalytic, precipitation, adsorption, evaporation, ion exchange, membrane technology, and electrochemical and biological processes toward the minimization of U from the environment. Finally, this chapter discusses the risk evaluation and progress of future strategies as well.

The presence of chromium (VI) in the environment poses significant threat to the world as it led mutagenic and carcinogenic effects to all living species. In addition, several health diseases associated with chromium (VI) have been reported. Through metal complexation mechanisms, the toxic chromium (VI) produces reactive oxygen species across the cell membranes, thereby damaging the cellular structures in living species. Therefore, it is important to control chromium (VI) pollution. Plenty of control methodologies have been reported for vanishing the chromium from the environment. In this context, Chap. 4 deals with the occurrence of chromium, chromium toxicity, and contamination, and chromium removal by different techniques. Moreover, this chapter highlights the chromium removal by various biological methods and mechanism of chromium reduction by microbes. Furthermore, it talks about the pilot-scale studies involved in chromium removal. Finally, this chapter also discusses the future perspectives in chromium removal as well.

Owing to various anthropogenic activities and natural processes, arsenic (As) is existing in ground water over the years. The presence of As in drinking water has been receiving special attention as they can lead to serious health issues to humans. Therefore, As removal from drinking water has become unavoidable. Until now, several point-of-use and community-based techniques have been implemented across the world with the aim of providing As-free drinking water to the affected population. Chapter 5 provides an overview of the various treatment technologies such as coagulation and filtration, electrocoagulation, adsorption, ion exchange, membrane filtration, capacitive deionization, and phytoremediation reported for the removal of As from ground water. In addition, this chapter also discusses the physical and chemical properties, occurrence, drinking water standards, and origin of As in groundwater.

Microplastics (MPs) are smaller plastic particles (<5 mm) and are abundant in the environment. Due to the steady increase in plastics usage, plastics production, and their accumulation in the environment over the years, MPs have become one of the major concerns to the ecosystem. Sewage treatment plants and surface runoff are regarded as the potential sources for the existence of MPs in ecosystem. In this context, Chap. 6 provides an overview of plastic pollution, various techniques involved during the MPs sampling, MPs extraction, and purification techniques, occurrence and morphology of MPs, fate, and removal of MPs at different treatment stages in sewage treatment plants, the release of MPs to the environment, and eventually advance techniques adopted for the removal of MPs.

To compensate the water scarcity across the world, along with emerging contaminants removal with the traditional treatment technology, desalination and wastewater reuse through membrane technology have been suggested. Several conventional water treatment technologies have been adopted for desalination and wastewater treatment. Nevertheless, the effectiveness of conventional treatment processes has become limited over the last few decades. Therefore, the use of membrane technology has been widely recommended in desalination and wastewater recycling to overcome the limitations of the conventional treatment processes and to meet stringent water quality regulations. However, the effective usage of membrane technology has been hindered due to many factors, including membrane fouling, reduced life span, etc. Chapter 7 discusses the role of membrane technology in desalination and wastewater treatment and describes the advantages of membrane technology over conventional treatment techniques. Moreover, it provides a brief overview of commonly used membrane processes, membrane fabrication techniques, development of new generation membrane materials, and membrane systems.

Forward osmosis (FO) membrane technology has the potential to treat the wastewater and to produce high-quality water. In spite of the wide range of applications, FO suffers from some major technological problems such as lack of proper membrane design, separation of water from draw solution, etc. Therefore, FO has been integrated with other membrane-based technologies to overcome its shortcomings. In this context, Chap. 8 provides an overview of the FO working principle, membrane type, and its orientation, factors that significantly affect the filtration process, membrane fouling, and FO integration with other techniques. Eventually, this chapter also describes the advantages and contributions of FO membrane technology in desalination and wastewater treatment.

A huge quantity of wastewater has been generated from various food industries, including dairy, meat processing facilities, alcoholic beverages, edible oil, confectionery and sugar industries, etc. The wastewater characteristics from those industries significantly differ from each other, and they must be treated properly. Chapter 9 summarizes relevant and recent literature gathered from various scientific databases regarding the wastewater from various food industries. Also, it describes the characteristics of the wastewater from the various food industries, including dairy, poultry, alcohol fermentation, edible oil, fruit juice, sugar, slaughterhouse, and cattle. Moreover, it provides information about the various treatment technologies reported for the treatment of wastewater from the above-mentioned food industries. Furthermore, this chapter identifies the key challenges associated with each treatment system and explores future opportunities in these wastewater treatment sectors.

Sustainable wastewater management systems (WWS) should be established not only to control water pollution but also to transform wastewater into the water with good quality and energy. But increased water and energy shortage have forced all wastewater treatment plants across the world to remodify its operations and functioning of the WWS. Nevertheless, efficient WWS requires huge financial and energy investment and skilled workers. In this aspect, Chap. 10 provides a comprehensive overview of overall energy analysis along with the evaluation of several approaches for assessing the total energy consumption and energy efficiency. Furthermore, it reveals that the energy efficiency indicators developed exclusively for WWS can bring a better solution to the process of energy evaluation and process benchmarking.

Epidemiological studies have been conducted to assess the adverse effects of particulate matter on human health, and the results showed that particulate matter pollution from certain sources caused more deaths in India. Moreover, many studies have incorporated time series analysis to investigate the influence of particulate matter pollution on mortality and morbidity using classic regression models. Chapter 11 describes the application of a random forest regression model to predict the number of patients who are registering at the respiratory outpatient department in a government hospital named Shyama Prasad Mukharji in Lucknow. Lucknow has been chosen for this investigation due to its variation in pollution levels throughout the year and its administrative and industrial importance. This random forest regression model considers the concentration of pollutants as a predictor variable and number of patients visits as a reasonable variable, while relative humidity and temperature are considered as confounders. Pollutant data were collected from the central pollution control board network of continuous ambient air quality monitoring stations in Lucknow. In this study, individual and synergistic effects of various air pollutants such as particulate matter ($PM_{2.5}$), sulfur dioxide

(SO_2), ozone (O_3), and nitrogen dioxide (NO_2) on human health have been evaluated and found that the synergistic effect of $PM_{2.5}$ and NO_2 showed significant harmful effects on human health. Moreover, the regression model predicted that the post-monsoon could be the most polluted season of the year with a higher number of hospital visits.

Owing to the rapid growth in industrialization and inadequate air pollution control measures, the emittance of particulate matter to the atmosphere is also rapidly increased. This has resulted in the deterioration of the air quality and thus showed significant adverse effects on human health and local and global climates. Several approaches have been developed and implemented to reduce the emission of hazardous pollutants from the emission source and to meet the given standards. In this aspect, Chap. 12 provides an overview of the techniques available for controlling particulate matter emission in industries and provides significant information about the recent developments of the emerging techniques to control particulate matter emission.

It is believed that this book chapter provides a comprehensive overview and discussion on several pollution control technologies available today to deal with the problems associated with water and air pollution.

Chapter 2
Remediation of Persistent Organic Pesticides from Wastewater Matrices—Present and Future Conceptions

Sivakumar Vigneshwaran, Palliyalil Sirajudheen, Perumal karthikeyan, Chettithodi Poovathumkuzhi Nabeena, and Sankaran Meenakshi

Abstract Water pollution is one of the significant environmental issues at present, which primarily originates from the discharge of industrial and agricultural wastewater back into water resources. Persistent organic compounds such as pesticides, dyes, and pharmaceutical wastes are the prominent sources of water pollution. Pesticides are the substances which increase agricultural productivity by destroying pests that are harmful to crops. However, the runoff and leaching of toxic pesticide residues lead to bioaccumulation, xenobiotic and persistence impacts, various health hazards, and severe ecological issues. Also, the industrial effluents containing the recalcitrant pesticide pollute water system by their odor, color, and formation of harmful or oncogenic by-products upon degradation. Due to the complex nature of chemical mixtures present in wastewater, conventional wastewater treatment processes are not always sufficed to remove the entire contaminant load. Therefore, the mineralization of persistent organics by the semiconductor-based photocatalytic process turns out to be a low-cost, environmentally responsive, and sustainable treatment technology which align with the 'zero' waste scheme. In this chapter, we discussed and detailed some data on persistent pollutants such as pesticides, their metabolic activities, and environmental impacts. The present perspective on their degradation using (p–n) heterojunction-based semiconductors has also been reviewed.

Keywords Pesticides · Classification · Removal methods · Nanocomposites

Abbreviations

2,4-D	2, 4-Dichlorophenoxyacetic acid
AHS	Agricultural health study

S. Vigneshwaran · P. Sirajudheen · P. karthikeyan · S. Meenakshi (✉)
Department of Chemistry, The Gandhigram Rural Institute—Deemed to Be University, Gandhigram, Tamil Nadu 624302, India

P. Sirajudheen · C. P. Nabeena
Pocker Sahib Memorial Orphanage College, Tirurangadi, Malappuram, Kerala 676306, India

S. P. Singh et al. (eds.), *Pollution Control Technologies*, Energy, Environment, and Sustainability, https://doi.org/10.1007/978-981-16-0858-2_2

AML	Acute myelocytic leukemia
AOPs	Advanced oxidation processes
CA	Chromosomal aberrations
CPF	Chlorpyrifos
CRC	Colorectal cancer
DCP	2,4-Dichlorophenol
DDD	Dichlorodiphenyldichloroethane
DDT	Dichlorodiphenyltrichloroethane
DHT	5α-Dihydrotestosterone
DSAs	Dimensionally stable anodes
EDCs	Endocrine disrupting chemicals
EPA	Environmental Protection Agency
HD	Hodgkin's disease
MN	Micronucleus
NHL	Non-Hodgkin lymphoma
OPs	Organophosphate pesticides
PCP	Pentachlorophenol
PPE	Personal protective equipment
RO	Reactive oxygen species
PD	Parkinson's disease
SD	Sprague Dawley
TCDD	2,3,7,8-Tetrachlorodibenzo-p-dioxin
TCP	3,5,6-Trichloro-2-pyridinol
TOC	Total organic carbon
XPS	X-ray photoelectron spectroscopy

2.1 Introduction

'Substances used for the destruction of insects or other organisms that are harmful to cultivated plants or livestock' are called as pesticides. In today's scenario, most of the chemical pesticides used worldwide are developed initially in the laboratory scale under an expert and are produced in industrial scale only after the positive results. Some pesticides are harmful, to humans and other organisms. The pesticide contamination of surface and groundwater poses a serious threat to neighboring environments. Pesticides find extensive agricultural and domestic usage. Environmental Protection Agency (EPA) describes pesticides as any substance or mixture of materials intended for preventing, destroying, repelling, or mitigating any pest. They are used for increasing the agricultural outcome (Sabarwal et al. 2018). In USA around the twentieth century, some of the pesticides were banned as they were not registered under EPA. Hence, banned pesticides were exported to other developing countries like India. As a result, the cases of pesticide poisoning increased 13 times as compared to USA (Swanimathan 1982). The usage of pesticides was presented as a means to meet the increased demand for food products owing to increase in population and an aid to the survival of many people who depend on agriculture for their livelihood. But their prolonged usage leads to adulteration of foodstuffs and making them unsafe for mankind or even oncogenic in nature. Besides the environmental issues and health hazards, the extensive usage of pesticides has affected biodiversity as well. Conventional techniques or technologies which were used to eliminate pesticides from the environment were flocculation, coagulation, photolysis, adsorption, etc. These methods also have an adverse effect on air, soil, and water quality as they lead to the formation of toxic products. High operational cost low efficiency is some of the major drawbacks. A promising technology which utilizes the ability of light as the source to remove pollutants from the environment is photodegradation. This relatively low-cost method is also eco-friendly and versatile. The methods and materials used for removal of pesticides ranged from adsorption with clay, activated carbon, etc., and treating with polymeric materials having semiconducting properties by oxidation. The photodegradation process is dependent on several factors such as pH, kind of matrix, temperature, and nature of pesticide.

2.2 Types of Pesticides

Pesticides can be grouped according to the type of target as insecticides, rodenticides, fungicides, herbicides, etc. Pesticides can also be classified as organic and inorganic pesticides depending on their chemical structure. Those which consist of carbon along with any groups like F, O, P, S, and Cl in their molecular structure are called as organic pesticides. Inorganic pesticides are comprised of groups like S and Cu in their structure. Some of the conventional pesticides used for agricultural, commercial, and residential purposes are listed in Table 2.1.

The major classifications of pesticides pertaining to their respective chemical nature are organophosphate pesticides (OPPs), organochlorine insecticides (OCPs), carbamate, and pyrethroids. The organochlorides are found to be most dangerous than the other pesticides, because of their stable nature and subsequent leaching into water and soil. It remains in the environment for more than three decades, polluting surface as well as groundwater, turning out to be a heavy burden on humankind.

2.3 Non-biodegradable and Biodegradable Pesticides

Some pesticides do not breakdown easily and persist in the applied field for a long time. These kinds of pesticides are used consistently due to their long-lasting actions toward the pests. These are also called as residual chemicals as it resides on the soil or on the treated surface for a prolonged period after its usage. Such residual

Table 2.1 Classification of pesticides with respect to their toxic and chemical behavior

Pesticides	Classification	Examples
Insecticides	• Organochlorines • Organophosphates • Carbamates • Pyrethroids	DDT, BHC, malathion, aldicarb, carbofuran, pyrethrum, allethrin, parathion, chlorpyrifos, diazinon, dichlorvos, bifenthrin, cyfluthrin, cypermethrin, carbaryl
Herbicides	• Triazines • Urea and substituted urea • Carbamates and thiocarbamates • Chloroacetanilides	Atrazine, simazine, diuron, linuron, oxamyl, methomyl, alachlor, acetochlor, metolachlor, dimethenamid, 2,4-D, glyphosate, cyanazine, pendimethalin, trifluralin, dicamba
Rodenticides		Warfarin, bromadiolone, maneb, difenacoum
Fungicides	• Phthalimide • Thiocarbamate	Captan, sulfur, mancozeb, mefenoxam
Larvicides		Temephos, methoprene, monomolecular films

Source Morillo and Villaverde (2017). Reproduced with permission from Copyright © 2017 Elsevier Ltd.

pesticides can enter into air and water also and remain poisonous and active for many years. These kinds of residual chemicals are called as **non-biodegradable pesticides**. DDT, BHC, and dieldrin are examples of **non-biodegradable pesticides**.

There is another class of pesticides developed by scientists, called **biodegradable pesticides** that stay active, last only long enough to do their activity, and then decompose into harmless and simple chemicals like water and carbon dioxide. Preparations of less toxic residual chemicals are also being tried by scientists from past years (Fig. 2.1).

2.4 Utilization of Pesticides All Over the World

Two million tonnes are the global usage of pesticides per year. The worldwide percentage of utilization of pesticides is listed below:

(1) Europe: 45%
(2) USA: 25%
(3) Other countries: 25% (India—3.75%, and Korea and Japan—6.6%).

In India, the usage is around 0.5 kg/ha, whereas in Korea and Japan, the usage around 12.0 kg/ha. In the cultivated land area, the pesticides cover only 25% globally. There are three most commonly used pesticides which account for about 70% of the total pesticide consumption; they are HCH (only gamma-HCH is allowed), DDT, and malathion. Since these pesticides are cost-effective and readily available and display a wide spectrum of bioactivity, it remains as the best choice of many farmers even though new pesticides are developed (Table 2.2).

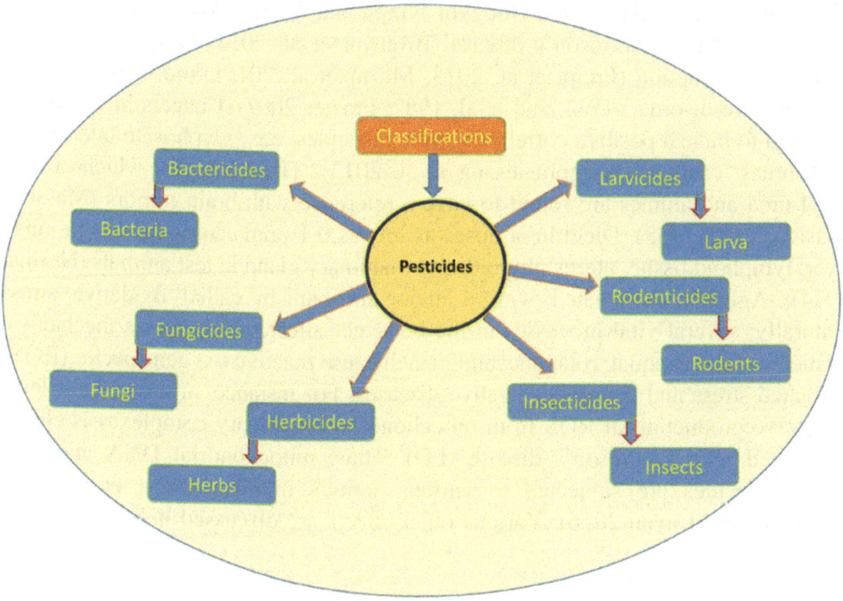

Fig. 2.1 Flowchart representation of different types of pesticides with its subclasses

Table 2.2 Current consumption percentage of various pesticides all over the world

S. no.	Consumption of pesticides	Indian scenario (%)	Worldwide scenario (%)
1	Fungicides	2	17.5
2	Herbicides	15	47.5
3	Insecticides	80	29.5
4	Other pesticides	3	5.5

In India, weed control is mainly done manually; thus, the consumption of herbicides is probably low. Pesticides also find their use in other sectors too, in addition to agricultural use (Nandipati and Litvan 2016).

2.5 Effects of Pesticides

Pesticides can enter the human body in three ways such as oral entry, respiratory entry, and dermal entry. Humans get exposed to pesticides in many ways. It causes various effects based on the exposure intensity of the pesticide. People working in pesticide industries, workers transporting hazardous chemicals, farmers, and fruits and vegetable vendors and consumers get exposed to various concentration levels of pesticides. Exposure to pesticides leads to numerous health disorders such as

Hodgkin's disease (HD), non-Hodgkin lymphoma (NHL) (Wiklund et al. 1987; Luo et al. 2016), Parkinson's disease (Brouwer et al. 2017; Wang et al. 2014), endocrine disruption (Freire et al. 2013; Mazur et al. 2015), and respiratory and reproductive disorders (Wiklund et al. 1987; Frazier 2007). Cancers in human are believed to have a positive correlation with pesticides, e.g., glyphosate interference with breast cancer (Thongprakaisang et al. 2013). The pesticides which contain alkyl urea and amines are found to have a reference with brain tumors (Massimo Musicco et al. 1988). Dieldrin at doses as low as 0.1 ppm causes tumors in lungs, liver, lymphoid tissue, uterus, thyroid, and mammary gland in test animals (Norman 1974). Aging is associated with a major mechanism called oxidative stress. Naturally, several vital processes in the body get affected due to this mechanism. Pesticides like paraquat, rotenone, and maneb cause reactive oxygen species (ROS)-mediated stress and neurodegenerative diseases. For instance, at the cellular level, excessive production of ROS from mitochondrial respiratory complexes is closely associated with Parkinson's disease (PD). Thus, mitochondrial DNA and other macromolecules are subjected to various serious injuries (Blesa et al. 2015; Nandipati and Litvan 2016). Various other effects are discussed below.

2.6 Exposures to Pesticides

Pesticides can cause active or passive health risk because the majority of the human population gets exposed to pesticides. The way we get exposed to pesticides is inevitable, but the dosage of exposure is more crucial. The pesticide exposure can be classified mainly into three categories as follows (Gunnell and Eddleston 2003).

(a) Short-term very high level of exposure (intentional and occupational)—11%
(b) Long-term high level of exposure (occupational)—33%
(c) Long-term shallow level of exposure (non-occupational)—56%.

The above-said percentage level describes the population being exposed to the risk factors of pesticides.

2.6.1 Intentional Exposure

In most of the developing countries, many people commit suicide using pesticides as it is readily available, especially in rural areas. In China and Southeast Asian countries, nearly 300,000 deaths occurred due to intentional ingestion of pesticides (Bertolote et al. 2006; Gunnell and Eddleston 2003). In Sri Lanka, a case study was performed to evaluate the intentional and accidental exposure to these pesticides. It was found that due to pesticide poisoning in a single year, around 13,000 patients were admitted in hospitals. Toxicity has occurred due to both intentional and unintentional exposures. In Sri Lanka, about a thousand people lost their life due to pesticide poisoning each year. Seventy-three percentage of patients consumed these chemicals intentionally; accidental or occupational exposures were approximately 24.9%. Organophosphorous pesticides were most commonly involved in such poisonings (Jeyaratnam et al. 1982).

2.6.2 *Occupational Exposure*

People employed in pesticide industries, in the transportation of pesticides, dealers, farmers, and sellers of fruits and vegetables in the markets, are the primary victims of occupational exposure. In Pakistan, one of the pesticide occupational outbreaks occurred, around 2800 workers involved to control malaria by spraying malathion for intoxication. From various studies, it was found that due to organophosphate poisoning, the patients had decreased cholinesterase activity in red blood cells (Baker et al. 1978). It is also surveyed that parents who are depended entirely on agriculture and are readily exposed to the hazardous effects of pesticides have a greater possibility in development of cancer in their children also (Fear et al. 1998).

2.6.3 *Non-occupational Exposure*

People who consume cultivated crops (fruits, vegetables, grains, etc.) are the non-occupational victims of pesticides. Majority of the human population and animals fall in the latter category. It is challenging to relate to the exposure of pesticides and their hazardous effects. Its effects are classified based on whether the toxicity is acute or chronic.

2.7 Health Effects Associated with Pesticides

Studies revealed that the exposure of pesticides leads to the development of cancers in both children and adults. Various malignancies such as leukemia, Burkitt lymphoma, neuroblastoma, Wilm's tumor, non-Hodgkin lymphoma, soft tissue sarcoma, ovarian cancer, cancers of lung, stomach, colon, bladder, and rectum were found in people who are closely encountered with the pesticide exposure (Brouwer et al. 2017; Jan Dich et al. 1997; Polanco Rodriguez et al. 2017; Schinasi and Leon 2014; Xu et al. 2010).

2.7.1 Cancer in Children

Cancers in children were mostly associated with pesticide exposure. In a population-based case–control study of acute myelocytic leukemia (AML), more convincing evidence was presented as compared to other cancer types. A comparison of 491 cases among an age group of 0–9 years was made for polymorphisms in CYP1A1, CYP2D6, GSTT1, and GSTM1 genes responsible for encoding enzymes that metabolize carcinogenic substances (Claire Infante-Rivard et al. 1999). The enzymes of cytochrome P-450 family are involved in the transformation of pro-carcinogenic compounds to reactive species which have genotoxic and cytotoxic effects. Glutathione S-transferases, the enzyme which gradually inactivates the reactive species of many pesticides, e.g., lindane, parathion, chlorophenol, and atrazine, use these enzymes as substrates (Guengerich 1998). There was also a significant association between maternal exposure of pesticides and childhood leukemia. Pregnant ladies working in agricultural lands for applying of fertilizers were also the primary victims of pesticide exposure, and its effects were later transferred to the newly born child in the form of cancers (Kumar et al. 2014).

2.7.2 Non-Hodgkin Lymphoma (NHL)

Lymph and immune system were affected by non-Hodgkin lymphoma, which is a diverse group of malignancies; it consists of more than 20 deferent malignancies. In the past few decades, this particular type of malignancy has been increased throughout the world (Alavanja et al. 2003; Brouwer et al. 2017). A meta-analysis was performed with organophosphate pesticides (OPs) which resulted in a positive association. Three non-halogenated OPs terbufos, malathion, and diazinon were selected from 1985 to 2017 for this study. Diazinon showed a significant positive relation with exposure and NHL. Immune dysfunction directly related to NHL was reported here. Immune cells are directly attacked by malathion, and disruption of a neuro-immune system was caused by diazinon, which involves a cholinergic system of lymphocytes (Wiklund et al. 1987).

2.7.3 Alzheimer's Disease (AD)

The most common cause of dementia in aged people is due to Alzheimer's disease (AD). The presence of extracellular amyloid beta (Aβ) plaques, neuronal death, and the loss of synapses are the characteristic features of the disease. Pathogenesis of AD is found to be positively associated with environmental contaminants. Individuals who are chronically exposed to pesticides have a high prevalence of cognitive, behavioral, and psychomotor dysfunction and Alzheimer's disease dementia which is proved by many studies (Zaganas et al. 2013). One of the prodromal stages of dementia is a mild cognitive impairment (MCI), and it is associated with the exposure of pesticides. Every year, nearly 5–15% of individuals are diagnosed with MCI, which otherwise develops AD (Petersen 2011). A case–control study was done at Andalusia (Spain), which involved 17,429 subjects, and data was collected during 1998–2005 using hospital records. Categorizations of the issues were done based on low agricultural practice/use of pesticides and high agricultural practice/high use of pesticides. Greater risk to PD, AD, and multiple sclerosis was shown by the subjects lived in the areas with a higher amount of pesticides (Parron et al. 2011).

2.7.4 Reproductive Disorders

Risk of sperm abnormalities, aberrant fertility abortions, defects in birth, and fetal growth retardation are due to the abnormal exposure to certain pesticides (Frazier 2007). An increase in chromosomal aberrations (CA), bone marrow micronucleus formation (MN), and sperm abnormality in mice are shown by carbosulfan, a carbamate pesticide. It acts as a potent genotoxic agent and could also act as a potent germ cell mutagen which was ingested at a rate of 1.25, 2.5, and 5 mg/kg (Sarbani Giri et al. 2002). Toxic potentials on reproductive and fertility parameters in the adult male Sprague Dawley rats were evaluated by cypermethrin, a pyrethroid insecticide. These animals were exposed to cypermethrin in ppm concentrations for 12 weeks through tap water containing different doses. In male rats, a significant decrease in fertility was recorded, which ingested cypermethrin at a concentration of 13.15 and 18.93 mg day. The treated male rats had a significant reduction in the number of viable fetuses of the impregnated females. In the treated animals, gain of weight in testis, seminal vesicles, and decrease in the epididymis and testicular sperm counts were also observed (Elbetieha et al. 2001). Genotoxicity in mouse somatic and germ cells is shown by paraquat (herbicide) which is extensively used. The mice were subjected to single or multiple exposures. Chromosomal aberration (CA) was not caused by single time exposure, but there was an increase in CA with multiple exposures. Spermatozoa, spermatid, and preleptotene spermatogonial cells are the three stages of cell development, and thus there is an increase of sperm shape abnormality when they are treated (Rios et al. 1995). In Andalusia (Spain), a case–control study was carried out on pregnant women and male children living in high-risk and low-risk areas. The areas were used pesticides in more elevated amounts; it was found that prevalence rates and risk of miscarriages, low birth weight, hypospadias, cryptorchidism, and micropenis were significantly high (Garcia et al. 2017).

2.7.5 *Respiratory Disorders*

Bernardino Ramazzini was one of the first researchers who informed that the risk of respiratory disorder is higher in farmers, in the 1700s (Hoppin et al. 2006). In developing countries like India, most of the pesticide applicators do not use personal protective equipment (PPE) like safety masks, gloves, etc., which brings them into direct contact and higher doses of pesticide exposure. As the pesticides sprayed, aerosols are generated in the fields, and thus it can directly enter into the respiratory system of the farmers. Pesticides cause diverse respiratory illnesses which include wheezing, dry cough, irritation of the respiratory tract, blood in sputum, etc. (Fareed et al. 2013). There is a decrease of red blood cell cholinesterase in 77% of pesticide applicators by organophosphate and carbamate exposure, while 27% had a decrease in plasma cholinesterase levels (Jaga and Rama 1992). A comparative study was conducted with the workers in a bottling plant and with the workers involved in the pesticide processing plant. The result revealed a much higher risk of inducing respiratory disorders, including chronic cough, dyspnea grades 3 and 4, throat irritation, nasal catarrh, and nasal dryness (Zuskin et al. 2008).

2.8 Various Methods Involved in Pesticide Removal

The removal of pesticides has been employed through several techniques, including the biological, chemical, physical, and physicochemical process of remediation from different types of matrices, such as water and soil. Various treatment methods are used for the removal of pesticides from the wastewater such as incineration, ozonation/UV irradiation, Fenton oxidation, phytoremediation, electrocoagulation, electro-oxidation, photocatalytic degradation, adsorption, reverse osmosis, nanofiltration, and electrodialysis. The removal method of specific pesticides depends on its physical and chemical characteristics. The use of hybrid removal techniques offers potential opportunities to develop innovative pesticide removal techniques.

2.8.1 *Physical Methods*

Physical method is the process of removing the pollutants by physical means. Various physical methods adopted for the remediation of pesticides are described below.

2.8.1.1 Adsorption

Biochar is a carbon-rich solid made from agricultural crop residues, wood, or waste via pyrolysis. Extensive and non-judicious use of pesticides over the last several decades resulted in severe soil and water contamination, which causes severe toxic effects on living organisms. Biochar has emerged as a promising material for adsorbing and thus decreasing the bioavailability of pesticides in polluted soils, due to its high porosity, surface area, pH, abundant functional groups, and highly aromatic structure, and mainly depends on the feedstock and pyrolysis temperature. Pyrolysis temperature may have a positive or negative effect on the sorption of pesticides on soil, one by large surface area and one by a large number of functional groups. The sorption and desorption mechanism pesticides in the soil by biochars are not well explored so far. Biochar, organic green wastes, and composites demonstrated proper physicochemical methods to immobilize pesticides. But its application was limited to certain classes of pesticides. Thus, the applications of biochar and organic green wastes for the remediation of contaminated soil have mainly been conducted in the laboratory, greenhouses, or small plot experiments. Large-scale field trials are needed before operational-scale remediation projects are implemented. Thus, biochar was preferred because it enhances the sorption of hydrophobic organic compounds.

Porosity and surface area are two primary properties which affect the sorption capacity of biochar for organic pollutants, including pesticides through pore filling and surface sorption, respectively. More porous structures and higher surface area will result in higher sorption capacities. Biochar porosity and surface area are varying substantially with pyrolysis temperature. Several previous studies showed that increased temperature results in larger pore volume and surface area. The pore volume of biochar increased from 0.056 to 0.099 $cm^3\ g^{-1}$ with the increase in temperature from 500 to 900 °C, while the surface area increased from 25.4 to 67.6 m^2g^{-1}. Apart from pyrolysis temperature, the composition of biochar feedstock is also an essential factor affecting its properties. Higher pH of biochar can accelerate the hydrolysis of organophosphorus and carbamate pesticides in the soil through alkali catalysis mechanism. The presence of surface functional groups including carboxylic (–COOH), hydroxyl (–OH), lactonic, amide and amine groups has some influence in the sorption capacity of biochar. Generally, pyrolysis temperature and raw material are the two crucial factors that influence the quantities of functional groups on the biochar surface.

There are mainly two different processes in biochar sorption, including surface adsorption and partition. The functional groups on the biochar form stable chemical bonds with ions or organic compounds, which leads to surface adsorption. The physical and chemical properties and structures of biochar surface are varied with pyrolysis temperatures. The surface adsorption capacity of biochar increases with increasing temperature, whereas partitioning plays a leading role in low-temperature pyrolyzed biochar. Pesticide sorption of biochar in soils depends on several factors including but not limited to biochar properties (e.g., surface area, porous structure, and aromaticity), pesticide characteristics (e.g., molecular dimensions, hydrophobicity), soil properties (e.g., soil pH, mineral content), and environmental factors. For example, the probability of getting adsorbed is lower for pesticides with macromolecular dimensions due to its difficulties in accessing the inner porous structure of biochar (Liu et al. 2018).

In contrast, polar compounds can be firmly adsorbed by biochar through interaction between electron donor and receptor. Biochar is one of the most efficient adsorbents for the removal of several groups of pesticides, including herbicides, insecticides, fungicides, rodenticides, and pediculicides. Biochars with lower surface area and higher dissolved organic carbon contents tend to decrease the sorption of herbicide in the soil. A significant quantity of biochar or some of its components do change with time, and the process commonly referred to as 'aging.' The aging process of biochar affects pesticide sorption–desorption in the soil due to the changes in its physicochemical properties. The sorption affinity of biochar could be slacked with aging or weathering in soils. The aging effects of biochar cause a decrease in its specific surface, which in turn decreases the potential sorption capacity of biochar. Sorption capacity is not as high as other types of sorbents, such as clays (e.g., stevensite, smectite), and activated carbon. Modifications of biochar in its essential properties and loading with organic functional groups, nanoparticles, and activation with alkali have been well reported in the previous literature. Biochar modification also includes loading biochar with different minerals such as hematite, magnetite, zero-valent Fe, hydrous Mn oxide, calcium oxide, and birnessite. The principal pathway of dissipation and decomposition for most of the pesticides such as isoproturon in the soil is biodegradation. Biochar is an effective remediation for pesticide pollution and decreases their biodegradation in soils due to sorption effect, and it was explained and indicated in many of the previous literatures. Besides, biochar may result in the microbial stimulation, which results in the higher microbial degradation of pesticides. Moreover, biochar promotes the growth of the plant, especially at the root which enhances the uptake and degradation of pesticides. Thus, more attention must be paid in the biochar soil root systems (Liu et al. 2018) (Fig. 2.2).

2.8.2 Chemical Methods

In chemical methods, the composites promote the removal of hazardous recalcitrant chemicals such as pesticides, convert toxic to non-toxic forms through a chemical process. In this process, the oxidation of organic pesticides carried out by various chemical techniques such as advanced oxidation processes (AOPs), photolysis, photo-Fenton, and photodegradation techniques is discussed below.

2.8.2.1 Degradation

The degradation of pesticides follows different mechanisms. Physical, chemical, and biological agents play significant roles in the transformation of insecticide, herbicide, and fungicide molecules into various degradation products. Oxidation, hydrolysis, reduction, hydration, conjugation, isomerization, and cyclization are transformation mechanisms. Resultant products are usually less bioactive than the

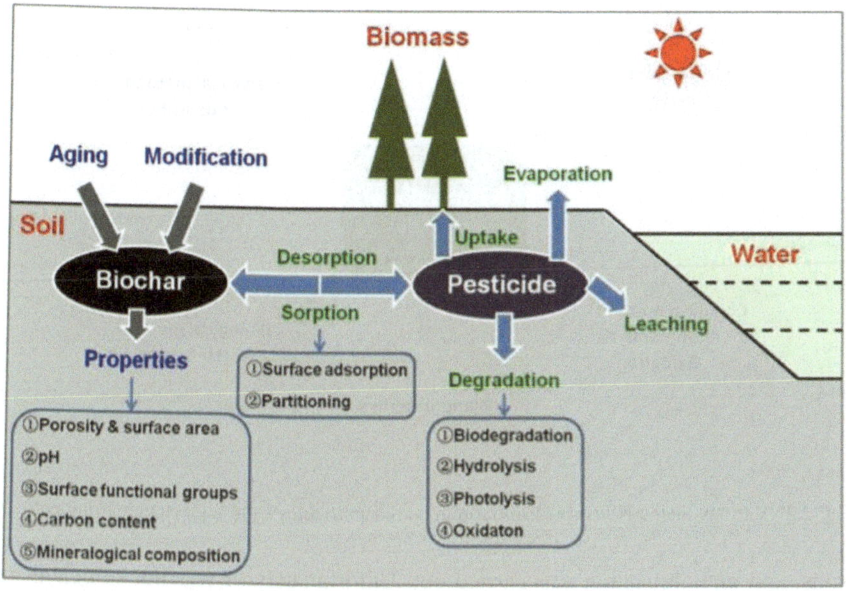

Fig. 2.2 Biochar properties and effects of biochar on pesticide behavior in the soil (Liu et al. 2018). *Source* Liu et al. (2018). Reproduced with permission from Copyright © 2018 Elsevier Ltd

parent pesticide molecule, but numerous cases have been documented of metabolites with higher bioactivity. The physical and chemical properties of the degradation products also vary from those of the parent compound, and their fate and significance in the environment also are altered with the structural changes (Felsot et al. 2003). A large percentage of pesticides applied in agriculture never reached their target organisms. Through dispersion, they are detected in soil, water, and air, and food for human consumption (Fig. 2.3).

Generally, three methods are adopted for the treatment of polluted soil: (I) the procedure in which contaminants are treated on the place where the contamination occurred without excavating the soil which is applied, (II) the polluted soil is excavated, treated on the site in which it exists, and deposited it on their original position, and (III) the final method involves the excavation of contaminated soil from its location and transported it into another site for treatment. However, processing of a large quantity of contaminated soil is rather costly and time-consuming. Usually, the above-said methods are not suitable for the eradication of pests in the agricultural fields due to the existence tribulations such as soil erosion, fertility loss, or nutrient leaching. A new and economical alternative for treating pesticide-contaminated soils is in demand (Morillo and Villaverde 2017). Washing of soil is another technique used for removing the chemical pollutants from soil using different kinds of aqueous solutions of extractants (organic compounds, acids, tensioactives, etc.). Soil flushing is another method adopted for immobilizing pesticides. In this process, the contaminated areas have been treated with injection

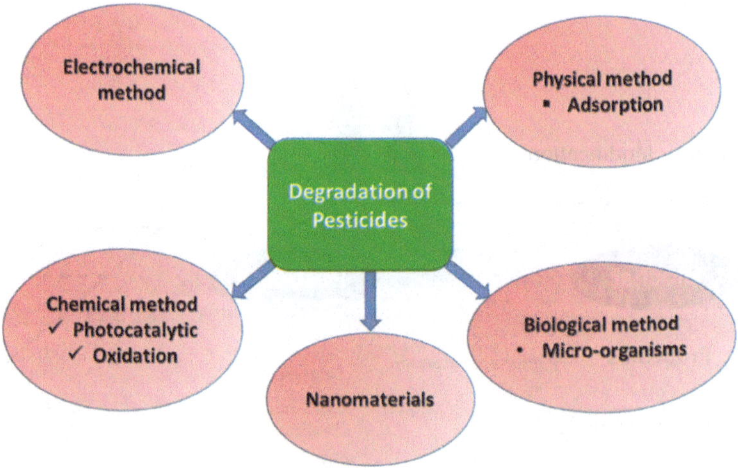

Fig. 2.3 Various methods involved in the removal of pesticides from aqueous environment

wells, and immobilization was carried out. Unfortunately, it was not sufficient for the remediation of pests. Nanofiltration and reverse osmosis membrane technologies were utilized at the earlier stages of water treatment to remove pesticides like atrazine and simazine. Chlorination is also another technique employed for removing pesticides in drinking water. Coagulation is a separation method used only for a particular group of pesticides (Ormad et al. 2008).

2.8.2.2 Photocatalytic Degradation

It is necessary to overlap the absorption spectrum of the pesticide with the emission spectrum of the light source for a molecular change to occur, and it is the most important condition for the photocatalytic process. For degradation, high energy lamps are used with a wavelength less than 254 nm, and direct environmental photolysis takes place since the energy of solar radiation is too low. Its wavelength at the earth's surface is greater than 280 nm (Sharma et al. 2016). Photosensitizers, such as rose bengal, methylene blue, or riboflavin, can absorb lower energy light and have been used to overcome this impediment, permitting the use of longer wavelength lamps. The sensitization processes involving humic substances in the soil which can absorb solar energy are similar to this (Jean Marie Herrmann et al. 1999). Thus, photolysis was the oldest method of degradation, and later degradation was followed with the help of photocatalytic methods which has a vast range of application in the degradation process. In order to study about photocatalytic degradation, there are many data available in the previous literatures. Photoreactor and photocatalyst are the primary requirement of this method. As a part of the study on pesticide-contaminated water, the effect of organophosphorus and phosphonoglycine is analyzed, and these kinds of pesticides are degraded with the help of

TiO_2 catalyst which is immobilized on a silica gel. The UV/TiO_2 process does not merely transfer pesticides from one phase into another, but it has an ability to destroy them into harmless final products including carbon dioxide, water, and inorganic ions (Echavia et al. 2009). It had recognized that when TiO_2 is irradiated with photons whose energy is equal or greater than its band gap energy (3.2 eV), the photons are absorbed. It results in the formation of electron–hole species (Jean Marie Herrmann et al. 1999). In most of the photocatalytic degradation process, TiO_2 is used in the form of powder suspensions or slurries (Han et al. 2009). The removal of pesticides such as organophosphate, acephate, dimethoate, phosphonoglycine herbicide glyphosate pesticides due to their widespread use, toxicity, water solubility, and their potential to contaminate the water resources under favorable conditions. In the photoreactor, 14.0 g of TiO_2 photocatalyst, 6 W blacklight UV fluorescent lamp, and pesticide solution were used. Chemically resistant Teflon tubes are used for connecting all the components of a photoreactor, and photocatalysis is conducted under ambient temperature of 22 ± 2 °C. This procedure was followed in dark condition also. Degradation of the pesticides was finalized by comparing the concentration in both untreated and irradiated samples with pure pesticide standards. Photocatalytic degradation of pesticides resulted in the breaking of the parent molecules as evidenced by the evolution of inorganic anions. SO^{2-} and NO^{-} which are released due to decomposition of acephate and dimethoate. Glyphosate liberated NO^{-} and PO^{3-}. Among the three pesticides, only dimethoate degradation showed the release of NO_2 as an intermediate of NO_3^{-}, which was demonstrated by the disappearance of NO_2 and concomitant formation of NO_3^-. Degradation pathway is explained below

$$TiO_2 + h\nu \rightarrow e^- + h^+ \tag{2.1}$$

$$Ti\,(IV) - OH + h^+ \rightarrow (O_2)_{ads} + e^- \tag{2.2}$$

$$Ti(IV) - H_2O + h^+ \rightarrow Ti(IV) - OH\bullet \tag{2.3}$$

$$Ti(IV) - OH\bullet + H^+ \rightarrow (O_2 \bullet -)_{ads} \tag{2.4}$$

$$\text{Compound} + h^+ \rightarrow \text{Compound oxidized} \tag{2.5}$$

List of by-products formed as a result of photocatalytic degradation:

- **Organophosphorus insecticides**: Trialkyl and dialkyl phosphorothioate or phosphate esters, carbamate insecticides: The breakdown of amide function C–N bonds and later hydroxylation of the products formed lead to more or less oxidized aliphatic compounds (aldehydes, acids, etc.); finally, carbon dioxide, nitrate, sulfate, and on the degradation of propoxur three intermediates were monitored, and they are O–isopropoxyphenol, isopropoxydihydroxyphenol, o-dihydroxybenzene.

- **Organochlorine pesticides**: Categories of chlorocyclohexanes, chlorocyclohexenes, chlorobenzenes, chlorophenols, chloropropanes, and the pentachloro cyclohexanone isomer and endosulfan sulfate are the degradation products of endosulfan.
- **Chlorophenol pesticides**: p-chloroaniline, tetrachlorohydroquinone, H_2O_2, tetrachlorocatechol, O-chloroaniline, 2,3,5,6-tetrachloro-1,4-benzoquinone,2,3,5,6-tetrachloro-1,4-hydroquinone, and 2,3,5,6-tetrachlorophenol.

For some kind of pesticides like chlorpyrifos, cypermethrin, and chlorothalonil, mineralization studies are done. Thus, an experiment was conducted under conditions (pesticide concentration 400 mg/L (COD 1130 mg/L), TiO_2 1.5 g/L, H_2O_2 concentration 100 mg/L, and pH 6) to study mineralization by UV/TiO_2/H_2O_2 photocatalysis. The concentration of ammonia–nitrogen (NH_3–N) decreased from 22 to 7.8 mg L^{-1}, and concentration of nitrate–nitrogen (NH–N) increased from 0.7 to 13.8 mg/L in 300 min, indicating mineralization. Complete degradation of the pesticides was achieved in 30 min. These kinds of pesticides degradation by UV/TiO_2/H_2O_2 are more effective than UV/TiO_2 (Affam and Chaudhuri 2013).

2.8.2.3 Oxidative Degradation

Chemical oxidation is the key technology to solve the pesticide contamination problems in both the water and wastewater treatments. For aqueous pesticide degradation, ozonation is considered as one of the most promising variations of chemical oxidation, and it has a long history of investigation. In addition to ozonation, various ozone-based advanced oxidation processes (AOPs), which utilize hydroxyl radicals for oxidation, were recently evaluated for the degradation of aqueous organic pollutants, including pesticides. In the water treatment industry, ozone has been used for disinfection, odor management, and color removal. Molecular ozone has a higher oxidation potential of 2.07 V (relative to the hydrogen electrode) than conventional chemical oxidants such as potassium permanganate and chlorine. Ozonation is proven to be effective in degrading several organic pollutants in water and wastewater. New water and wastewater treatment technologies are subjected to AOPs, which are characterized by the production of hydroxyl radicals through various chemical and photochemical reactions. The hydroxyl radicals have a higher oxidation potential of 2.8 V produced by AOPs and can attack organic and inorganic molecules non-selectively with very high reaction rates (Andreozzi et al. 1999). It should be noted that ozonation at high pH (>8) is also regarded as an AOP because hydroxyl radicals are produced at such conditions, and the reactions between the radicals and organic molecules take place. Oxidative degradation by ozonation and ozone-based AOPs takes place with the help of O_3/H_2O_2, O_3/UV, and O_3/H_2O_2/UV. Rapid and complete conversion of certain compounds has often been demonstrated. Ozonation alone or even with ozone-based AOPs, the total mineralization of pesticides is likely to be impossible.

However it is not practical to apply more intensive ozone-based treatment or any other chemical oxidation techniques to achieve the total decomposition and mineralization of the compounds. Several ozonation by-products, as well as intermediates, were identified, and for some of the significant pesticides, possible degradation pathways were proposed such as 2,4-D, pentachlorophenol (PCP), and diazinon. However, for many pesticides degradation by-products are mostly unknown because the relative importance of such pesticides might not be as high as that of the major pesticides or the toxicity and/or environmental impact of such pesticides may be relatively low (Ikehata and Gamal El-Din 2005).

2.8.2.4 Electrochemical Method

Since the 1990s, a technique has widely been studied for the removal of organic substances which was known as electrochemical combustion (total oxidation). Advantages of electrochemical treatment methods are:

The catalyst/electrode is immobilized (thus reducing the need to separate the catalyst from the reaction mixture).

- The variables (i.e., current, potential) are easily controlled and facilitate automation of a process.
- The cost of the equipment is generally low (Juttner et al. 2000).
- Electrochemical processes are easily adapted for use in flow systems, for the feasibility of pesticide treatment systems, thus is considered to be an essential factor (Felsot et al. 2003).

Oxidation of various pesticides in brine solutions using Ti/Pt anodes was studied. The electrochemical method can be used as a pre-treatment step in pesticide waste disposal because of its reasonable removal rates (Vlyssides et al. 2005). Pt-based electrodes are not ideal because it is expensive, especially with organic waste. Thus, it leads to the blocking of the electrode surface. Dimensionally stable anodes (DSAs) are suggested as possible alternative electrode materials which have received considerable attention in recent year. DSAs are promising materials for many electro-organic applications, and depending on the chemical nature, it is classified as active or non-active (Malpass et al. 2006). Active electrodes mediate the oxidation of organic species via the formation of higher oxidation state oxides of the metal, whenever the metal oxide can reach a higher oxidation state (e.g., RuO_2 or IrO_2) and it leads to selective oxidation. Non-active electrodes present no higher oxidation state to be available, and the organic species is directly oxidized by an adsorbed hydroxyl radical and results from incomplete combustion of the organic molecule (e.g., SnO_2 or PbO_2). With the help of oxide electrodes, a wide range of treatments is done on wastewater management including humic acid (Pinhedo et al. 2005), dyes (Xu et al. 2020), and pesticide removal (Malpass et al. 2006).

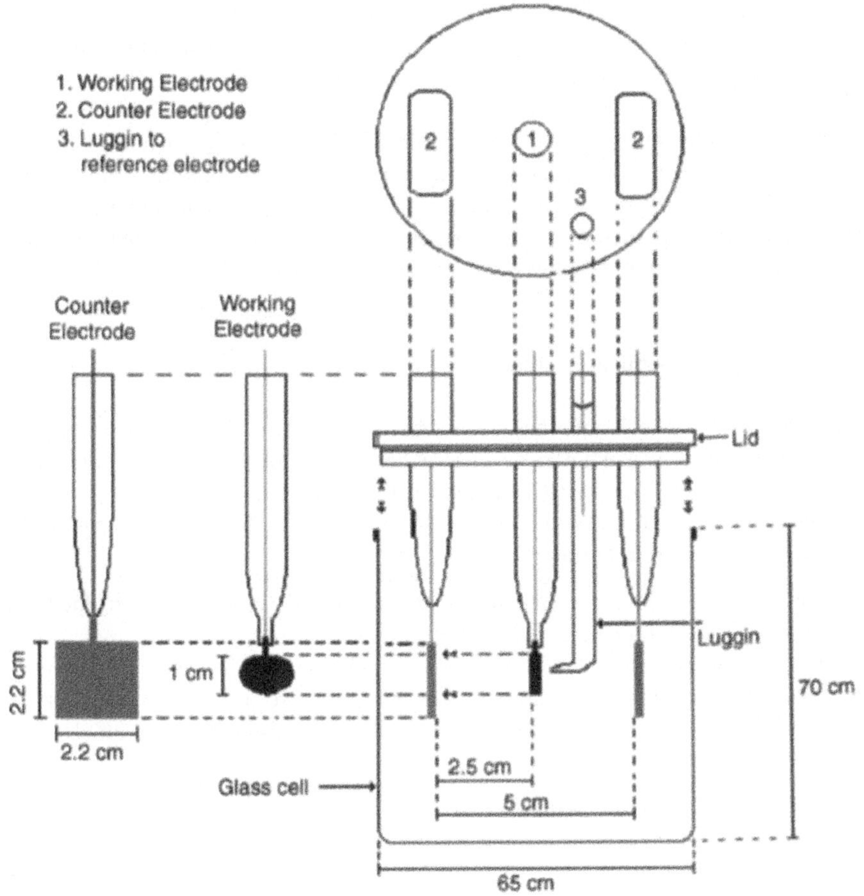

Fig. 2.4 Single compartment electrochemical cell used in the oxidation assays (Miwa et al 2006). *Source* Miwa et al. (2006). Reproduced with permission from Copyright © 2006 Elsevier Ltd

Carbaryl, a carbamate insecticide, is considered as an example to explain electrochemical degradation. $Ti/Ru_{0.3}Ti_{0.7}O_2$, $Ti/Ru_{0.3}Sn_{0.7}O_2$, and $Ti/Ir_{0.3}Ti_{0.7}O_2$ are the three different types of DSAs electrode materials employed. The following effects such as of current density (10, 20, 40, and 60 $mAcm^2$) and supporting electrolyte (0.1 mol/L NaCl and 0.033 mol/L H_2SO_4), upon the rate of carbaryl and total organic carbon (TOC) removal, are investigated. Apparatus consisting of a single compartment electrochemical cell made of glass was employed using a typical three-electrode arrangement (Fig. 2.4).

The working electrodes were DSAs disks (1 cm), and the counter electrodes were two platinum foils (2.2 cm × 2.2 cm), which is positioned parallel to the working electrode. All potentials are referred to as the reversible hydrogen electrode (RHE). Using a potentiostat/galvanostat (Autolab, model SPGSTAT30), all electrochemical measurements (cyclic voltammetry and galvanostatic electrolysis) were performed. And the results obtained indicated that the electrodes employed in this study have the following order of activity toward carbaryl oxidation in H_2SO_4 as $Ti/Ru_{0.3}Sn_{0.7}O_2 > Ti/Ir_{0.3}Ti_{0.7}O_2 > Ti/Ru_{0.3}Ti_{0.7}O_2$. Carbaryl and TOC removal in H_2SO_4 is limited, indicating that the oxidizing species formed on the electrode surface are not robust and abundant enough to result in rapid oxidation of the pesticide and its products. Carbaryl and TOC content is removed rapidly with the help of electrodes in the presence of NaCl. With all the electrodes, the removal of carbaryl is complete, and for $Ti/Ir_{0.3}Ti_{0.7}O_2$ electrode current densities and TOC removal is at least 80% to approximately 100%. Energy consumption factor in H_2SO_4 is lower for $Ti/Ru_{0.3}Sn_{0.7}O_2$ electrode. Since all the electrodes show similar efficiencies, the efficiency of an electrode in the presence of NaCl is most excellent. Thus, the degradation of pesticides is done in electrochemical method (Miwa et al. 2006).

2.8.3 *Biological Method*

2.8.3.1 Microorganisms

Bioremediation is one of the environmentally safe and cost-effective methods for decontamination and detoxification of pesticide-contaminated soil (Deo et al. 1994) (Table 2.3).

The use of the microorganisms for bioremediation requires an understanding of all the physiological, microbiological, and biochemical aspects involved in pollutant transformation. Considerable progress is made in understanding the mechanism of microbial degradation of pesticides. Under aerobic conditions, degradation mechanisms are mostly understood, and metabolism under anaerobiosis is studied. Oxygen is both the terminal electron acceptor and a reactant in the initial reactions, in the case of aerobic environment. Nitrate, sulfate, or carbonate may function as alternate electron acceptors in the microbial degradation of aromatic compounds in anaerobic conditions. To identify and isolate microbes with degradation in accordance with their potential toward pesticides and related compounds like 4-chlorobiphenyl (Steffan et al. 1989), polychlorinated biphenyl (Walia et al. 1990), 2,4-D (Amy et al. 1990), and 2,4,5-T (Steffan et al. 1989), molecular probes have been used from environmental samples. Genes encoding enzyme(s) have been used to detect the strain of interest from the environment for pesticide degradation. The application of molecular probes for detection and isolation of pesticide degrading microorganisms in the environment holds great potential. To detect specific microbes in the environment, polymerase chain reaction (PCR) and gene probe have been extensively used among the available methods (Christine Picard and Nesme 1992) Two most important classes of enzymes are hydrolases and oxygenases, and are involved in catalyzing the conversion of pesticides. Hydrolases include helidohydrolases, esterases, and amidases. Examples:

Table 2.3 Microorganisms responsible for pesticide degradation are listed below (Brajesh Kumar Singh et al. 1999). *Source* Brajesh Kumar Singh et al. (1999). Reproduced with permission from Copyright © 1999 Elsevier Ltd.

Pesticide	Microorganisms
Chlorophenoxy acids 2,4-D	Bacteria – *Alcaligenes eutrophus* – *Alcaligenes xylosoxidans* – *Flavobacterium* sp. 50001 – *Pseudomonas putida* – *Pseudomonas cepacia* – *Comamonas* sp.
2,4,5-T	– *Pseudomonas cepacia*
DPA	– *Flavobacterium* sp.
Mecoprop	– *Sphingomonas herbicidivorans MH*
Mecocarp	– *Alcaligenes denitrificans*
Organochlorines DDT	Bacteria – *Aerobacter aerogenes* – *Alcaligenes eutrophus A5* – *Agrobacterium tumefaciens* + *Arthrobacter* sp. + *Bacillus cereus* + *Bacillus coagulans* + *Bacillus megaterium* + *Bacillus subtilis* + *Clostridium pasteurianum* – *Clostridium michiganense* – *Enterobacter aerogenes* – *Erwinia amylovora* – *Escherichia coli* – *Hydrogenomonas* sp. + *Klebsiella pneumoniae* + *Kurthia zopfii; Johnson et al.* + *Micrococcus sp.* + *Nocardia* sp.; *Chacko et al.* – *Serratia marcescens* + *Streptomyces annomoneus* + *Streptomyces aureofaciens* + *Streptomyces viridochromogenes* + *Xanthomonas* sp. Fungi *Phanerochaete chrysosporium Trichoderma viride*
γ-HCH	Bacteria – *Aerobacter aerogenes* + *Bacillus cereus*
	+ *Bacillus megaterium* – *Citrobacter freundii* + *Clostridium rectum* – *E. coli* – *Pseudomonas fluorescens* – *Pseudomonas putida*

(continued)

Table 2.3 (continued)

Pesticide	Microorganisms
	– *Pseudomonas paucimobilis* – *Pseudomonas* sp. Cyanobacteria *Anabaena* sp. *Nostoc ellipssosum* Fungi *Phanerochaete chrysosporium, Trametes versicolor, Phanerochaete sordida, Cyathus bulleri*
Organophosphates Parathion	Bacteria – *Flavobacterium* sp. – *Pseudomonas aeruginosa* – *Pseudomonas diminuta* – *Pseudomonas melophthara* – *Pseudomonas stutzeri*
Carbamates Carbaryl	Bacteria – *Pseudomonas cepacia* – *Pseudomonas melophthora* – *Pseudomonas aeruginosa* Fungi *Gliocladium roseum Aspergillus flavus Aspergillus terreus Culcitalna* sp. *Halosphaeria* sp. *Fusarium solani Rhizopus* sp. *Penicillium* sp.
Carbofuran	Bacteria – *Achromobacter* sp. – *Arthrobacter* sp. – *Flavobacterium* sp. – *Pseudomonas cepacia* – *Pseudomonas stutzeri* – *Bacillus pumilus*
s-Triazines	Bacteria – *Pseudomonas* sp. + *Klebsiella pneumoniae* –*Rhodococcus corallinus* – *Rhizobium* sp. Fungi *Phanerochaete chrysosporium*

1. Halidohydrolases dehalogenate many halogenated aliphatic and aromatic compounds.
2. Esterases like parathion hydrolase attack phosphodiester bonds of organophosphates.
3. Amidase degrades propanil, etc. Oxygenases are less stable than hydrolases and require molecular oxygen as substrates.

Table 2.4 Enzymes responsible for microbial degradation of pesticides are listed below (Brajesh Kumar Singh et al 1999). *Source* Brajesh Kumar Singh et al. (1999). Reproduced with permission from Copyright © 1999 Elsevier Ltd.

Enzymes	Target pesticides
Esterases	Organophosphates, phenyl carbamates
Lyases	Organophosphates
Phosphatases	Organophosphates Dithioates
Acylamidases	Phenylamide
Oxygenases	Chlorinated phenols 2,4-D DDT HCH
Hydrolases	2,4-D, carbamates, chlorinated phenols s-triazines
Dehydrogenases	DDT, HCH
Cytochrome P-450	DDT
Dehalogenases	HCH, DDT

They can be divided further into two subgroups since they are more complex enzymes:

1. Monooxygenases require reduced pyridine nucleotides as cofactors.
2. Dioxygenases do not require a reduced compound as a cofactor (Table 2.4).

2.8.4 *Material Used for the Removal of Pesticides*

2.8.4.1 Nanomaterials

Nanostructured materials possess unique physical and chemical properties; they not only are useful for pesticide detection but also act as a promising tool for pesticide removal and degradation strategies to remediate environmental pollution. Some of the strategies in this area reported for real samples are summarized in the form of a table below; data such as degradation percentages is also included (Gemma Aragay and Merkoci 2012). Application of advanced oxidation processes (AOPs) is a promising method for mineralization of pesticides (von Sonntag 2008). Depending on the oxidants added to accelerate the photodegradation process, such as hydrogen peroxide and/or ozone, metallic salts, or semiconductors (e.g., TiO_2), various AOPs can be employed (Liu et al. 2011). For pesticide degradation, photocatalytic oxidation (e.g., TiO_2/UV) and photo-Fenton and Fenton-like systems (H_2O_2/UV/Fe^{3+}) are the most widely used AOPs (Gemma Aragay and Merkoci 2012). Photocatalytic oxidation is based on the use of UV irradiation of semiconductors (usually TiO_2). Electron–hole pairs are created on irradiation of TiO_2 with photons (e.g., at $\lambda = 390$ nm) whose energy is equal to or greater than its band gap energy (EG = 3.2 eV). These holes react with H_2O or OH^- adsorbed at the surface of the semiconductor to produce OH˙ radicals, which are the most potent oxidants in this process, in the aqueous system (Ahmed et al. 2011). Fe^{3+} ions are added to acidic H_2O_2 medium to form the complex $Fe(OH)^{2+}$, which, upon UV irradiation, decomposes to generate Fe^{2+} ions and OH˙ radicals, the latter of which act as oxidant, in photo-Fenton and Fenton-like systems. TiO_2 and ZnO NPs are the most widely studied nanoparticles for the pesticide degradation, as they function at room temperature and pressure with low-energy photons and do not require any chemical reagents except the oxygen. Much researches using TiO_2 NPs as nanostructures for photocatalytic degradation of either a single pesticide or mixed pesticides have been reported (Gupta and Tripathi 2011). The catalytic reaction involves the following steps:

(i) Adsorption of pollutant molecules onto the surface of titanium nanoparticles
(ii) Breakage of the chemical bonds
(iii) The formation and subsequent release of the mineralized products.

The unique surface area and surface activity of nanoparticles are crucial for catalytic reactions. TiO_2 nanowires simultaneously provide mechanical filtration and photocatalytic degradation of organic pollutants (e.g., atrazine) in UV irradiation (Hu et al. 2011). Photocatalytic degradation of atrazine with the combination of titanium nanoparticles with multiwalled carbon nanotubes (MWCNTs) is most recently reported (Chen et al. 2011). TiO_2 nanotubes were fabricated with diverse surface modifications for various functions. For example, metallic inclusions can trap photo-induced charge carriers to improve charge separation and enhance light absorption in semiconductor oxides (Gemma Aragay and Merkoci 2012). Au–Pd-modified TiO_2(Au–Pd–TiO_2) nanotube film photocatalyst was designed for the degradation of malathion and proposed a corresponding photocatalytic mechanism (Yu et al. 2010) (Fig. 2.5; Table 2.5).

Some of the organophosphorus (OP) pesticides were degraded with the help of ZnO nanoparticle (35.9 nm) in UV-radiated photodegradation system. In order to protect the external light source, UV irradiation was conducted in the dark. The UV visible absorbance spectra of OP pesticides are in the range 240–430 nm, and it was recorded up to 300 min with 10 min intervals. ZnO acts as a catalyst and enhances the rate of the reaction.

$$\mathrm{ZnO} + hn \rightarrow \mathrm{ZnO}\left(\mathrm{e}_{\mathrm{cb}}^{-} + h_{\mathrm{vb}}^{+}\right) \quad (2.6)$$

$$hn^{+} + \text{pesticides} \rightarrow \text{oxidation of pesticides} \quad (2.7)$$

$$h_{\mathrm{vb}}^{+} + \mathrm{H_2O} \rightarrow \mathrm{H}^{+} + \mathrm{OH}\bullet \quad (2.8)$$

$$h_{\mathrm{vb}}^{+}\mathrm{OH}^{-} \rightarrow \mathrm{OH}\bullet \quad (2.9)$$

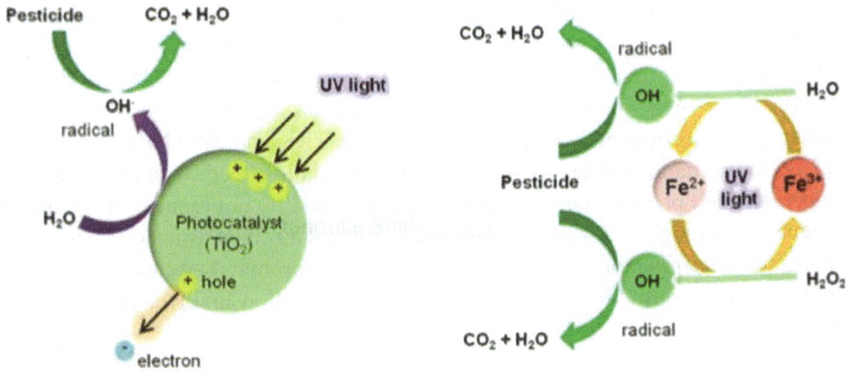

(A) Photocatalyticoxidation (B) Photo -Fentondegradation

Fig. 2.5 Schematic of pesticide degradation mechanism. **a** Photocatalytic oxidation. **b** Photo-Fenton degradation (Gemma Aragay and Merkoci 2012). *Source* Gemma Aragay and Merkoci (2012). Reproduced with permission from Copyright © 2012 Elsevier Ltd.

Table 2.5 List of nanomaterials used for degradation (Gemma Aragay and Merkoci 2012)

Pesticide	Nanomaterial	Strategy	Degradation efficiency (%)
Alachlor/fenitrothion	TiO_2 NPs	Photocatalysis	100
Atrazine	TiO_2 NWs	Photocatalysis	–
Atrazine	TiO_2 NPs-coated MWCNTs	Microwave-assisted photocatalysis	100
Methyl parathion/ 2,4-D	AuNP's-coated TiO_2NT array	Photocatalysis	100
Malathion	Au–Pd–TiO_2NT film	Photocatalysis	>96
Propachlor	TiO_2NPs/γ-Fe_2O_3	Photocatalysis	–
Pentachlorophenol	Nanoporous Ti-doped βBi_2O_3	Photocatalysis	–
2-CP/DCP/TCP	Ag-AgBr NP/MAP	Photocatalysis	71–80
4-CP	ZnO/Au NRs	Photocatalysis	91
Phenol	MNPs	Fenton-like catalysis	85
Metalaxyl	Bi-doped goethite–hematite NS	Photo-Fenton process	99
Atrazine	NZVI/CTMA-Bent	Chemical reduction	64
Lindane	Pd NPs	Catalytic dechlorination	75
Methylparathion/ Lindane/dichlorvos	Nano-TiO_2 immobilized on Pyrex glass	Photocatalysis	100

CP chlorophenol, *DCP* dichlorophenol, *TCP* 2,4,6, trichlorophenol, *NP* nanoparticle, *NW* nanowire, *NT* nanotube, *NR* nanorods, *NS* nanostructures, *MWCNTs* multiwalled carbon nanotubes
Source Gemma Aragay and Merkoci (2012). Reproduced with permission from Copyright © 2012 Elsevier Ltd.

$$e^-_{cb} + O_2 \rightarrow \bullet O^- \quad (2.10)$$

$$\bullet O_2^- + \text{pesticides} \rightarrow \text{organic peroxides} \quad (2.11)$$

$$\bullet O_2 + HO_2 + H^+ \rightarrow H_2O_2 + O_2 \quad (2.12)$$

$$OH\bullet + \text{organic peroxides} \rightarrow \text{degradation of pesticides} \quad (2.13)$$

(1) Photoexcitation of OP pesticides takes place followed by the formation of an electron–hole ($ZnO(e^-_{cb} + h^+_{vb})$) pair on the surface of the catalyst.
(2) Since ZnO is a UV-sensitive semiconductor, it permits the direct oxidation of pesticide to reactive intermediate due to high oxidative potential of the hole.
(3) According to Eq. 2.8, the formation of free radicals (i.e., hydroxyl radicals) in the photocatalytic system is ascribed.
(4) The free radicals are responsible for the degradation of OP pesticides due to their stable nonselective behavior. Electrons in the conduction band e^-_{cb} on the nanocatalyst surface can reduce molecular oxygen to superoxide anion according to Eq. 2.10.
(5) This radical, in the presence of organic scavengers, may form organic peroxides or hydrogen peroxide according to Eqs. 2.11 and 2.12.

The electrons produce hydroxyl radicals in the conduction band since they are responsible for it, which have been indicated as the primary cause of organic matter mineralization (Daneshvar et al. 2007).

Silica nanoparticles are used for the detection and degradation of pesticides. Properties of silica nanoparticles comprise sorbent properties, on account of their porosity, mono-dispersity, stability, low toxicity, and increased surface area which allow effective extraction of pesticides. They can be modified (with ionic liquids, silanes, or amines), coated with molecularly imprinted polymers or magnetized to improve the extraction of pesticide process. For detoxification of pesticides, enzymes like organophosphate hydrolases, carboxylesterases, laccases, and bacteria expressing recombinant proteins have been immobilized on silica nanostructures (Bapat et al. 2016).

Nanotechnology is essential in the context of environmental remediation (Warner et al. 2010). Noble metal nanoparticles cause dehalogenation of several halocarbons, and the chemistry and application can be extended to halocarbon pesticides also. Silver and gold nanoparticles are used to degrade chlorpyrifos (CP), belonging to the organophosphorothioate group. This pesticides decomposed to 3,5,6-trichloro-2-pyridinol (TCP) and diethyl thiophosphate at room temperature. Absorption spectroscopy and electrospray ionization mass spectrometry (ESI–MS) were used to characterize the degraded products. ESI tandem mass spectrometry was used for further confirmation. Transmission electron microscopy, energy-dispersive analysis of X-rays, Raman spectroscopy, and X-ray photoelectron spectroscopy (XPS) are used to determine the interaction of CP with nanoparticle surfaces. There is no change in the oxidation state of silver after the

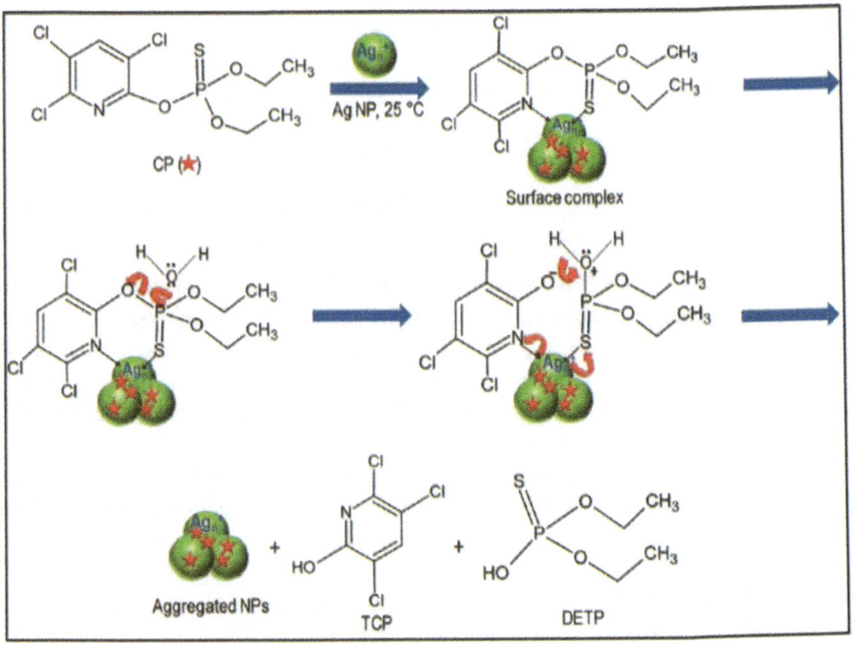

Fig. 2.6 Mechanism of decomposition of CP on NPs (Bootharaju and Pradeep 2012). *Adsorbed CP molecules are shown in red stars. *Source* Bootharaju and Pradeep (2012). Reproduced with permission from Copyright © 2012 Elsevier Ltd.

degradation of CP, and XPS reveals it. Raman spectroscopy confirms that the degradation of CP proceeds through the formation of AgNP–S surface complex. In this complex, a stable aromatic species TCP is formed by the breaking up of P–O bond, and with an increase of temperature and pH, the rate of degradation of CP increases. On the laboratory scale, complete degradation of 10 mL of 2 ppm CP solution is achieved in 3 h using 100 mg of supported (on neutral alumina) Ag@citrate NPs on neutral alumina at room temperature at a loading of ~0.5 wt%. The rate of degradation of CP by Au NPs is lesser than that of Ag NPs. The preparation of Ag and Au nanoparticle-supported alumina is easy, and degradation of CP in aqueous solutions occurs at room temperature by merely stirring without any additional irradiation of light. The application of noble metal nanoparticles in water purification applications is sustainable. It is proved that unsupported Ag and Au@citrate NPs show aggregation after adsorption of CP and it leads to the red shift of surface plasmon of NPs. Monolayered, unprotected, and protected Ag NPs are also found to show the degradation of CP. In these kinds of applications, there must be a post-degradation step to remove the mineralized products by adsorbents such as activated carbon (Bootharaju and Pradeep Bootharaju and Pradeep 2012) (Fig. 2.6; Table 2.6).

Table 2.6 Comparative studies on degradation efficiency of pesticides by various methods

Material	Pesticides	Removal methods	Efficiency (%)	References
NF-270 membrane	Malathion	Photo-Fenton process	73	Zhang and Pagilla (2010)
Fe_3O_4-@mTiO_2	Monocrotophos	Photodegradation	100	Lingling Zheng et al. (2016)
Fe_3O_4/rGO nanocomposite	Ametryn	Adsorption	93.61	Boruah et al. (2017)
CM-MgO	Fenitrothion	Adsorption	47	Armaghan and Amini (2017)
CMC-CTA	Chloroacetic acid	Flocculation	96.5	Agbovi and Wilson (2019)
CdS/BiOBr/ $Bi_2O_2CO_3$	Atrazine	Photodegradation	>95	Majhi et al. (2020)
Bi_2MoO_6/Ti_3C_2	Tetracycline hydrochloride	Photodegradation	>99	Zhao and Cai (2020)

2.9 Conclusion

The use of 2D-based nanomaterials as photocatalysts has been increased in the area of novel sustained composite formulations. These nanocatalysts are having a unique nature of forming composites. Nanocomposites have gained significant attention because of ecological issues and the decline in petroleum-based resources. Various 2D nanomaterials based on biological materials, including cellulose, chitin and chitosan, amino acids, etc., were often used to make materials and devices for industries. From this new knowledge, research has moved forward, and many attempts are made for functionalizing the nanomaterials specifically for other biomedical applications, wastewater treatments, pharmacological field, etc. The nanocomposite possesses significant flexibility advantages, compatibility, and also tunable degradability, which often make them useful for a wide range of industrial applications. This paper critically explored the various types of imprinted TiO_2 composites and their catalytic behaviors toward adsorption as well as degradation on different pollutant molecules. The exceptional physical and chemical properties of these polymers encouraged the researcher's to develop and utilization of these materials in the purification of water technologies in industrial scales.

References

Affam AC, Chaudhuri M (2013) Degradation of pesticides chlorpyrifos, cypermethrin and chlorothalonil in aqueous solution by TiO_2 photocatalysis. J Environ Manage 130:160–165

Agbovi HK, Wilson LD (2019) Optimisation of orthophosphate and turbidity removal using an amphoteric chitosan based flocculant–ferric chloride coagulant system. Environ Chem 16

Ahmed S, Rasul MG, Brown R, Hashib MA (2011) Influence of parameters on the heterogeneous photocatalytic degradation of pesticides and phenolic contaminants in wastewater: a short review. J Environ Manage 92:311–330

Alavanja MC, Samanic C, Dosemeci M, Lubin J, Tarone R, Lynch CF, Knott C, Thomas K, Hoppin JA, Barker J, Coble J, Sandler DP, Blair A (2003) Use of agricultural pesticides and prostate cancer risk in the Agricultural Health Study cohort. Am J Epidemiol 157:800–814

Amy PS, Staudaher MV, Seidler RJ (1990) Comparison of a Gene Probe with Classical Methods for detecting 2,4-dichlorophenoxyacetic acid (2,4-d)-biodegrading bacteria in natural waters. Curr Microbiol 21:95–101

Andreozzi R, Caprio V, Insola A, Marotta R (1999) Advanced oxidation processes (AOP) for water purification and recovery. Catal Today 53:51–59

Armaghan M, Amini MM (2017) Adsorption of diazinon and fenitrothion on nanocrystalline magnesium oxides. Arab J Chem 10:91–99

Baker EL, Warren M, Zack M, Dobbin RD, Miles JW, Miller W, Alderman S, Teeters WR (1978) Epidemic malathion poisoning in Pakistan malaria workers. Lancet 1:31–34

Bapat G, Labade C, Chaudhari A, Zinjarde S (2016) Silica nanoparticle based techniques for extraction, detection, and degradation of pesticides. Adv Colloid Interface Sci 237:1–14

Bertolote M, Fleischmann A, Eddleston M, Gunnell D (2006) Deaths from pesticide poisoning: a global response. Br J Psychiatry 189:201–203

Blesa J, Trigo-Damas I, Quiroga-Varela A, Jackson-Lewis VR (2015) Oxidative stress and Parkinson's disease. Front Neuroanat 9:91

Bootharaju MS, Pradeep T (2012) Understanding the degradation pathway of the pesticide, chlorpyrifos by noble metal nanoparticles. Langmuir 28:2671–2679

Boruah PK, Sharma B, Hussain N, Das MR (2017) Magnetically recoverable Fe_3O_4/graphene nanocomposite towards efficient removal of triazine pesticides from aqueous solution: Investigation of the adsorption phenomenon and specific ion effect. Chemosphere 168:1058–1067

Brajesh Kumar Singh RCK, Singh Ajay, Lal Rup, Tripathi KK (1999) Biochemical and molecular basis of pesticide degradation by microorganisms. Crit Rev Biotechnol 19(3):197–225

Brouwer M, Huss A, van der Mark M, Nijssen PCG, Mulleners WM, Sas AMG, van Laar T, de Snoo GR, Kromhout H, Vermeulen RCH (2017) Environmental exposure to pesticides and the risk of Parkinson's disease in the Netherlands. Environ Int 107:100–110

Chen H, Yang S, Yu K, Ju Y, Sun C (2011) Effective photocatalytic degradation of atrazine over titania-coated carbon nanotubes (CNTs) coupled with microwave energy. J Phys Chem A 115:3034–3041

Christine Picard CPEP, Nesme Xavier (1992) Detection and enumeration of bacteria in soil by direct DNA extraction and polymerase chain reaction. Appl Environ Microbiol 58:2717–2722

Claire Infante-Rivard DL, Krajinovic Maja, Sinnett Daniel (1999) Risk of childhood leukemiaassociated with exposure to pesticides and with gene polymorphisms. Epidemiology 10 (5):481–487

Daneshvar N, Aber S, Seyeddorraji M, Khataee A, Rasoulifard M (2007) Photocatalytic degradation of the insecticide diazinon in the presence of prepared nanocrystalline ZnO powders under irradiation of UV-C light. Sep Purif Technol 58:91–98

Deo PG, Ganesh Karanth N, Gopalakrishna N, Karanth K (1994) Biodegradation of hexachlorocyclohexane isomers in soil and food environment. Crit Rev Microbiol 20(1):57–78

Echavia GR, Matzusawa F, Negishi N (2009) Photocatalytic degradation of organophosphate and phosphonoglycine pesticides using TiO_2 immobilized on silica gel. Chemosphere 76:595–600

Elbetieha A, Da'as SI, Khamas W, Darmani H (2001) Evaluation of the toxic potentials of cypermethrin pesticide on some reproductive and fertility parameters in the male rats. Arch Environ Contam Toxicol 41:522–528

Fareed M, Pathak MK, Bihari V, Kamal R, Srivastava AK, Kesavachandran CN (2013) Adverse respiratory health and hematological alterations among agricultural workers occupationally exposed to organophosphate pesticides: a cross-sectional study in North India. PLoS One 8:

Fear NT, Roman R, Reeves G, Pannett B (1998) Childhood cancer and paternal employment in agriculture: the role of pesticides. Br J Cancer 77(5):825–829

Felsot AS, Racke KD, Hamilton DJ (2003) Disposal and degradation of pesticide waste. Rev Environ Contam Toxicol 177:123–200

Frazier LM (2007) Reproductive disorders associated with pesticide exposure. J Agromedicine 12:27–37

Freire C, Koifman RJ, Sarcinelli PN, Simoes Rosa AC, Clapauch R, Koifman S (2013) Long-term exposure to organochlorine pesticides and thyroid status in adults in a heavily contaminated area in Brazil. Environ Res 127:7–15

Garcia J, Ventura MI, Requena M, Hernandez AF, Parron T, Alarcon R (2017) Association of reproductive disorders and male congenital anomalies with environmental exposure to endocrine active pesticides. Reprod Toxicol 71:95–100

Gemma Aragay FP, Merkoci Arben (2012) Nanomaterials for sensing and destroying pesticides. Chem Rev 112:5317–5338

Guengerich FP (1998) Activation of procarcinogens by human cytochrome P450 enzymes. Mutat Res 400:201–213

Gunnell D, Eddleston M (2003) Suicide by intentional ingestion of pesticides: a continuing tragedy in developing countries. Int J Epidemiol 32:902–909

Gupta SM, Tripathi M (2011) A review of TiO_2 nanoparticles. Chin Sci Bull 56:1639–1657

Han S, Li J, Xi H, Xu D, Zuo Y, Zhang J (2009) Photocatalytic decomposition of acephate in irradiated TiO_2 suspensions. J Hazard Mater 163:1165–1172

Hoppin JA, Umbach DM, London SJ, Lynch CF, Alavanja MC, Sandler DP (2006) Pesticides and adult respiratory outcomes in the agricultural health study. Ann N Y Acad Sci 1076:343–354

Hu A, Zhang X, Oakes KD, Peng P, Zhou YN, Servos MR (2011) Hydrothermal growth of free standing TiO_2 nanowire membranes for photocatalytic degradation of pharmaceuticals. J Hazard Mater 189:278–285

Ikehata K, Gamal El-Din M (2005) Aqueous pesticide degradation by ozonation and ozone-based advanced oxidation processes: a review (Part I). Ozone: Sci Eng 27:83–114

Jaga K, Rama DB (1992) Pesticide exposure and cholinesterase levels among farm workers in the Republic of South Africa. Sci Total Environ 122:315–319

Jan Dich SHZ, Hanberg Annika, Adami Hans-Olov (1997) Pesticides and cancer. Cancer Causes Control 8:420–443

Jean Marie Herrmann CG, Arguello M, Agüera Ana, Tejedor Ana, Piedra Luis, Fernández-Alba Amadeo (1999) Photocatalytic degradation of pesticide pirimiphos-methyl Determination of the reaction pathway and identification of intermediate products by various analytical methods. Catal Today 54:353–367

Jeyaratnam J, de Alwis Seneviratne RS, Copplestone JF (1982): Survey of pesticide poisoning in Sri Lanka. Bull World Health Organ 60(4):615–619

Juttner K, Galla U, Schmieder H (2000) Electrochemical approaches to environmental problems in the process industry. Electrochim Acta 45:2575–2594

Kumar A, Vashist M, Rathee R (2014) Maternal factors and risk of childhood leukemia. Asian Pac J Cancer Prev 15:781–784

Lingling Zheng HX, Pi Fuwei, Zhang Yinzhi, Sun Xiulan (2016) Synthesis of Fe_3O_4@mTiO_2 nanocomposites for the photocatalytic degradation of monocrotophos under UV illumination. RSC Adv 6:87273

Liu Y, Su G, Zhang B, Jiang G, Yan B (2011) Nanoparticle-based strategies for detection and remediation of environmental pollutants. Analyst 136:872–877

Liu Y, Lonappan L, Brar SK, Yang S (2018) Impact of biochar amendment in agricultural soils on the sorption, desorption, and degradation of pesticides: a review. Sci Total Environ 645:60–70

Luo D, Zhou T, Tao Y, Feng Y, Shen X, Mei S (2016) Exposure to organochlorine pesticides and non-Hodgkin lymphoma: a meta-analysis of observational studies. Sci Rep 6:25768

Majhi D, Das K, Mishra A, Dhiman R, Mishra BG (2020): One pot synthesis of CdS/BiOBr/$Bi_2O_2CO_3$: A novel ternary double Z-scheme heterostructure photocatalyst for efficient degradation of atrazine. Appl Catal B: Environmental 260:118222

Malpass GR, Miwa DW, Machado SA, Olivi P, Motheo AJ (2006) Oxidation of the pesticide atrazine at DSA electrodes. J Hazard Mater 137:565–572

Massimo Musicco MS, Molinaki Silvia, Fllippini Graziella, Gatta Gemma, Berrino Franco (1988) A case-control study of brain glimos and occupational exposure to chemical carcinogens: the risk to farmers. Am J Epidemiol 128:778–785

Mazur CS, Marchitti SA, Zastre J (2015) P-glycoprotein inhibition by the agricultural pesticide propiconazole and its hydroxylated metabolites: implications for pesticide-drug interactions. Toxicol Lett 232:37–45

Miwa DW, Malpass GR, Machado SA, Motheo AJ (2006) Electrochemical degradation of carbaryl on oxide electrodes. Water Res 40:3281–3289

Morillo E, Villaverde J (2017) Advanced technologies for the remediation of pesticide-contaminated soils. Sci Total Environ 586:576–597

Nandipati S, Litvan I (2016) Environmental exposures and Parkinson's disease. Int J Environ Res Public Health 13

Norman C (1974) EPA halts dieldrin production. Nature 250:528

Ormad MP, Miguel N, Claver A, Matesanz JM, Ovelleiro JL (2008) Pesticides removal in the process of drinking water production. Chemosphere 71:97–106

Parron T, Requena M, Hernandez AF, Alarcon R (2011) Association between environmental exposure to pesticides and neurodegenerative diseases. Toxicol Appl Pharmacol 256:379–385

Petersen RC (2011) Mild cognitive impairment. N Engl J Med 364:2227–2234

Pinhedo L, Pelegrini R, Bertazzoli R, Motheo AJ (2005) Photoelectrochemical degradation of humic acid on a $(TiO_2)_{0.7}(RuO_2)_{0.3}$ dimensionally stable anode. Appl Catal B 57:75–81

Polanco Rodriguez AG, Riba Lopez MI, DelValls Casillas TA, Araujo Leon JA, Mahjoub O, Prusty AK (2017) Monitoring of organochlorine pesticides in blood of women with uterine cervix cancer. Environ Pollut 220:853–862

Rios ACC, Salvadori DMFS, Oliveira SV, Ribeiro LR (1995) The action of the herbicide paraquat on somatic and germ cells of mice. Mutat Res 328:113–118

Sabarwal A, Kumar K, Singh RP (2018) Hazardous effects of chemical pesticides on human health-Cancer and other associated disorders. Environ Toxicol Pharmacol 63:103–114

Sarbani Giri AG, Sharma Gouri Dutt, Prasad Surya Bali (2002) Mutagenic effects of carbosulfan, a carbamate pesticide. Mutat Res 519:75–82

Schinasi L, Leon ME (2014) Non-Hodgkin lymphoma and occupational exposure to agricultural pesticide chemical groups and active ingredients: a systematic review and meta-analysis. Int J Environ Res Public Health 11:4449–4527

Sharma AK, Tiwari RK, Gaur MS (2016) Nanophotocatalytic UV degradation system for organophosphorus pesticides in water samples and analysis by Kubista model. Arab J Chem 9: S1755–S1764

Steffan RJ, Breen A, Atlas RM, Sayler GS (1989) Application of gene probe methods for monitoring specific microbial populations in freshwater ecosystems. Can J Microbiol 35:681–685

Swanimathan MS (1982) Politics of pesticides. Nature 296:521–522

Thongprakaisang S, Thiantanawat A, Rangkadilok N, Suriyo T, Satayavivad J (2013) Glyphosate induces human breast cancer cells growth via estrogen receptors. Food Chem Toxicol 59:129–136

Vlyssides A, Arapoglou D, Mai S, Barampouti EM (2005) Electrochemical detoxification of four phosphorothioate obsolete pesticides stocks. Chemosphere 58:439–447

von Sonntag C (2008) Advanced oxidation processes: mechanistic aspects. Water Sci Technol 58:1015–1021

Walia S, Khan A, Rosenthal N (1990) Construction and applications of DNA probes for detection of polychlorinated biphenyl-degrading genotypes in toxic organiccontaminated soil environments. Appl Environ Microbiol 56:254–259

Wang A, Cockburn M, Ly TT, Bronstein JM, Ritz B (2014) The association between ambient exposure to organophosphates and Parkinson's disease risk. Occup Environ Med 71:275–281

Warner CL, Addleman RS, Cinson AD, Droubay TC, Engelhard MH, Nash MA, Yantasee W, Warner MG (2010) High-performance, superparamagnetic, nanoparticle-based heavy metal sorbents for removal of contaminants from natural waters. Chemsuschem 3:749–757

Wiklund K, Dich J, Holm LE (1987) Risk of malignant lymphoma in Swedish pesticide appliers. Br J Cancer 56:505–508

Xu X, Dailey AB, Talbott EO, Ilacqua VA, Kearney G, Asal NR (2010) Associations of serum concentrations of organochlorine pesticides with breast cancer and prostate cancer in U.S. adults. Environ Health Perspect 118:60–66

Xu Y, Shi X, Hua R, Zhang R, Yao Y, Zhao B, Liu T, Zheng J, Lu G (2020) Remarkably catalytic activity in reduction of 4-nitrophenol and methylene blue by Fe_3O_4@COF supported noble metal nanoparticles. Appl Catal B: Environ 260

Yu H, Wang X, Sun H, Huo M (2010) Photocatalytic degradation of malathion in aqueous solution using an Au-Pd-TiO_2 nanotube film. J Hazard Mater 184:753–758

Zaganas I, Kapetanaki S, Mastorodemos V, Kanavouras K, Colosio C, Wilks MF, Tsatsakis AM (2013) Linking pesticide exposure and dementia: what is the evidence? Toxicology 307:3–11

Zhang Y, Pagilla K (2010) Treatment of malathion pesticide wastewater with nanofiltration and photo-Fenton oxidation. Desalination 263:36–44

Zhao D, Cai C (2020) Preparation of Bi_2MoO_6/Ti_3C_2 MXene heterojunction photocatalysts for fast tetracycline degradation and Cr(vi) reduction. Inorg Chem Frontiers 7(15)

Zuskin E, Mustajbegovic J, Schachter EN, Kern J, Deckovic-Vukres V, Trosic I, Chiarelli A (2008) Respiratory function in pesticide workers. J Occup Environ Med 50:1299–1305

Chapter 3
Uranium(VI) Ions Removal from the Aqueous Environment: Current Treatment Technologies and Its Future Prospects

Perumal Karthikeyan, Palliyalil Sirajudheen, Vigneshwaran Sivakumar, and Sankaran Meenakshi

Abstract Pollution causes a significant change in the physico-chemical properties of air, water, and soil that may induce harmful impact and make potential hazards to all living beings. Pollution is one of the severe problems observed throughout the world. Water pollution has ascended due to the nearness of toxic environment, which arisen by human activities such as increased agricultural, industrial, and urban area developments. According to the thriving of the human population, the need for energy is increasing for the development of the world and its economy. Coal, oil, and other nonrenewable fossil are the prime sources of energy which have regularly consumed. Now, the world is facing an unexpected energy emergency. Due to the utilization of radioactive nuclides with the progress in human civilization, the atmosphere is facing cumulative radioactive pollution. Natural and anthropogenic radionuclides and its effect on humans, animals, and all other living beings have become a major environmental problem because of their prevalent occurrence in the atmosphere with concentrations that exceed the World Health Organization (WHO) recommended maximum levels. Operative, suitable, effective, and environmental friendly techniques are essential to mitigate this problem. This chapter describes the development of different conventional techniques and their removal performance of uranium(VI) ions from aqueous solution. Moreover, the chapter also investigates the various adsorbent materials used for the removal of radioactive uranium(VI) ions from water.

Keywords Uranium(VI) · Pollution · Adsorption · Adsorbent · Prospects

---

P. Karthikeyan · P. Sirajudheen · V. Sivakumar · S. Meenakshi (✉)
Department of Chemistry, The Gandhigram Rural Institute (deemed to be University), Gandhigram, Dindigul 624302, Tamil Nadu, India

S. Singh et al. (eds.), *Pollution Control Technologies*, Energy, Environment, and Sustainability, https://doi.org/10.1007/978-981-16-0858-2_3

3.1 Introduction

Population explosion and industrial developments increased the energy requirement of the world. As the out-dated oil, coal, and other nonrenewable fossil fuels have frequently utilized, the biosphere is facing an unpredicted energy emergency (Bleise et al. 2003). Expansion and utilization of nuclear power is one of the solutions for this problematic task. However, processing and mining of radionuclides resources will cause radioactive pollutions. In recent years, radionuclides have been used frequently in nuclear fuels, nuclear reactors, and nuclear weapons. Besides, uranium is a naturally existing primal and omnipresent heavy metal ions, originate in numerous forms in the atmosphere. From its discovery in 1789, it has been an interesting material because of its favorable claims in producing nuclear energy. In nature, uranium exists as ^{238}U (99.27–99.28%), ^{235}U (0.71–0.72%), and in trace amount as ^{234}U (0.005–0.005%). Uranium exists in the atmosphere by discharge from uranium products, minings, natural cycles, and radiations from the nuclear manufacturing industries and usage of uranium-containing fertilizers (Krestou et al. 2003; Meinrath et al. 2003; Mitsakou et al. 2003). The harmful form of uranium, like ^{235}U and ^{238}U, has a half-life of 7×10^8 and 4.4×10^9 years, respectively, and they are spread in the atmosphere and eventually stores in the ecology system. Henry Becquerel revealed the radioactive behavior of uranium in 1896, which reduced it as the supreme energetic component for nuclear power stations, serving as an important and viable energy source for power generation (Andrade Martins 1997). About 14% of the global electricity is produced from uranium through nuclear power stations (Key World Energy Statistics 2009; World Nuclear Association 2013; Bhalara et al. 2014) (Fig. 3.1). South Korea, Bulgaria, France, Hungary, Belgium, Czech Republic, Slovakia, Finland, Lithuania, Sweden, Slovenia, Switzerland, and Ukraine obtain >30% of their electricity from nuclear sources. Japan and Germany develop more than a quarter of their electricity from uranium ions. The United States of America has 100 reactors, producing 20% of electric power. Nations like Australia, Kazakhstan, and Canada account for nearly 63.5% of the world's uranium manufacture (Bhalara et al. 2014).

Uranium can exist as mostly two different oxidation states, which are U(IV) and U(VI), and their two parallel forms are uranium trioxide (UO_3) and uranium dioxide (UO_2) (Seaborg 1968). The noxiousness of uranium has thoroughly linked with its high solubility (Craig 2001). U(VI), has high water solubility in the environmental contexts as compared to U(IV) (Coetze et al. 2006). Considering the case of radioactive pollution of soil and water, the poisonousness of elemental uranium produces severe health problems (Rollett 2008). Uranium having 1–10 mm in average size are unsolvable and when they reached in the lungs, may cause radiation injuries and over a tenacious phase, prominent to lung cancer. Uranium, due to its enhanced coordination efficiency toward biomolecules, acts as a potential

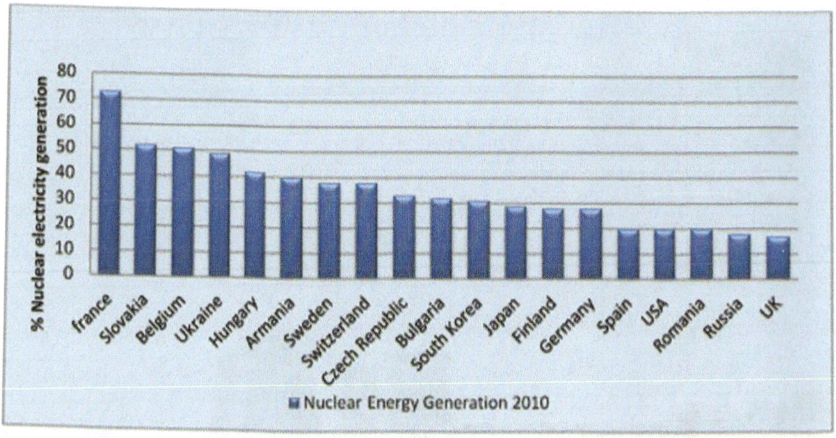

Fig. 3.1 Nuclear energy generation-2010 around the globe. *Source* Bhalara et al. (2014). Reproduced with permission from Copyright © 2018 Elsevier Ltd.

mutagen and carcinogen. As per the World Health Organization (WHO), the maximum tolerable concentration of U(VI) ions in water is 50 mg/L. The United States Environmental Protection Agency (USEPA) has set a maximum contaminant level for uranium of 30 μg/L (WHO 1998, 2003).

Environmental adulteration affected by uranium ions is being an important problem in the world, which has advanced in route with the expansion of nuclear technology over the last few decades. The average uranium concentration in earth crust is about 2.8 mg/kg (World-Nuclear.org 2013; McGraw-Hill 2004). Launching of nuclear fuels, nuclear reactors and nuclear weapon results in the contamination of radionuclides especially in water bodies and adjoining soil surface (Fig. 3.2).

Natural and anthropogenic radionuclides are major contaminants of water sources. The occurrence of these radionuclides, particularly uranium, and its decay products in the environment often threats the environment stability and possess a severe hazard to human and all other living beings, and they can easily be integrated into the food chain and ingested by human beings. Due to its long half-life, high biological activity and capability to be reused as an energy resource, need some significant attention in eliminating and recovering these pollutants from the environment (Bhalara et al. 2014). Several methods are adopted for the removal of radioactive pollutants such as physical, chemical, and biological strategies. Each strategy has both disadvantages and advantages and the suitable restoration methods would be based on the environments. This chapter covers the information about the development of adsorbents and their properties for the remediation of radionuclides in water. It gives an insight into the principal adsorption characteristics and efficiency of many adsorbents based on their physico-chemical modification concerning uranium ions from water.

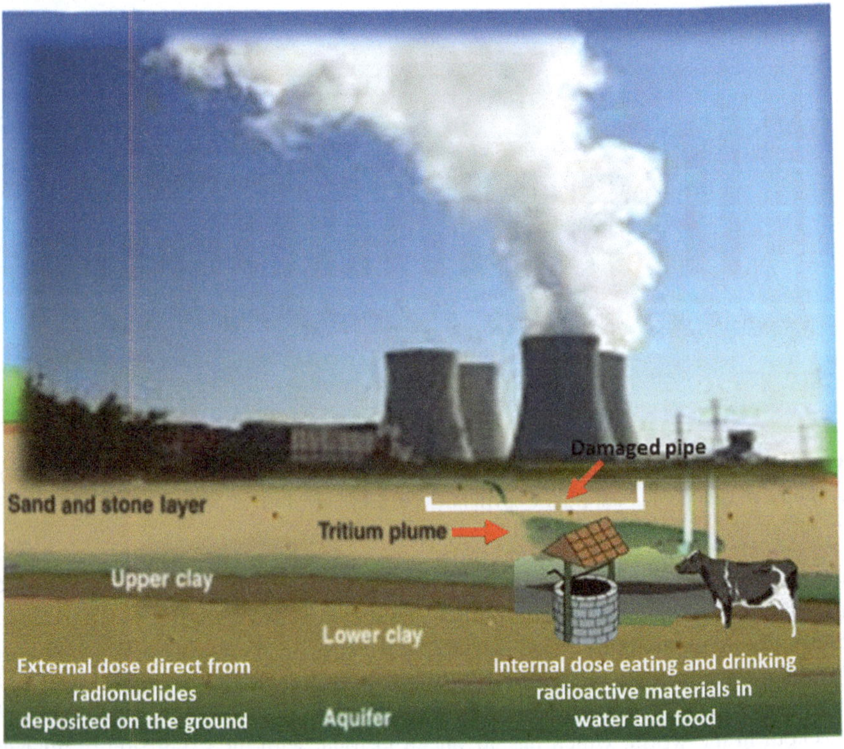

Fig. 3.2 Outline of radioactive pollution

3.2 Radioactive Pollution

Radioactive pollution induces severe adulteration in the environment, which will adversely affect community health. There are two types of radionuclides exist in the environment, which are natural and anthropogenic. Agreeing to the endorsement of the International Commission on Radiological Protection (ICRP 2007), there having some methodologies to regulate the radioactivity in the atmosphere for both artificial and natural radionuclides. The natural radionuclides existing in the atmosphere as long-lived primeval radionuclides (^{40}K, ^{87}Rb, ^{238}U, ^{235}U, ^{232}Th and some others) and cosmogenous radionuclides (^{3}H, ^{14}C, ^{7}Be). However, thorium, uranium, and their offspring nuclides (^{226}Ra, ^{228}Ra, ^{210}Po, and ^{210}Pb) contribute more than half of yearly internal dose of human and all other living beings. The radionuclides present in the environment contain fusion-fission yields (^{129}, ^{131}I, ^{90}Sr–^{90}Y, ^{134}, ^{137}Cs, ^{106}Ru, ^{99}Tc), transuranium elements (long-lived elements of Pu, Am, Np, and in a less grade heavier metals) and initiation products (^{3}H, ^{14}C, ^{60}Co, ^{55}Fe, ^{56}Mn, etc.) (Gupta and Walther 2014; Lauria and Godoy 2002; Povinec et al. 2004). The Chernobyl disaster occurred in 1986 is the most affected nuclear accident in the world and resulted in water pollution of seas, rivers, and groundwater. The atmospheric release of xenon-133 is another disaster, which has a $t_{1/2}$ of five days, projected at 5200 PBq 50–60% of all radioiodine in the container, about 1760 PBq (1760 × 1015 becquerels), or about 0.4 kilograms (0.88 lb), has been discharged as a mixture of sublimed steam, solid, and organic compounds. Iodine-131 has a $t_{1/2}$ of eight days, 20–40% of all core cesium-137 are released, 85 PBq in all, cesium has released in aerosol form; cesium-137, along with isotopes of strontium, is the two primary elements preventing the Chernobyl exclusion zone being reinhabited. 8.5 × 1016 Bq equals 24 kilograms of cesium-137. Cs-137 has a half-life of 30 years. Tellurium-132, having a half-life of 78 h, an estimated 1150 PBq was released (Marples David 1988; Mould 2000). Another nuclear accident has taken place at the Fukushima Daiichi Nuclear Power Plant in Ōkuma, Fukushima Prefecture. This disaster was the most severe nuclear accident since the 26 April 1986 Chernobyl disaster and the only other disaster that had occurred at the Level 7 of event classification of the International Nuclear Event Scale. As of October 2019, 1.17 million cubic meters of polluted water deposited in the plant area. The water had been treated by the decontamination process to eliminate contaminants, except tritium, near a level that Japanese strategies allow to discharging into the sea (Amano 2015; Aliyu et al. 2015). Besides, radioactive pollutants may be removed from water bodies through some natural and artificial strategies such as adsorption, co-precipitation, evaporation, membrane technique, etc. Briefly, the radionuclides existing in water and maybe expended by humans affecting the contact of interior irradiation quantities. Thus, the removal of radioactive pollutants from water/wastewater and hindering their movement through the food cycles is a critical job for radiation safety and public health.

3.2.1 Classification of Techniques

The objective of this radioactive pollution control treatments is to minimize the capacity and proper management of radionuclide waste. Radioactive pollutants removal depends on its quantity, physico-chemical and radiological properties. This unit aims to convey a brief outline of frequently used techniques for the removal of radionuclides from water. Despite numerous methods employed for the remediation of uranium ions from water, it is essential to quote the choice of the most appropriate methods depending on the initial concentration, composition of wastewater, operational cost, reliability, and environmental influence. Figure 3.3 summarizes various techniques which have been employed for the removal of (U(VI) ions from aqueous environment.

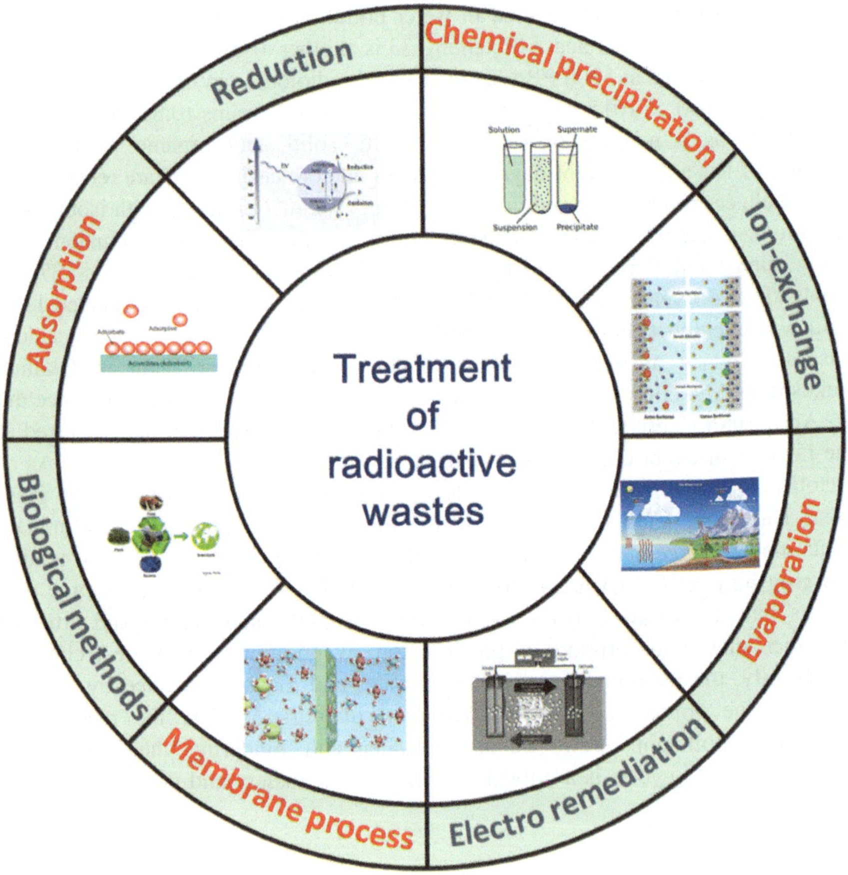

Fig. 3.3 Different strategies involved for the removal of U(VI) ions from aqueous environment

3.2.1.1 Adsorption

The significance of the adsorption process is based on its higher reliability with effectiveness, regeneration capacity, convenience, easy operation, cost-effectiveness, and low secondary products. In the adsorption process, the solution phase of adsorbate containing a dissolved species gets adhered on the solid adsorbent phase. The process of adsorption has influenced by some mechanistic approaches involving physico-chemical interactions on the surface such as pores, complexation, ion exchange, chelation, and setup in capillaries.

Figure 3.4 summarizes different adsorbent materials which have been used so far for the removal of U(VI) ions. Agricultural waste materials have reflected to be a suitable choice for U(VI) adsorption due to their commercial, eco-friendly, and unique properties along with renewable nature, low-cost abundance, and more efficacy (Bailey et al. 1999; Orhan and Bujukgungor 1993). Besides, the reactive groups existing in the molecules such as structural polysaccharides, phenolic, amino, carboxyl, carbonyl, esters, alcohols, facilitate the retrieval of U(VI) ions from water/wastewater. Bioadsorption has been considered as an innovative strategy for remediation of U(VI) ions from water because of its obvious advantages like high aptness, selectivity for absorbing radionuclides at low concentrations with wide range of solution pHs (Hashem et al. 2005a, b; Gupta and Ali 2000). Rice husk and its hydrochloric acid, sodium carbonate, sodium hydroxide, tartaric acid, and epichlorohydrin treated modifications have commonly used as bioadsorbent. The maximum adsorption efficiency has achieved when using the tartaric acid-modified rice husk. In strongly acidic conditions, carbon-based adsorbents have the advantage of higher mechanochemical resistance than organic resins and superior physico-chemical behavior than inorganic materials in treating nuclear wastewater. Nowadays, the carbon constituents such as carbon nanotubes, activated carbon, mesoporous carbon, and graphene compounds have been used for metal separation. Activated carbon is one of the most often used materials as an adsorbent in worldwide due to its most operative and cost-effective nature. Activated carbon has fabricated by carbonization at high temperatures. The extreme removal percentage of U(VI) >98% was detected at the pH of 3.0 with a concentration of 100 mg/L and the amount of adsorbent used as 100 mg. The pseudo-second-order kinetics defines the best-suited model. Maximum adsorption efficiency of activated carbon was found to be 28.30 mg/g (Mellah et al. 2006). Coconut shells are initially preserved by 20% phosphoric acid to enhance the surface and texture property. The carbonization is achieved by heating at 500 °C for 3 h with a rate of 10 °C per min. The carbon shows high adsorption efficiency for Eu^{3+}, Ce^{3+}, but lower adsorption efficiency for Sr^{2+} and Cs^{+} in the aqueous medium (Moloukhia et al. 2016). Ershov et al. (1993a, b) studied the physico-chemical performance of sorption-active materials, and the authors explained the constituent of the ion-exchange process. They examined the adsorption of cesium, thorium, uranium, strontium, americium, technetium, and plutonium from several aqueous forms. Significantly, much research interest motivated on the utilization of iron-based adsorbents for the adsorption behavior of radionuclides from water (Noubactep et al. 2006). However, understanding the remediation mechanism of U(VI) ions onto nanoscale zero-valent iron (nZVI) is vital to progress new strategies for polluted water. Till date, various types of nZVI were employed to eliminate U(VI) ions from water, with scrap iron achieved from a diverse source of materials like reactive iron foil, Fe-based sludge, as well as prepared nZVI, etc. The concentration of U in a range of 50-2500 mg/L, the effective dose of nZVI was found to be less than 0.5 g/L with elimination efficiency from 85 to 99%, respectively, and the adsorption process was completed within 24 h (Crane and Scott 2014; Dickinson and Scott 2010; Li et al. 2013). The removal efficiency of U by nZVI series from 0.43 to 8170 mg U/g with identical experimental environments (Klimkova et al. 2011; Riba et al. 2008). The nZVI fabricated at laboratory conditions was more effective and selectively removed the U than commercially available ZVI. Chen et al. examined the effectiveness of synthetic hydroxyapatite for the removal U from wastewater and also calculated the influencing factors of hydroxyapatite for the removal of U(VI) ions (Chen et al. 2009). The results showed that 95% of U(VI) ions could be removed by hydroxyapatite from aqueous medium (Chen et al. 2009). Philiips et al. examined the removal of U from artificial groundwater using Purolite A-520E and Dowex1-x8 anion exchange resin which was more operative than metal-binding resins (Philiips et al. 2008). Chitosan, a biopolymer composed of glucosamine derived from chitin by the deacetylation process, employed as an adsorbent for U adsorption in strongly alkaline medium. The removal efficiency of U(VI) ions onto cross-linked chitosan increases with rising contact time and attains saturation within 120 min (Wang et al. 2009). A resin was fabricated by poly(amido)amine (PAMAMG3) dendron on the surface of styrene-divinylbenzene (SDB) (Ilaiyaraja et al. 2013) acted as a potential adsorbent. It was found that U(VI) ions adsorption on the resin follows PSO kinetic and saturation parameters well associated with the Langmuir isotherm. Besides, PAMAM possesses a high content of nitrogen and oxygens of amide and amino-functional moieties on the surface, which can show strong binding behavior toward U(VI) ions from aqueous solution. Talc is a layered material with the formula of $Mg_3Si_4O_{10}(OH)_2$. The Langmuir isotherm was established to be the superior fit for the removal of U(VI) ions by Talc (Charnay et al. 2001; Meldon et al. 2011) 58.3 m^2/g of talc surface area was engaged by an extreme amount of 41.6 mg/g uranium during adsorption. $Ti_3C_2(OH)_2$ was used as another material for the removal of U(VI) and it shows a maximum removal efficiency of 595.3 mg/g (Zhang et al. 2016). The adsorption of U(VI) ions from water using two-dimensional (2D) multilayered V_2CT_X material showed a maximum adsorption capability of 174 mg/g, which was higher than the adsorption efficiency of many other adsorbents (Wang et al. 2016). Also, the existence of multilayered active groups such as –O, –F, and –OH in MXene were accountable for U(VI) adsorption. Wang et al. (2017) (Wang et al. 2017) reported a novel method to extend the interlayer space of $Ti_3C_2T_x$, complexed to the fabrication of pristine and intercalated MXenes in identical conditions. Layered double hydroxides (LDHs) are highly tunable brucite-like adsorbents, in which a segment of divalent ions are exchanged by trivalent ions, creating positively charged species that are recompensed by multivalent anions intercalated into the interlayer distance. Kulyukhin's et al. reported uranium enrichment studies for U(VI) remediation by various LDHs (Kulyukhin et al. 2010, 2011; Kulyukhin and Krasavina 2016). Characteristically, they conducted a study on the removal of U(VI) by various compositions (Mg, Al and Nd) of LDHs. Figure 3.5 shows the recycling process of the adsorption technique.

3.2.1.2 Reduction

For the environmental remediation of uranium, the TiC_2T_x condensed the extremely toxic U(VI) ions to the less toxic U(IV) and revealed the removal efficiency of 470 mg/g at pH 3.0 via sorption-reduction method. A facile $Ti_3C_2/SrTiO_3$ heterocatalyst was prepared through partial oxidation of a multilayered Ti_3C_2 in simple hydrothermal strategy (Deng et al. 2019). The $Ti_3C_2/SrTiO_3$ employed to remove U (IV) by photocatalytic reduction conversion. The study on 2D MXene revealed that

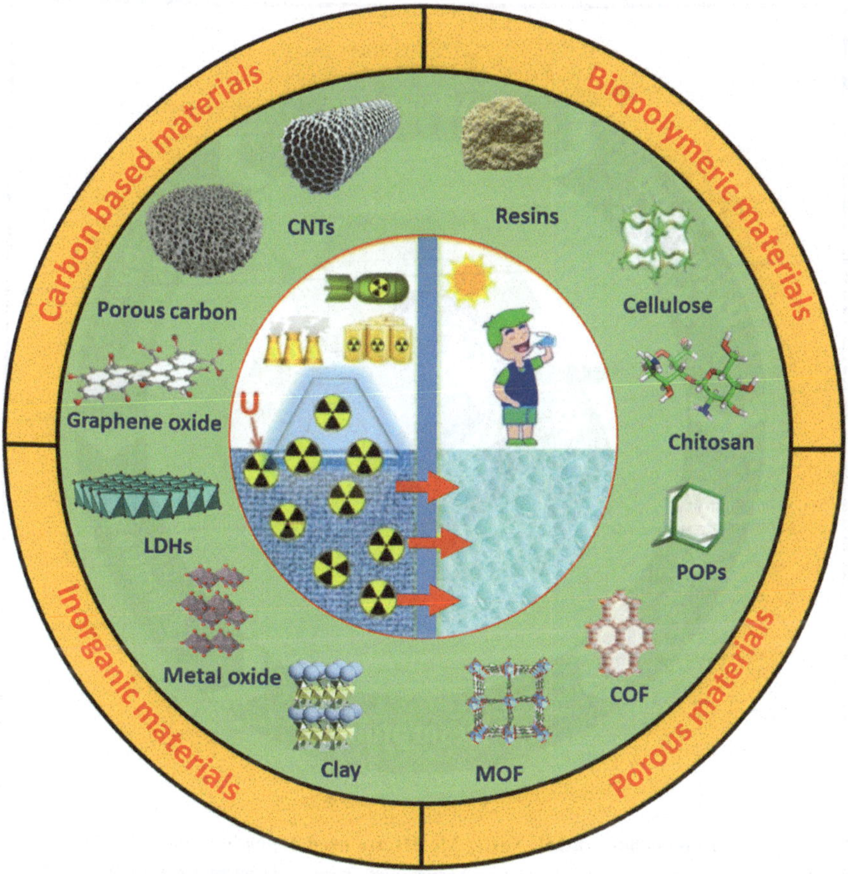

Fig. 3.4 Different adsorbent materials used for the removal of U(VI) ions from water

the amidoxime functionalized $Ti_3C_2T_x$ loaded carbon cloth can exhibit an enhanced U(VI) ions removal efficiency of 626 mg/g in the presence of electric field (Zhang et al. 2020). Uranium flexibility is contingent heavily on its oxidation states. U(VI) is highly water soluble and portable, and can be condensed to less soluble complexes (U_3O_8/U_4O_9/UO_2), when several reluctant are existing in the reduction process (Lovley and Phillips 1992; Lovley et al. 1993). The reduction of U(VI) has become the primary technique for removing U(VI) adulteration and several reductants have examined for the removal of uranium contamination. Fe(0), Fe(II), (FeS), (FeS_2), ($FeCO_3$), and magnetite (Fe_3O_4) are robust reductants and are being used widely to condense U(VI). Also, Fe(II) can reduce soluble U(VI) through a rapid reaction when the thermodynamic system is favorable. The outcome of findings provided new probable ways for U(VI) reduction process (Regenspurg et al. 2009; Du et al. 2011).

Fig. 3.5 Recycling process for the adsorption technique

3.2.1.3 Chemical Precipitation

Precipitation methods have been utilized commonly for eliminating radioactive materials from low and in-between levels through fuel recycling stations, power stations, and energy production places. These techniques are knowingly multipurpose, moderately low-cost, smooth operative and may react with sewages that containing moderately low concentrations of target pollutants among those covering vast quantities of contaminants or great concentrations of in-active species. However, in some cases, pretreatment phases such as the reaction of inorganic pollutants, solution pH, the variation of the valence states, should have some influence in the formation of the precipitate in the system. As a precipitating agent, iron oxide has some impact in the precipitation of uranium due to the reactivity and exclusive properties of the material and not of the specific attraction of the contaminant with the sorbent in the aqueous medium. Zero-valent iron shows much more precipitating power than the other sorbents in eliminating U(VI) ions from water. Precipitation process over iron oxide is more preferred when the external surface of the sorbent has not shielded by corrosion yields (Noubactep et al. 2005; Gu et al. 1998; Zhao et al. 2012).

3.2.1.4 Ion Exchange

Ion-exchange techniques have a widespread prospect in eliminating soluble and insoluble radionuclides from the aquatic environment, which has formed as a byproduct in nuclear fuel systems. It is instrumental in transferring a solution of the large volume of harmful contents into a smaller amount of solid (International Atomic Energy Agency 1970).

The ion-exchange method contains the exchange of anions or cations of a solution using an insoluble solid surface holding polar functional groups which support the exchange of more objectionable materials with another nonobjectionable, or less objectionable ionic substance. When the groups are negatively charged, involve the replacement of cations and when they are positively charged, involve the replacement of anions. So, ion exchanger can be 'restored,' radioactive waste can be separated with high activity, and if the ion exchangers become 'drained,' they can be eliminated and treated as harmful sewage.

Different types of ion exchangers have been employed for the removal of radioactive contaminations such as (a) synthetic materials like hydrous oxide gels of metals, zeolites, or resins designed by polymerized cross-linked hydrocarbons holding ionic functional groups (carboxylic acid, amino groups, sulfonic acid, etc.) and (b) natural ion exchangers (clays, zeolites, cellulose, charcoals, collagen). Ion-exchange processes can be accomplished in continuous and batch methods. If the wastes are having high contamination of salts, radionuclide, inorganic and organic contaminants, and suspended solids, the ionic form is not appropriate. The effluents will have to be pretreated before proceeding to the ion-exchange method (International Atomic Energy 1992; Adenot et al. 2005).

3.2.1.5 Evaporation Process

Evaporation technique is an operative and active method for concentrating inorganic salts, metal ions, and various harmful substances from wastewater effluent and decreases the high content of wastes with great decontamination factor. The method is employed for the intermediate, high and low-level effluents, in specific, for the highly reactive wastes and may be achieved through the generally available evaporation apparatus. However, this technique has some critical restrictions: inappropriate for waste sewages containing a large number of active salts, high cost because of its massive energy consumption, and the presence of some inorganic and organic compounds can result from explosions during evaporation (International Atomic Energy 1992; Adenot et al. 2005).

3.2.1.6 Electroremediation Technique

The electrochemical process and electroremediation, also recognized as the electrokinetic system, are categorized as a physico-chemical strategy and can also be used for the treatment of uranium ions. The electrochemical treatment of organic and inorganic compounds involves the direct use of electric current for the removal of contaminants. This process is informal to control and contains the setup of electrodes into the inorganic or organic effluents. This technique is proficient for mineralizing the organic compounds into CO_2 and water molecules directly, without the emission of any toxic compounds. Different catalytic agents such as Ce(IV)/Ce(III), Co(III)/Co(II), Ag(II)/Ag(I), have been recognized with this remediation system (Prabhakaran et al. 2009; Farmer et al. 1991; Chang 2001).

3.2.1.7 Membrane Process

Membrane technologies with several modifications have been surprisingly applied in U(VI) ions removal from water due to their high efficacy, easy process, and friendly handling instruments (Van der Bruggen and Vandecasteele 2003; Bian et al. 2000; Bacchin et al. 2002; Bhattacharjee et al. 1999). The commonly used membrane processes for removing the U(VI) ions from water/wastewater are ultrafiltration, nanofiltration, and reverse osmosis techniques.

Nanofiltration has an advantage comparing with the other membrane techniques, as it can be able to perform at lower operative pressures, and hence, it consumes little energy. Also, the nanofiltration is very selective and will not intercalate with the highly charged U(VI) ions from water during the treatment process.. Raff and Wilken were established that 81–99% of U(VI) ions could be rejected using nanofiltration technique (Raff and Wilken 1999). Ultrafiltration is a unique kind of membrane process; hydrostatic compression is the energetic force for wastewater rejection. The uranium typically exists in different forms; the studies were performed at pH 4. The rate of flux changed from 20 mg/s at the pH 4.0 and the lower level of 18 mg/s at the of pH 5.0 (Schaep et al. 1998; Favre-Reguillon et al. 2003; Bellona et al. 2004; Raff and Wilken 1999; Hilal et al. 2004). The reverse osmosis technique is semipermeable in nature. In this process, the feed solution was passed through it and eliminating the pollutants from aqueous solution. However, its difficulties like complexity, membrane fouling expensiveness, and a small quantity of permeate nature have restricted their use in U(VI) ions remediation.

3.2.1.8 Biological Process

The biological process exhibits extracellular bindings and the solution pH will change the biological actions of microorganisms and plants. Through varying the form and biodegradability of U(VI) ions, it is possible to minimize the transmission of uranium ions to the food cycles, which may help to prevent the environmental adulteration of uranium ions in water/wastewater. U(VI) removal using the biological technique may be more operative in the remediation of U(VI) ions from HCO_3^- solution than other conventional methods like ion exchange and biosorption process (Kalin et al. 2005). Various physico-chemical strategies and biological methods are enough to achieve the operative costs, and the biological material sources are economical and easily accessible (Loviey and Phillips 1992; Veglio et al. 2003). Pseudomonas MGF-48, which assembles U(VI) ions with high efficacy, is a catalase-positive, gram-negative, yellow-pigmented bacterium, motile derived from electroplating effluents. This bacterium can quickly remove U(VI) ions from water.

3.3 Risk Evaluation and Progress of Future Strategies

Nuclear waste processing and effective site repairing depend on the accessibility of appropriate waste administrative strategies and facilities. Operative administration of both radioactive and nonradioactive waste is an important for wastewater management. Strategic resolutions about the wastewater management are learned by the subsequent key ethics: remediation is import, integrated, and multisite attitudes would be reflected where it may be beneficial, waste should be reduced, and the waste order should be used as the outline for waste management choice creating and permits an active balance of significances containing the cost-effective, affordability and the health security, safety and the atmospheric balance. The management of radioactive waste in worldwide includes the categorization of the waste from its generation to final disposals such as classification, treatment, transport, pretreatment, storage, and removal. All of these processes will reduce the volume of radioactive wastes from the aqueous environment.

3.4 Conclusion

The problem of U(VI) ions removal from large volumes of natural aquatic systems could not be resolved using the synthetic materials because of their high cost and low attraction of the pollutants from the aqueous environment. The radioactive contamination is one of the ultimate imperative ecological problems throughout the world. The reduction of radioactive uranium becomes a concern for investigators. In this point of view, numerous researches have been made to study the effective remediation of U(VI) from water. This study critically demonstrated the various strategies for U(VI) ions elimination from the aqueous environment which has been examined extensively by various researchers through different studies. An extensive range of treatment strategies like adsorption, precipitation, ion exchange, reduction, biological process, and membrane filtration have been established for removing U(VI) ions from water/wastewater. It was obvious from the literature that adsorption, ion exchange, and membrane processess are the most frequently used methods for the removal of U(VI) ions from water. Adsorption using cost-effective sorbents and bioadsorbents was well-known, effective, and economically sound technique for the removal of low concentration of radioactive pollutants. Activated carbon is an alternative for the removal of high concentration of pollutants since the carbonaceous materials have an enhanced sorption efficiency. The cost of compounds utilized and approaches of modification also have to be taken into consideration to make 'low-cost' adsorbent materials. The presence of a flexible polymer chain structure, high hydrophobicity, and a high number of reactive groups will induce the adsorption of the pollutants. The membrane process has shown high rejection abilities but the operative cost is still challenging. Bioremediation has accomplished via the biomineralization of U(VI) ions from water. The bioremediation acts in the same way to that of the more extensively examined U(VI) ions bioreduction. It is necessary to develop some sophisticated methodologies by conjoining the organic and inorganic materials for the fabrication of adsorbents for the treatment of radioactive pollutants, which will refine the quality of the environment by reducing the quantity of radioactive pollutants from water/wastewater.

References

Adenot F, Advocat T, Bisel I, Broudic JC, Brun P, Cau-Dit-Coumes C, et al (2005) Le, traitement et le conditionnement des desechts. Rapport technique DTCD/SCD/2005/001, CEA, France

Aliyu AS, Evangeliou N, Mousseau TA, Wu J, Ramlib AT (2015) An overview of current knowledge concerning the health and environmental consequences of the Fukushima Daiichi Nuclear Power Plant (FDNPP) accident. Environ Int. 85:213–28

Amano Y (2015) The Fukushima Daiichi Accident (PDF). IAEA—International Atomic Energy Agency, Wienna. August 31, 2015. pp 131–132. ISBN 978-92-0-107015-9

Andrade Martins R (1997) Becquerel and the choice of uranium compounds. Arch Hist Exact Sci 51:67–81

Bacchin P, Si-Hassen D, Starov V, Clifton M, Aimar P (2002) A unifying model for concentration polarization, gel-layer formation and particle deposition in crossflowmembranefiltration of colloidal suspensions. Chem Eng Sci 57:77–91

Bailey SE, Olin TJ, Bricka RM, Adrian DD (1999) A review of potentially low-cost sorbents for heavy metals. Water Res 33:2469–2479

Bellona C, Drewes JE, Xu P, Amy G (2004) Factors affecting the rejection of organic solutes during NF/RO treatment—a literature review. Water Res 38:2795–2809

Bhalara PD, Punetha D, Balasubramanian K (2014a) A review of potential remediation techniques for uranium(VI) ion retrieval from contaminated aqueous environment. Environ Chem Eng 2:1621–1634

Bhalara PD, Punetha D, Balasubramanian K (2014b) A review of potential remediation techniques for uranium(VI) ion retrieval from contaminated aqueous environment. J Environ Chem Eng 2:1621–1634
Bhattacharjee S, Kim A, Elimelech M (1999) Concentration polarization of interacting solute particles in cross-flow membrane filtration. J Colloid Interface Sci 212:81–99
Bian R, Yamamoto K, Watanabe Y (2000) The effect of shear rate on controlling the concentration polarization and membrane fouling. Desalination 131:225–236
Bleise PR, Danesi W, Burkart B (2003) Properties, use and health effects of depleted uranium (DU): a general overview. J Environ Radioact 64:93–112
Chang HO (2001) Hazardous and Radioactive Waste Treatment Technologies Handbook. CRP Press, New York
Charnay C, Lagerge S, Partyka S (2001) Assessment of the surface heterogeneity of talc materials. J Colloid Interface Sci 233:250–258
Chen CM, Zeng GM, Tang C (2009) Research of hydroxyapatite Adsorption to uranium wastewater. Metal Mine 5:135–137
Coetze H, Winde F, Wade PW (2006) An assessment of sources, pathways, mechanisms and risks of current and potential future pollution of water and sediments in goldmining areas of the Wonderfonteinspruit catchment Report to the Water Research Commission, Research Report No.1214/1/06
Craig DK (2001) Chemical and Radiological Toxicity of Uranium and its Compounds. Washington, DC, Westinghouse Savannah River Company
Crane RA, Scott TB (2014) The removal of uranium onto nanoscale zero-valent iron articles in anoxic batch systems. J Nanomater 956360
Deng ZJ, Li L, Wang LY, Yuan JH, Lan ZY, Chang ZF, Chai WQ (2019) Nanolayered Ti_3C_2 and $SrTiO_3$ composites for photocatalytic reduction and removal of uranium (VI). ACS Appl. Nano Mater. 2(4):2283–2294
Dickinson M, Scott TB (2010) The application of zero-valent iron nanoparticles for the remediation of a uranium-contaminated waste effluent. J Hazard Mater 178:171e179
Du X, Boonchayaanant B, Wu WM, Fendorf S, Bargar J, Criddle CS (2011) Reduction of uranium (VI) by soluble iron(II) conforms with thermodynamic predictions. Environ Sci Technol 45 (11):4718–4725
Ershov BG, Bykov GL, Seliverstov AF, Milyutin VV, Gelis VM (1993a) The study of radionuclides sorption on oxidized wood carbons. Radiochemistry 6:100–110
Ershov BG, Bykov GL, Seliverstov AF, Milyutin VV, Gelis VM (1993b) Synthesis and sorption properties of oxidized wood carbons. Russ J Appl Chem 5:1074–1078
Farmer JC, Hickman RG, Wang FT, Lewis PR, Summers LJ (1991) Initial study of the complete mediated electrochemical oxidation of ethylene glycol, Report No. UCRL-LR-106479, Lawrence Livermore National Laboratory, Livermore
Favre-Reguillon A, Lebuzit G, Foos J, Guy A, Draye M, Lemaire M (2003) Selective concentration of uranium from seawater by nanofiltration. Ind Eng Chem Res 42:5900–5904
Gu B, Liang L, Dickey MJ, Yin X, Dai S (1998) Reductive precipitation of uranium(VI) by zero-valent iron. Environ Sci Technol 32:3366–3373
Gupta VK, Ali I (2000) Utilization of bagasse fly ash (a sugar industry waste) for the removal of copper and zinc from wastewater. Sep Purif Technol 18:131–140
Gupta DK, Walther C (2014) Radionuclide contamination and remediation through plants. Springer, Heidelberg
Hashem A, Abdel-Halim ES, El-Tahlawy KF, Hebeish A (2005a) Enhancement of adsorption of Co (II) and Ni (II) ions onto peanut hulls though esterification using citric acid. Adsorpt Sci Technol 23:367–380
Hashem A, Akasha RA, Ghith A, Hussein DA (2005b) Adsorbent based on agricultural wastes for heavy metal and dye removal: a review. Energy Edu Sci Technol 19:69–86
Hilal N, Al-Zoubi H, Darwish NA, Mohamma AW, Abu Arabi M (2004) A comprehensive review of nanofiltration membranes: treatment, pretreatment, modelling, and atomic force microscopy, Desalination 170:281–308

Ilaiyaraja P, Deb AKS, Sivasubramanian K, Ponraju D, Venkatraman B (2013) Adsorption of uranium from aqueous solution by PAMAM dendron functionalized styrene divinylbenzene. J Hazard Mater 250–251:155–166

International Atomic Energy (1992) AgencyIAEA Treatment and conditioning of radioactive organic liquids, IAEA-TEC-DOC-656. International Atomic Energy Agency (IAEA), Vienna

International Atomic Energy Agency (1970)IAEA Standardization of Radioactive Waste Categories, Technical Reports Series No. 101, International Atomic Energy Agency (IAEA), Vienna

Kalin M, Wheeler WN, Meinrath G (2005) The removal of uranium from mining waste water using algal/microbial biomass. J Environ Radioact 78:151–177

Key World Energy Statistics (2009) OECD International Energy Agency, 9 rue de la Fe´de´ ration, 75739 Paris Cedex 15, France

Klimkova S, Cernik M, Lacinova L, Filip J, Jancik D, Zboril R (2011) Zero-valent iron nanoparticles in treatment of acid mine water from in situ uranium leaching. Chemosphere 82:1178e1184

Krestou A, Xenidis A, Panias D (2003) Mechanism of aqueous uranium (VI) uptake by natural zeolitic tuff. Miner Eng 16:1363–1370

Kulyukhin SA, Krasavina EP (2016) Sorption of U(VI) onto layered double hydroxides and oxides of Mg and Al, prepared using microwave radiation. Radiochemistry 58:405–408

Kulyukhin SA, Krasavina EP, Gredina IV, Mizina LV (2010) Sorption of U(VI) from aqueous solutions on layered double hydroxides of Mg, Al, and Nd. Radiochemistry 52:653–661

Kulyukhin SA, Krasavina EP, Gredina IV, Mizina LV, Konovalova NA, Gredina IV (2011) Effect of complexing anions on sorption of U(VI), 90Sr, and 90Y from aqueous solutions on layered double hydroxides of Mg. Al, and Nd, Radiochemistry 53:504–509

Lauria DC, Godoy JM (2002) Abnormal high natural radium concentration in surface waters. J Environ Radioact 61:159–168

Li XY, Zhang M, Liu YB, Li X, Liu YH, Hua R, He CT (2013) Removal of U(VI) in aqueous solution by nanoscale zero-Valent iron(nZVI). Water Qual Expo Health 5:31e40

Loviey DR, Phillips EJP (1992) Bioremediation of uranium contamination with enzymatic uranium reduction. Environ Sci Technol 26:2228–2234

Lovley DR, Phillips EJP (1992) Bioremediation of uranium contamination with enzymatic uranium reduction. Environ Sci Technol 26(11):2228–2234

Lovley DR, Roden EE, Phillips EJP, Woodward JC (1993) Enzymatic iron and uranium reduction by sulfate-reducing bacteria. Mar Geol 113(1–2):41–53

Marples David R (1988) The Social Impact of the Chernobyl Disaster. New York: St. Martin's Press. ISBN 978-0-312-02432-1

McGraw-Hill (2005) McGraw-Hill Concise Encyclopedia of Science and Technology, 5th edition, The McGraw-Hill Companies, McGraw-Hill Professional, New York, ISBN: 978-0-07-142957-3

Meinrath A, Schneider P, Meinrath G (2003) Uranium ores and depleted uranium in the environment, with a reference to uranium in the biosphere from the Erzgebirge/Sachsen, Germany. J Environ Radioact 64:175–193

Meldon JH, Sprynskyy M, Kowalkowski T, Tutu H, Cukrowska EM, Buszewski B (2011) Adsorption performance of talc for uranium removal from aqueous solution. Chem Eng J 171:1185–1193

Mellah A, Chegrouche S, Barkat M (2006) The removal of uranium(VI) from aqueous solutions onto activated carbon: kinetic and thermodynamic investigations. J Colloid Interface Sci 296:434–441

Mitsakou C, Eleftheriadis K, Housiadas C, Lazaridis M (2003) Modeling of the dispersion of depleted uranium aerosol. Health Phys 84:538–544

Moloukhia H, Hegazy WS, Abdel-Galil EA, Mahrous SS (2016) Removal of Eu^{3+}, Ce^{3+}, Sr^{2+}, and Cs^{+} ions from radioactive waste solutions by modified activated carbon prepared from coconut shells. Chem Ecol 32:324–345

Mould RF (2000) Chernobyl Record: The Definitive History of the Chernobyl Catastrophe. Institute of Physics Publishing. Bristol, UK. ISBN 978-0-7503-0670-6

Noubactep C, Meinrath G, Merkel BJ (2005) Investigating the mechanism of uranium removal by Zerovalent Iron. Environ Chem 2:235–242

Noubactep C, Schoner A, Meinrath G (2006) Mechanism of uranium removal from the aqueous solution by elemental iron. J Hazard Mater 132:202e212

Orhan Y, Bujukgungor H (1993) The removal of heavy metals by using agricultural wastes. Water Sci Technol 28:247–255

Philiips DH, Gu B, Watson DB (2008) Uranium removal from contaminated groundwater by synthetic resins. Water Res 42:260–268

Povinec PP, Hirose K, Honda T, Ito T, Scott EM, Togawa O (2004) Spatial distribution of ^{3}H, ^{90}Sr, ^{137}Cs and $^{239,240}Pu$ in surface waters of the Pacific and Indian Oceans-GLOMARD database. J Environ Radioact 76:113–137

Prabhakaran D, Kannadasan T, Basha CA (2009) Mediated electrochemical oxidation process for destruction of TOC in a batch recirculation reactor. Inter J Chem Technol Res 1(4):962–969

Raff O, Wilken RD (1999) Removal of dissolved uranium by nanofiltration. Desalination 122:147–150

Regenspurg S, Schild D, Schäfer T, Huber F, Malmström ME (2009) Removal of uranium(VI) from the aqueous phase by iron(II) minerals in presence of bicarbonate. Appl Geochem 24 (9):1617–1625

Riba O, Scott TB, Ragnarsdottir KV, Allen GC (2008) Reaction mechanism of ranyl in the presence of zero-valent iron nanoparticles. Geochim. Cosmochim. Acta 72:4047e4057

Rollett AD (2008)Applications of Texture Analysis, Wiley, p 108, ISBN: 0-470-40835-9

Schaep J, der Bruggen B, Vandecasteele C, Wilms D (1998) Retention mechanisms in nanofiltration. Environ Sci Res 55:117–126

Seaborg GT (1968) Uranium, The Encyclopedia of the Chemical Elements. Reinhold Book Corporation, Skokie, IL, pp 773–786

Van der Bruggen B, Vandecasteele C (2003) Removal of pollutants from surface water and groundwater by nanofiltration: overview of possible applications in the drinking water industry. Environ Pollut 122:435–445

Veglio F, Esposito A, Reverberi AP (2003) Standardisation of heavy metal biosorption tests: equilibrium and modeling study. Process Biochem 38:953–961

Wang G, Liu J, Wang X, Xie Z, Deng N (2009) Adsorption of uranium (VI) from aqueous solution onto cross-linked chitosan. J Hazard Mater 168:1053–1058

Wang L, Yuan L, Chen K, Zhang Y, Deng Q, Du S, Huang Q, Zheng L, Zhang J, Chai Z, Barsoum MW, Wang X, Shi W (2016) Loading actinides in multilayered structures for nuclear waste treatment: the first case study of uranium capture withvanadium carbide MXene. ACS Appl Mater Interfaces 8(25):16396–16403

Wang W, Tao L, Yuan Z, Liu Q, Huang Z, Chai JK, Gibson W (2017) Rational control of the interlayer space inside two-dimensional titanium carbides for highly efficient uranium removal and imprisonment. Chem Commun 53(89):12084–12087

WHO (1998) Guidelines for Drinking Water Quality, 2nd edition, Addendum to vol. 2. Health Criteria and Other Supporting Information, WHO/EOS/98.1, Geneva, 283

WHO (2003)Guidelines for Drinking Water Quality, 3rd edition

World Nuclear Association (2013) Table of Nuclear Share Figures, IAEA

World-Nuclear.org. http://www.world-nuclear.org/info/Nuclear-Fuel-Cycle/Uranium-ResourcesSupply-of-Uranium/. Accessed 7 July 2013

Zhang YJ, Lan JH, Wang L, Wu QY, Wang CZ, Bo T, Chai ZF, Shi WQ (2016) Adsorption of uranyl species on hydroxylated titanium carbide nanosheet: a firstprinciples study. J Hazard Mater 308:402–410

Zhang P, Wang L, Huang Z, Yu J, Li Z, Deng H, Yin T, Yuan L, Gison JK, Mei L, Zheng L, Wang H, Chai Z, Shi W (2020) Aryl diazonium-assisted amidoximation of MXene for boosting water stability and uranyl sequestration via electrochemical sorption. ACS Appl Mater Interfaces 12:11587–15579

Zhao G, Wen T, Yang X, Yang S, Liao J, Hu J et al (2012) Preconcentration of U(VI) ions on few-layered graphene oxide nanosheets from aqueous solutions. Dalton Trans 41:6182–6188

Chapter 4
A Study on Different Bioremediation Approaches to Hexavalent Chromium

Shivangi Upadhyay and Alok Sinha

Abstract Hexavalent chromium is a geochemical element and designated as priority pollutant. It has mutagenic and carcinogenic property and poses a serious threat to both humanity and ecosystem. Despite of toxicity, little dose of chromium acts as micronutrient in the diets of animals and humans and also helps in sugar, protein, and lipid metabolism in mammals. Chromium speciation exists in two states: hexavalent chromium and trivalent chromium, out of which the latter is nontoxic. Health problems associated with high dose of chromium are ulcers, diarrhea, irritation of skin, eye and lung carcinoma, dysfunction of kidney, birth defects, and reduced reproductive health. The lethal dose $(LD)_{50}$ value for oral toxicity in rats is 50–100 mg kg^{-1} and 1900–3000 mg kg^{-1} for Cr (VI) and Cr (III), respectively. Due to high toxicity of Cr (VI) compounds, there is multiplicity of treatment technologies including physico-chemical and biological methods. Physico-chemical methods are high energy demanding, have high operational cost, generate secondary pollutant, and sometimes have lesser efficiency due to high metal concentration and interferences. In contrast to physciochemical method, bioremediation of Cr (VI) reduction is operated at low cost, and less energy is required with high efficiency of reduction, no health and environmental hazards. Microorganisms involved in remediation metabolize the chemicals via enzyme-catalyzed pathway converting into harmless compounds and often use compounds as a source of their growth. Despite of all these methods, some green technologies and modification in these techniques also proved to be effective in chromium reduction. This chapter deals with occurrence and fate of chromium, speciation, various treatment technologies, mechanism of reduction and their advantages-disadvantages, pilot-scale studies, and future perspectives in remediating toxic hexavalent chromium.

Keywords Bioreduction · Biotransformation · Constructed wetlands

S. Upadhyay · A. Sinha (✉)
Department of Environmental Science and Engineering, Indian Institute of Technology (Indian School of Mines), Dhanbad 826004, India
e-mail: alok@iitism.ac.in

S. P. Singh et al. (eds.), *Pollution Control Technologies*, Energy, Environment, and Sustainability, https://doi.org/10.1007/978-981-16-0858-2_4

4.1 Introduction

The seventh most abundant element and metal found in earth's crust is chromium (Cr) which is excavated as chromite ($FeCr_2O_4$) (Sultana et al. 2014). Being a geochemical element, its presence is found in rocks, fresh water, and mineral soils. Despite of several oxidation states, it is the most stable form, and trivalent chromium Cr (III) and hexavalent chromium Cr (VI) are prevalent in environment (Sultana et al. 2014; Fernandez et al. 2018). Chromium cycle mainly follows oxidation and reduction of its different form. Oxidation of Cr (III) into hexavalent form in sediments and soils occurs by manganese oxide while Cr (VI) is reduced to its trivalent form by soil compounds that are in reduced form. Hexavalent chromium has wide range of industrial application like chrome plating, leather tanning, electroplating, stainless steel industries, and wood preservation (Jobby et al. 2018). United States Environmental Protection Agency-USEPA has designated this element as priority pollutant as it becomes easy for the metal complexes to cross the membrane, thereby generating reactive oxygen species (ROS) which in turn alters cellular structure (Fernandez et al. 2018; Fedorovych et al. 2009; Juvera-Espinosa et al. 2006). Owing to toxicity of chromium element, hexavalent chromium is 1000 times mutagenic and 100 times lethal than trivalent state (Chojnacka 2010). Organisms come into contact with chromium via inhalation, oral digestion, or dermal contact. Cr (III) enters through digestive system, and if there enters Cr (VI), simultaneously most of them gets reduced to Cr (III) (Hamilton et al. 2018; Pechancova et al. 2019). Absorption of Cr in the gastrointestinal tract follows the unsaturated passive transport and is severely influenced by substances present in diet. The Cr species then travels in the bloodstream and gets accumulated in the deep organs like liver, kidney, and spleen and then excreted out via urine and negligible amount by bile or hair (Finley et al. 2017). Cr (III) is helpful in human metabolism like cholesterol and triglyceride levels, stability and amalgamation of proteins, nucleic acids, glucose maintenance, and stability of cell membrane (Di Bona et al. 2011; Frois et al. 2011; Fernandez et al. 2014). A study compared rats supplied with less content of Cr (III) and another provided with heavy amount of Cr (III) confirms that those having little amount of trivalent chromium had no adverse consequences and are not used as nutritional supplement. Higher dosing of trivalent chromium forms composite with organic compounds which interferes with metalloenzymatic process (Poljsak et al. 2010) and may cause lung cancers, decrease in reproductive health and birth deficiency (Fernandez et al. 2018).

These polluting agents are directly thrown into water and soil from various sources generates pollution, contamination and demolition of the ecosystem. For the treatment of chromium ions, various technologies have been developed till date from water, wastewater, and soil. Physico-chemical methods like use of activated carbon, chemical precipitation, reverse osmosis, ion exchange, membrane technologies, and adsorption (Krowiak 2013) have been extensively used but sometimes these processes are exceptionally expensive when the metal concentration varies from 1 to 100 mg/L. Another drawback of these techniques is they generate huge quantity of toxic chemical sludge which creates a serious concern for disposal. On the other side, microorganisms interact with chromium via biosorption, enzymatic reduction, and bioaccumulation is gaining grounds due to its low operationality and minimum chemical requirement. Use of scrap iron in reducing Cr (VI) is a promising and emerging technology because of its easy availability, faster reaction rate, and cost-effectiveness.

4.2 Chromium Toxicity and Contamination

Chromium is an essential micronutrient in the diets of humans and animals which is helpful in lipid, sugar and proteins metabolism, but in case of plants and microorganisms, there is no such known necessity of chromium in their metabolic pathways. However, chromium is toxic at high level depending upon its oxidation state. In between its two oxidation states (III) and (VI), hexavalent state is highly lethal, carcinogenic, and mutagenic and has effects on lowering reproductive capacity and birth defects as well. The casualty may occur due to large dosage of Cr (VI). The lethal dose $(LD)_{50}$ value for oral toxicity in rats is 50–100 mg kg^{-1} for Cr (VI), and for Cr (III), it is 1900–3000 mg kg^{-1} (Jobby et al. 2018). The other toxicity effects associated with chromium are that it reduces the plant's capacity to grow by decreasing uptake of nutrients and photosynthesis. The morphological, physiological, and biochemical processes of plants are rigorously affected by high dose of chromium which induces formation of reactive oxygen species. In plants, a phenomenon called chlorosis and necrosis indicates its toxicity.

Chromium has corrosion-resistant quality due to which it has been widely used in industrial processes like manufacturing of stainless steel, metallurgical, tanning, wood preservation, electroplating, pulp and paper, and production of paints. These industries generate huge amount of wastes in the environment. In many countries, the chromium contamination in surface water has crossed its permissible limits approximately 0.5–2 mg/L. (US Environmental Protection Agency 1987) and (Guidelines for drinking-water quality 1996) reported that the total chromium content of 84 mg/L and 0.2–44 mg/L has been found in the surface water of Central Canada and USA, respectively. The contamination to groundwater by Cr (VI) is due to leaching and seeping of dumped wastes as a filling material possess a great threat to health. Major source of Cr (VI) pollution are tannery industries where chromium compound has been used to tan hides. However, Cr is not completely used in the tanning process of leather, a large portion of it is discharged as it is in the effluent. Around the world, approximately 40 million tonnes of chromium waste is produced by tanning industries which is directly released into water and disposed of on land. It has been detected in India that around 2000–32,000 tonnes of elemental chromium are discharged annually by tannery industries in the environment having chromium concentration ranging between 2 and 5 g/L which is higher than the permissible limit.

4.3 Chromium Reduction by Different Methods

4.3.1 Physical and Chemical Methods

Physical methods of Cr (VI) reduction include membrane filtration, granular activated carbon, adsorption, photocatalysis, electrodialysis, soil washing (Wang et al. 2008), and chemical methods include use of chemicals like ferrous sulfate, sulfur dioxide, barium sulfite, sodium metabisulfite, lime and limestone, and sodium sulfite, for reduction of Cr(VI) to Cr(III). The disadvantages of these methods are high energy demand, high operational cost, generation of secondary pollutant and sometimes lesser efficiency due to high metal concentration and interferences (Zouboulis et al. 2004). A list of physico-chemical treatment technologies is discussed in Table 4.1.

4.3.2 Biological Methods

Bioremediation is one of the best approachs toward remediating heavy metal pollution. It is a phenomenon of transforming harmful pollutants into nontoxic compounds by involving living organisms (fungi, bacteria, plants, yeast, and algae). The advantage of this method is that there is low operational cost, less energy requirement, high efficiency, no health and environmental hazards, metal recovery and possibility to reuse. Microorganisms involved in remediation metabolize the chemicals via enzyme-catalyzed pathway converting into harmless compounds and often uses compounds as a source of their growth. Some of the parameters that affect the efficiency of the process are chemical nature of pollutants, structure of the compound, pH and temperature of the system, nutritional state, presence of microbial community and hydrogeology. Bioremediation approaches to heavy metal treatment are bioaccumulation, biosorption, and biotransformation.

4.3.2.1 Bioremediation by Fungi

Fungi has been well known for biosorption of Cr (VI). Several genera of fungi had been discovered in biosorption process such as *Aspergillus oryzae*, *Trichoderma* sp., *A. niger*, *Fusarium oxysporum*, *Trichoderma inhamatum*, *Hypocrea tawa*, *Fusarium oxysporum* NCBT-156, *Saccharomyces cerevisiae*, *Penicillium griseofulvum* MSR1, and *Acremonium* sp.. Cr (VI) biosorption by fungi can be accomplished by metabolism-dependent and independent pathway. The mechanism for Cr (VI) sorption involves adsorption of Cr (VI) on the cell surface of fungi by formation of a chemical bond that have some functional groups present. The presence of hydroxyl, carboxyl, amino, and carbonyl groups on the cell surface helps in attachement of Cr (VI) on the wall of fungal cell. Involvement of different

Table 4.1 Different physico-chemical methods of Cr (VI) removal

Materials applied	Mechanism of treatment	Cr (VI) concentration (mg/L)	Efficiency (%)	Advantages	Disadvantages	References
Mixed maghemite–magnetite		0.5–3.5	95	–	–	Chowdhury et al. (2012)
Activated carbon-coated α-Fe2O3 nanoparticles	Adsorption	25	94	Faster removal rate	Inhibition of Cr removal due to deprotonation of functional groups at high pH	Lia et al. (2019)
Nanoscale zero-valent metal by sodium borohydride reduction	Reductive immobilization	60	99.99	Faster reaction rate	Reduction efficiency decreased while increasing Cr concentration	Fang et al. (2011)
FeS-coated iron (Fe/FeS) magnetic nanoparticle	Adsorption, oxidation-redcution, precipitation	25	82.1	Aggregation of particles was prevented providing larger BET surface area and higher removal efficiency	Limitation of reactive sites with increase in Cr concentration	Gong et al. (2017)
Carboxymethyl cellulose (CMC) stabilized microscale iron sulfide (FeS)	Reduction	1407 mg/kg Cr (VI) spiked in soil	98	Higher removal and immobilization efficiency	–	Li et al. (2017)
Biochar-supported zero-valent iron nanoparticles		320 mg/kg Cr (VI) spiked soil	100	Enhanced immobilization and decreased migration of Cr, reduced bioavailability and bioaccumulation, improved soil fertility	–	Su et al. (2016)

(continued)

Table 4.1 (continued)

Materials applied	Mechanism of treatment	Cr (VI) concentration (mg/L)	Efficiency (%)	Advantages	Disadvantages	References
Graphene oxide incorporate mixed matrix membrane (MMM) hollow fibers	Adsorption	10, 25, 50, 100, 250, 500	88	–	–	Mukherjee et al. (2014)
Iron electrodes	Electrocoagulation	49.96	100	–	–	Khan et al. (2019)
MIL-53(Fe) as positive photoelectrode and active carbon as negative electrode	Photocatalysis and capacitive deionization	50	81.6	Visible light absorption capacity was good, bigger surface area, rich metal-containing catalytically active sites, and desirable pores	MIL-53(Fe) does not fit for practical application	Houa et al. (2018)
TiO2 P25	Photocatalysis	10–200	79	–	–	Sanea et al. (2018)
Powdered activated carbon (PAC)	Adsorption	50	>80	High adsorption rate	Quick saturation with time	Wang (2018)
Graphene-coated iron oxide nanoparticles	Adsorption	25–175	74	–	–	Khare et al. (2018)

fungal species in Cr (VI) reduction exerts different results such as *Saccharomyces cerevisiae* from culture collection bank reduced 200 mg/L of Cr (VI) in 24 h with 85% efficiency (Mahmoud and Mohamed 2017), *A. niger* was capable of reducing 18.125 mg/L having efficiency 96.3 and operating duration was seven days (Sivakumar 2016), *Paecilomyces lilacinus* isolated from tannery effluent reduced 200 mg/L Cr in 120 h with 100% removal (Sharma and Adholeya 2011).

4.3.2.2 Bioremediation by Bacteria

Cr (VI) remediation using bacteria proved to be a very efficient, cheaper, no chemical input, less energy requiring method that converts it in less toxic Cr (III) form. Both gram positive and gram negative bacteria, living and dead cells are efficient in remediation approaches. A work reported ability of chromium biosorption in *B. circulans* (34.5 mg Cr g^{-1} of dry weight), *B. megaterium* (32.0 mg Cr g^{-1} of dry weight) and *B. coagulans* (39.9 mg Cr g^{-1} of dry weight) and it was found out that the biosorption ability of living and dead cells of *B. coagulans* and *B. megaterium* were compared and dead cells were found to be more effective for chromium biosorption. Some of the bacterial species applied for the treatment of hexavalent chromium reduction are provided in Table 4.2.

4.3.2.3 Bioremediation Using Algae

Biosorption by algae is another remediation method for Cr (VI) which involves adsorption of metal on the algal cell surface, and may further follows accumulation inside the cell. Specific molecules like phytochelatins, metallothioneins, guluronic acid, alginates, sulfated polysaccharides with hydroxyl, amino, carboxyl, and sulfate as functional groups trigger Cr (VI) remediation. Variety of algal species has been used in Cr (VI) removal like, *Euglena*, *Scenedesmus*, *Cladophora* sp. *Selenastrum*, *Ceramium virgatum*, *Spirulina* sp. *Nostoc linckia* and *Chlorella vulgaris*. It was studied by (Pradhan et al. 2017) using *Chlorella vulgaris* for 3.22 mg/L of hexavalent chromium. At 28 °C, complete reduction was observed within 12 days. Furthermore, organelles (Chloroplasts) were extracted from *Chlorella vulgaris* for reduction and the results stated that it adsorbed total chromium (21%) and reduced 70% of Cr (VI). An algal species, *Sargassum cymosum*, has been used in Cr (VI) reduction as an electron donor and the reduction was due to acidic carboxylic group associates with the surface of the biomass which mediates sequestration of trivalent chromium. During the reduction process, the oxidation of biomass forms the binding sites on the surface.

Table 4.2 Different bacterial species in Cr (VI) reduction

Species	Cr (VI) concentration (mg/L)	Duration of reaction (h)	% efficiency	Growth condition	References
Cellulosimicrobium sp.	50, 100, 200, and 300	24 and 96	99.33 and 96.8	–	Bharagava and Mishra (2018)
Bacillus circulans BWL1061	50	–	100	–	Liu et al. (2017)
Serratia sp.	4, 8, 12, 16, 20	36 and 48	100	Aerobic	Upadhyay et al. (2018)
Bacillus sp. (CSB-4)	10–500	144	90	–	Dhal et al. (2010)
Ochrobactrum sp.	300	–	96.5	–	Chen et al. (2016)
Acinetobacter baumannii L2	1000	24	99.58	–	Sathishkumar et al. (2016)

4.4 Mechanism of Cr Reduction by Microbes

Microbial remediation is the process of quickly degrading the lethal pollutants to naturally safer limit in water, sludge, soil, residues, and subsurface materials. (Asha Latha and Sandeep Reddy 2013). The different remediation mechanisms followed by microbes are biosorption, bioaccumulation, and biotransformation. Biosorption is a reversible, passive, and rapid process that involves a biosorbent and a sorbate. (Ahluwalia and Goyal 2007). Enormous variety of microorganisms have been applied for biosorption activity such as cyanobacteria, algae, microalgae (Khoubestani et al. 2015; Kwak et al. 2015; Nemr et al. 2015), yeast (Fernandez et al. 2013; Farina 2012; Khani et al. 2012) fungi (Huang et al. 2016), and bacteria (Wu et al. 2015; Bahafid et al. 2013). Structural integrity and many functional groups like hydroxyl, amino, carboxylate, and phosphate are provided by microbial cell wall that helps in binding of heavy metal ions. The sorption of metal ions requires energy consumption which is provided by citoplasmatic metal binding proteins. A study by (Thatoi et al. 2014) found out that hexavalent chromium removal involves adsorption on functional groups like polysaccharides, amide I, amide II, amide III, carboxyl, and sulfonate and which further accumulates within the cell and thus biotransforming hexavalent chromium to its trivalent form. Biotransformation of Cr (VI) to Cr (III) is mediated by soluble cytosolic proteins or insoluble cell membrane enzymes (Viti et al. 2014; Kadlec and Wallace 2009). Biotransformation in plants is mediated by chemical or enzymatic process. The reduction of Cr (VI) by chemically induced mechanism is mediated by cysteine, sulfite, thiosulfates, and glutathione that are present in the plant cell. The enzymatic method of reduction is carried out by diverse group of bacteria such as *Bacillus* sp., *Pseudomonas* sp., *Staphylococuusarlettae* sp. etc. The presence of soluble and membrane-bound reductases such as flavin reductase, cytochromes, and hydrogenases is used by the bacteria that can use chromate as the terminal electron acceptor in electron transport system. Also, presence of different chromate reductase such as YieF, LpDH ChrR, and NemA, and that are present in section of cytoplasm else are membrane bounded helps in transforming activity. The mechanism of chromium reduction varies with different microorganisms. In yeast, the detoxification occurs indirectly by riboflavin and sulfate and is released to the extracellular medium by the yeast cells. (Fedorovych et al. 2009). A pictorial representation is depicting the reduction of Cr (VI) by microorganism and their effects associated with the process in Fig. 4.1. A flowchart showing different scale-up approach for Cr (VI) reduction is depicted in Fig. 4.2.

4.5 Cr Removal by Constructed Wetlands

Constructed wetlands (CW) play an important role in chromium removal process involving a combination of biological and physico-chemical processes which includes sedimentation, attachment to porous media, plant uptake, and precipitation as insoluble forms (mainly sulfides and (oxy-) (hydroxides) (Maine et al. 2009). Rhizosphere is the most efficient reaction zone where both biological and physico-chemical processes and interaction of microorganisms, plants, and pollutants takes place. CW vegetation in Cr removal follows release of root exudates which impacts metal toxicity and their mobility and provides surface area for microbial growth to occur, and the tissues accumulates Cr in themselves. Root zone accumulates metals and inhibits metal mobility from roots to shoots in vascular

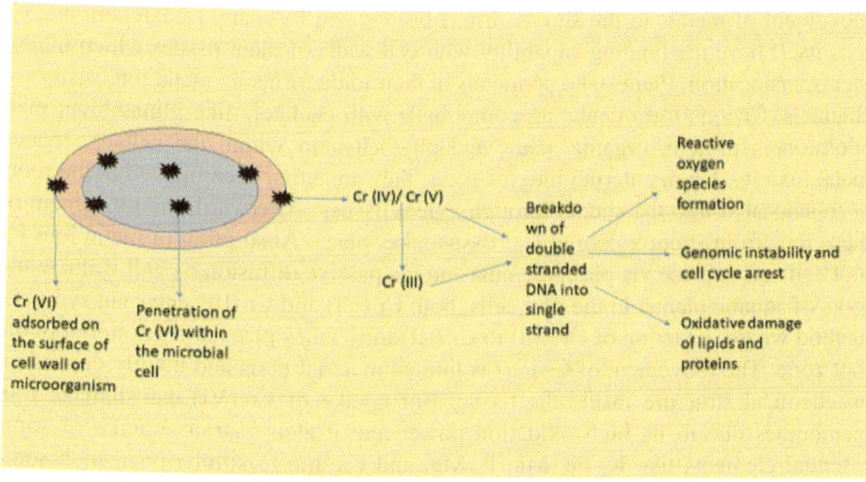

Fig. 4.1 Chromium reduction by microbes and the possible effects associated with cell

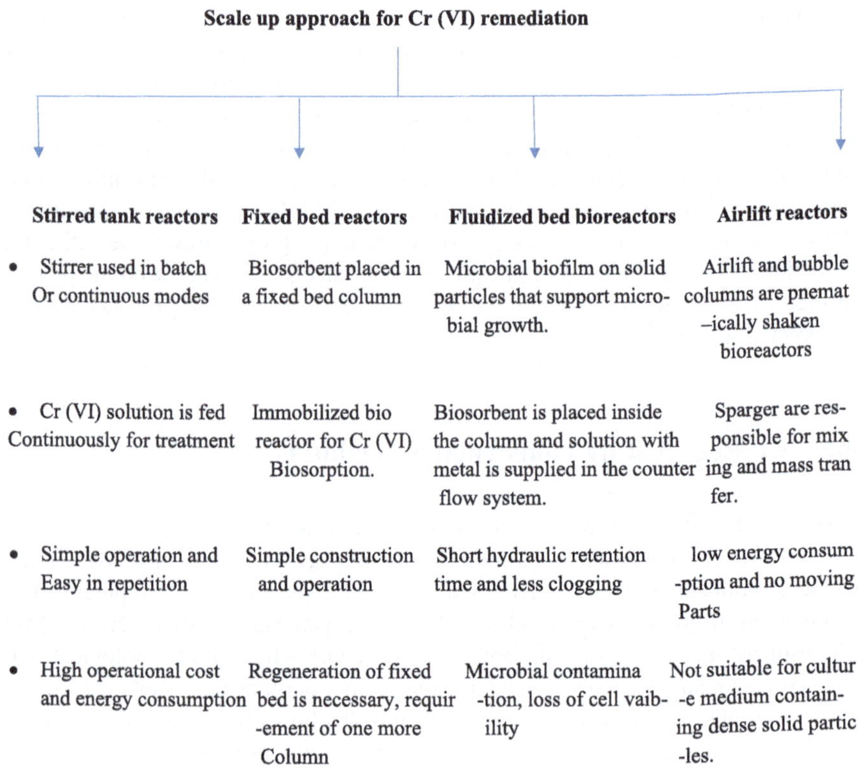

Fig. 4.2 Flowchart showing different scale-up approach for Cr (VI) reduction

plants, and also the complex compounds formed with carboxyl group prohibits movement of metals to the shoots. It has been stated by many researchers that Cr ions itself have the binding capability with cell walls of plant tissues which inhibit their translocation. Plants take up metals in their ionic forms by metal ion carriers or channels. Cr ions form chelating compounds with chelators like glutathione, metallothionein protein, organic acids, and phytochelatin within the cells to reduce metal toxicity. Upon entering into the roots, they are either accumulated in the roots or translocated into the shoots through xylem tissues. Afterward, the movement of these ions from shoot xylem to leaf tissue takes place. Absorption of metal ions by root cells takes place via plasmalemma and via passive diffusion by cell walls in the roots of aquatic plants. In the root cells, both Cr (VI) and Cr (III) enter via symplast method where reduction of Cr (VI) to Cr (III) form takes place and get stored in the root zone. The movement of Cr ions is limited in aerial parts and mainly dependent on chemical structure inside the tissue. But uptake of Cr (VI) can damage root membranes due to its high oxidation power and it also restricts uptake of some essential elements like K, Fe, Mn, P, Mg, and Ca due to similarity in their ionic forms.

Microorganisms too have vital role in CW function as they involve metal reduction mainly by an energy-dependent active process called as bioaccumulation and another one is nonenergy-dependent passive process known as biosorption. In CW, microorganisms influence metals by biosorption, metal speciation, methylation of heavy metals and precipitation by sulfate reduction. Metal sorption by some bacteria occurs by formation of amorphous mineral inclusions. Mycorrhizae forms a symbiotic connection between roots and soil and provides adsorptive surface area for metals present in toxic form in the soil (Table 4.3).

4.6 Biostimulation

Biostimulation is the modification of the process to enhance the growth of existing bacteria during course of bioremediation. Various nutrients and electron accepters or donors like lactate, molasses, or acetate are required in the modification process. Based on the physico-chemical properties and indigenous microbial communities, the efficiency of each electron donors in this process depends. It helps in heavy metal reduction and microbial growth kinetics in a specified environmental condition. A study conducted by (Brodie et al. 2011) stated that in presence of acetate as an electron donor, 16 mM Cr (VI) was reduced to zero within 25 days. Some more experiments were conducted by (Varadharajan et al. 2015) using variety of electron donors for chromium reduction based on lactate polymerization. The experiments involved treatment of Cr (VI) contaminated groundwater using different electron donors such as polylactate cysteine, primer hydrogen release compound (HRC), and extended HRC. Both the electron donors proved to be effective in overall reduction process with enhanced biomass and their activity.

4.7 Pilot-Scale Studies

Generally, the findings obtained from laboratory experimental setup do not essentially equate to the results from large scale on-site operating conditions. A very restricted number of pilot-scale experiments have been implemented for Cr (VI) remediation (Table 4.4).

4.8 Future Perspectives in Chromium Removal

The development and certain modification in the reduction processes offer great opportunities for the ongoing heavy metal pollution problem. Fungi and bacteria have been used recently in the proteomic or transcriptomic studies on their response to hexavalent chromium. Certain modifications such as use of immobilized

Table 4.3 Different types of constructed wetlands operational for Cr (VI) reduction

Types of constructed wetlands (CW)	Cr concentration (mg/L)	Porous media used in CW	Time (days)	zTypes of plants used	Removal efficiency (%)	Effect on plants	References
Free water surface CW	0.018	Sediment	7–12	*T. domingensis* *E. crassipes* *E. crassipes* + *T. domingensis*	88 7 30		Hadad et al. (2010)
	0.033	Muddy sediment	7–12	*Typha domingensis*	85	Biomass and plant height was higher than natural wetlands	Di Luca et al. (2011)
	0.4	Sediment	7–12	*Typha domingensis*	90		Fibbi et al. (2012)
Horizontal surface flow CW	0.0008–0.0158	Gravel	2.1	*Phragmites australis*	72		Dotro et al. (2012)
	1.1, 0.08–5.9	Granitic rock	5.2.4	*Typha latifolia*	50–95		Kelvin and Tole (2011)
	0.5	Gravel	11	Macro-hydroplants	60		Yadav et al. (2010)
Vertical flow CW	10, 15, 20	Gravel	6–48 h	*C. indica* L.	98.3	Continuous plant growth and production of new biomass	Mant et al. (2006)
	10–20	Gravel	6 h	*Penisetum purpureum* *Brachiaria decumbens* *Phragmites australis*	78.1 68.5 56.7		Kongroy et al. (2012)
Hybrid CW systems	0.016	Gravel and soil	–	*Cyperus alternifolius*, SSF—*Phragmites australis*, *Vetiveria zizanioides*, *Typha orientalis*, FWS-*P. australis*, *T. orientalis*, *Eleocharis dulcis*,	44		Lara et al. (2017)

Table 4.4 Pilot-scale studies for Cr (VI) reduction

Type of reactor used	Method of reduction	Cr (VI) concentration	Operating time (days)	Efficiency (%)	References
Arthrobacter viscosus biofilm supported on granular activated carbon	Biosorption	10 and 100 mg/L	30	99.9 and 72, respectively	Quintelas et al. (2009)
Biobarrier	–	50 mg/L	130	–	Jeyasingh et al. (2011)
Acinetobacter haemolyticus in ChromeBac™ system	–	17–81 mg/L	60	100	Ahmad et al. (2010)
Aerobic packed-bed mixed cultures	–	5.5	3 h	100	Tziotzios et al. (2008)
Aerobic trickling filter mixed cultures	–	5–100	0.3–6 h	100	Dermou and Vayenas (2007)
Desulfomicrobium norvegicum in fixed bed column	–	15 mg/L	18	100	Battaglia-Brunet et al. (2006)

microbial cells and enzymes in combination with nanotechnology like infusion of carbon nanotubes into calcium alginate beads have better efficiency in Cr (VI) reduction. Application of nanomaterials with metal reducing bacteria can be efficient as they act as an electron donor, thus accelerating Cr (VI) reduction into Cr (III) (Gutierrez-Corona et al. 2016; Seo and Roh 2015). Genetically, engineered bacteria have the great adaptability and treatment efficiency for the removal of chromium compounds. Development of technologies like combining electrochemical and biological processes together may prove to be helpful in treatment of chromium released from tannery effluent in future. Bioaugmented microorganisms have diverse metabolic pathways and robustness which is a must requirement for high-scale application (He et al. 2014).

4.9 Conclusion

Diversity in anthropogenic activities and industrialization has increased the risk of life due to heavy metal pollution. Among the heavy metal, Cr (VI) possesses a great threat to environment as well as life of living beings due to its mutagenic, carcinogenic, and teratogenic behavior. Different Cr (VI) remediation techniques like biosorption and biotransformation involving variety of microorganisms have been implemented and proven to be cost-effective, eco-friendly and efficient. Presence of functional groups like polysaccharides, amide I, amide II, amide III, carboxyl, and sulfonate mediates the Cr (VI) accumulation inside the cell, thereby transforming Cr (VI) into Cr (III) form. Use of different types of constructed wetlands has been also proven to be effective in treatment process. Despite of all these above-explained treatment technology, there is a gap between laboratory outcomes and pilot-scale studies of Cr contaminated sites. A suitable operational strategy may fulfill the gap and can be applied to Cr contaminated sites.

References

Ahluwalia SS, Goyal D (2007) Microbial and plant derived biomass for removal of heavy metals from wastewater. Bioresour Technol 98:2243–2257. https://doi.org/10.1016/j.biortech.2005.12.006

Ahmad WA, Zakaria ZA, Khasim AR, Alias MA, Ismail SMHS (2010) Pilot-scale removal of chromium from industrial wastewater using the ChromeBac™ system. Biores Technol 101 (12):4371–4378. https://doi.org/10.1016/j.biortech.2010.01.106

Asha Latha P, Sandeep Reddy S (2013) Review on bioremediation potential tool for removing environmental pollution. Int J Basic Appl Chem Sci 3:21–33

Bahafid W, Joutey NT, Sayel H, Boularab I, El Ghachtouli N (2013) Bioaugmentation of chromium-polluted soil microcosms with Candida tropicalis diminishes phytoavailable chromium. J Appl Microbiol 115:727–734. https://doi.org/10.1111/jam.12282

Battaglia-Brunet F, Touze S, Michel C, Ignatiadis I (2006) Treatment of chromate polluted groundwater in a 200 dm3 pilot bioreactor fed with hydrogen. J Chem Technol Biotechnol 81:1506–1513. https://doi.org/10.1002/jctb.1551

Bharagava RN, Mishra S (2018) Hexavalent chromium reduction potential of *Cellulosimicrobium* sp. isolated from common effluent treatment plant of tannery industries. Ecotoxicol Environ Saf 147:102–109. https://doi.org/10.1016/j.ecoenv.2017.08.040

Brodie EL, Joyner DC, Faybishenko B, Conrad ME, Rios-Velazquez C, Malave J, Martinez R, Mork B, Willett A, Koenigsberg S, Herman DJ (2011) Microbial community response to addition of polylactate compounds to stimulate hexavalent chromium reduction in groundwater. Chemosphere 85(4):660–665. https://doi.org/10.1016/j.chemosphere.2011.07.021

Chen CY, Cheng CY, Chen CK, Hsieh MC, Lin ST, Ho KY, Li JW, Lin CP, Chung YC (2016) Hexavalent chromium removal and bioelectricity generation by *Ochrobactrum* sp. YC211 under different oxygen conditions. J Environ Sci Health 51:502–508. https://doi.org/10.1080/10934529.2015.1128731

Chojnacka K (2010) Biosorption and bioaccumulation-the prospects for practical applications. Environ Int 36:299–307. https://doi.org/10.1016/j.envint.2009.12.001

Chowdhury SR, Yanful EK, Pratt AR (2012) Chemical states in XPS and Raman analysis during removal of Cr(VI) from contaminated water by mixed maghemite-magnetite nanoparticles. J Hazard Mater 235:246–256

Dermou E, Vayenas DV (2007) A kinetic study of biological Cr (VI) reduction in trickling filters with different filter media types. J Hazard Mater 145:256–262. https://doi.org/10.1016/j.jhazmat.2006.11.017

Dhal B, Thatoi H, Das N, Pandey BD (2010) Reduction of hexavalent chromium by *Bacillus* sp. isolated from chromite mine soils and characterization of reduced product. J Chem Technol Biotechnol 85:1471–1479. https://doi.org/10.1002/jctb.2451

Di Bona KR, Love S, Rhodes NR, McAdory D, Sinha SH, Kern N, Kent J, Strickland J, Wilson A, Beaird J, Ramage J, Rasco LF, Vincent JB (2011) Chromium is not an essential trace element for mammals: effects of a “lowchromium” diet. J Biol Inorg Chem 16:381–390. https://doi.org/10.1007/s00775-010-0734-y

Di Luca GA, Maine MA, Mufarrege MM, Hadad HR, Sanchez GC, Bonetto CA (2011) Metal retention and distribution in the sediment of a constructed wetland for industrial wastewater treatment. Ecol Eng 37:1267–1275. https://doi.org/10.1016/j.ecoleng.2011.03.003

Dotro G, Castro S, Tujchneider O, Piovano N, Paris M, Faggi A, Palazolo P, Larsen D, Fitch M (2012) Performance of pilot-scale constructed wetlands for secondary treatment of chromium-bearing tannery wastewaters. J Hazard Mater 239–240:142–151. https://doi.org/10.1016/j.jhazmat.2012.08.050

Fang Z, Qiu X, Huang R, Qiu X, Li M (2011) Removal of chromium in electroplating wastewater by nanoscale zero-valent metal with synergistic effect of reduction and immobilization. Desalination 280:1–3, 224–231

Fedorovych DV, Gonchar MV, Ksheminska HP, Prokopiv TM, Nechay HI, Kaszycki P, Koloczek H, Sibirny AA (2009) Mechanisms of chromate detoxification in yeasts. Microbiol Biotechnol 3:15–21

Fernandez PM, Cabral ME, Delgado OD, Farina JI, Figueroa LIC (2013) Textile dye polluted waters as an unusual source for selecting chromate-reducing yeasts through Cr(VI)-enriched microcosms. Int Biodeterior Biodegrad 79:28–35. https://doi.org/10.1016/j.ibiod.2013.01.009

Fernandez PM, Cruz EL, Figueroa LI (2014) Perspective in bioremediation: enhancing the hexavalent chromium removal using native yeasts from Tucuman, Argentina. In: Alejandra P, Alvarez A (eds), Current Status of Bioremediation in Latin American Countries: Perspective of a Region in Progress. Springer, pp 209–222 (ISBN 978-3-319-05738-5)

Fernandez PM, Vinarta SC, Bernal AR, Cruz EL, Figueroa LIC (2018) Bioremediation strategies for chromium removal: Current research, scale-up approach and future perspectives. Chemosphere 208:139–148. https://doi.org/10.1016/j.chemosphere.2018.05.166

Fibbi D, Doumett S, Lepri L, Checchini L, Gonnelli C, Coppini E, Del Bubba M (2012) Distribution and mass balance of hexavalent and trivalent chromium a subsurface, horizontal

flow (SF-h) constructed wetland operating as posttreatment of textile wastewater for water reuse. J Hazard Mater 199–200:209–216. https://doi.org/10.1016/j.jhazmat.2011.10.089
Finley B, Scott PK, Glynn ME, Paustenbach D, Donovan E, Thuett KA (2017) Chromium speciation in the blood of metal-on-metal hip implant patients. Toxicol Environ Chem 99:48–64. https://doi.org/10.1080/02772248.2016
Frois SR, Grassi MT, Fernandes TC, Barreto RAS, Abate G (2011) Preconcentration of Cr (III) and speciation analysis of chromium employing montmorillonite saturated with potassium ions. Quím Nova 34:462–467. https://doi.org/10.1590/S0100-40422011000300018
Gong Y, Gai L, Tang J, Fu J, Wang Q, Zeng EY (2017) Reduction of Cr(VI) in simulated groundwater by FeS-coated iron magnetic nanoparticles. Sci Total Environ 595:743–751
Guidelines for drinking-water quality (1996) Health Criteria and Other Supporting Information, 2nd ed, World Health Organization, Geneva
Gutierrez-Corona JF, Romo-Rodríguez P, Santos-Escobar F, Espino-Saldaña AE, Hernández-Escoto H (2016) Microbial interactions with chromium: basic biological processes and applications in environmental biotechnology. World J Microbiol Biotechnol 32(12):191. https://doi.org/10.1007/s11274-016-2150-0
Hadad HR, Mufarrege MM, Pinciroli M, Di Luca GA, Maine MA (2010) Morphological response of Typha domingensis to an industrial effluent containing heavy metals in a constructed wetland. Arch Environ Contam Toxicol 58:666–675. https://doi.org/10.1007/s00244-009-9454-0
Hamilton EM, Young SD, Bailey EH, Watts MJ (2018) Chromium speciation in foodstuffs: a review. Food Chem 250:105–112. https://doi.org/10.1016/j.foodchem.2018.01.016
He Z, Yao Y, Lu Z, Ye Y (2014) Dynamic metabolic and transcriptional profiling of *Rhodococcus* sp. strain YYL during the degradation of tetrahydrofuran. Appl Environ Microbiol 80:2656–2664. https://doi.org/10.1128/AEM.04131-13
Houa S, Xua X, Wanga M, Lua T, Sunb CQ, Pan L (2018) Synergistic conversion and removal of total Cr from aqueous solution by photocatalysis and capacitive deionization Chemical Engineering Journal. 337:398–404
Huang H, Wu K, Khan A, Jiang Y, Ling Z, Liu P, Chen Y, Tao X, Li X (2016) A novel Pseudomonas gessardii strain LZ-E simultaneously degrades naphthalene and reduces hexavalent chromium. Bioresour Technol 207:370–378. https://doi.org/10.1016/j.biortech.2016.02.015
Jeyasingh J, Somasundaram V, Philip L, Bhallamudi SM (2011) Pilot scale studies on the remediation of chromium contaminated aquifer using bio-barrier and reactive zone technologies. Chem Eng J 167:206–214. https://doi.org/10.1016/j.cej.2010.12.024
Jobby R, Jha P, Yadav AK, Desai N (2018) Biosorption and biotransformation of hexavalent chromium [Cr(VI)]: A comprehensive review. Chemosphere 207:255–266. https://doi.org/10.1016/j.chemosphere.2018.05.050
Juvera-Espinosa J, Morales-Barrera L, Cristiani-Urbina E (2006) Isolation and characterization of a yeast strain capable of removing Cr(VI). Enzym Microb Technol 40:114–121. https://doi.org/10.1016/j.enzmictec.2005.10.045
Kadlec RH, Wallace SD (2009) Treatment Wetlands, 2nd edn. CRC Press, Boca Raton, Florida
Kelvin K, Tole M (2011) The efficacy of a tropical constructed wetland for treating wastewater during the dry season: the Kenyan experience. Water Air Soil Pollut 215:137–143
Khan SU, Islam DT, Farooqi IH, Ayub S, Basheer F (2019) Hexavalent chromium removal in an electrocoagulation column reactor: Process optimization using CCD, adsorption kinetics and pH modulated sludge formation. Process Saf Environ Prot 122:118–130
Khani MH, Pahlavanzadek H, Alizadeh K (2012) Biosorption of strontium from aqueous solution by fungus Aspergillus terreus. Environ Sci Pollut Res Int 19:2408–2418. https://doi.org/10.1007/s11356-012-0753-z
Khare N, Bajpai J, Bajpai AK (2018) Graphene coated iron oxide (GCIO) nanoparticles as efficient adsorbent for removal of chromium ions: Preparation, characterization and batch adsorption studies Environmental Nanotechnology. Monitoring & Manag 10:148–162
Khoubestani RS, Mirghaffari N, Farhadian O (2015) Removal of three and hexavalent chromium from aqueous solutions using a microalgae biomass-derived biosorbent. Environ. Prog. Sustain. Energy 34:949–956. https://doi.org/10.1002/ep.12071

Kongroy P, Tantemsapya N, Lin YF, Jing SR, Wirojanagud W (2012) Spatial distribution of metals in the constructed wetlands. Int J Phytoremediat 14:128–141. https://doi.org/10.1080/15226514.2011.573825

Krowiak W (2013) Kinetics and equilibrium of copper and chromium ions removal from aqueous solutions using sawdust. Environ Eng Manag J 12:2125–2135. https://doi.org/10.30638/eemj.2013.264

Kwak HW, Kim MK, Lee JY, Yun H, Kim MH, Park YH, Lee KH (2015) Preparation of bead-type biosorbent from water-soluble Spirulina platensis extracts for chromium (VI) removal. Algal Res 7:92–99. https://doi.org/10.1016/j.algal.2014.12.006

Lara P, Morett E, Juarez K (2017) Acetate biostimulation as an effective treatment for cleaning up alkaline soil highly contaminated with Cr(VI). Environ Sci Pollut Res 24:25513–25521. https://doi.org/10.1007/s11356-016-7191-2

Li Y, Wang W, Zhou L, Liu Y, Mirza ZA, Lin X (2017) Remediation of hexavalent chromium spiked soil by using synthesized iron sulfide particles. Chemosphere 169:131–138

Lia B, Yin W, Xu M, Tana X, Lia P, Gua J, Chianga P, Wua J (2019) Facile modification of activated carbon with highly dispersed nano-1 sized α-Fe2O3 for enhanced removal of hexavalent chromium from aqueous solutions. Chemosphere. https://doi.org/10.1016/j.chemosphere.2019.02.121

Liu W, Liu C, Liu L, You Y, Jiang J, Zhou Z, Dong Z (2017) Simultaneous decolorization of sulfonated azo dyes and reduction of hexavalent chromium under high salt condition by a newly isolated salt-tolerant strain *Bacillus circulans* BWL1061. Ecotoxicol Environ Saf 141:9–16. https://doi.org/10.1016/j.ecoenv.2017.03.005

Mahmoud MS, Mohamed SA (2017) Calcium alginate as an eco-friendly supporting material for baker's yeast strain in chromium bioremediation. Housing Build Nat Res Center J 13:245–254. https://doi.org/10.1016/j.hbrcj.2015.06.003

Maine MA, Sune N, Hadad H, Sanchez G, Bonetto C (2009) Influence of the vegetation on the removal of heavy metals and nutrients in a constructed wetland. J Environ Manag 90:355–363. https://doi.org/10.1016/j.jenvman.2007.10.004

Mant C, Costa S, Williams J, Tambourgi E (2006) Phytoremediation of chromium by model constructed wetland. Bioresour Technol 97:1767–1772. https://doi.org/10.1016/j.biortech.2005.09.010

Martorell MM, Fernandez PM, Farina, JI, Figueroa LIC (2012) Cr(VI) reduction by cell-free extracts of Pichia jadinii and Pichia anomala isolated from textile dye factory effluents. Int Biodeterior Biodegrad 71:80–85. https://doi.org/10.1016/j.ibiod.2012.04.007

Mukherjee R, Bhunia P, De S (2014) Long term filtration modelling and scaling up of mixed matrix ultrafiltration hollow fiber membrane: a case study of chromium (VI) removal. J Mem Sci 570–571:204–214

Nemr AE, El-Sikaily A, Khaled A, Abdelwahab O (2015) Removal of toxic chromium from aqueous solution, wastewater and saline water by marine red alga *Pterocladia capillacea* and its activated carbon. Arab J Chem 8:105–117. https://doi.org/10.1016/j.arabjc.2011.01.016

Pechancova R, Pluhaček T, Milde D (2019) Recent advances in chromium speciation in biological samples. Spectrochimica Acta Part B. 152:109–122. https://doi.org/10.1016/j.sab.2018.12.008

Poljsak B, Pocsi I, Raspor P, Pesti M (2010) Interference of chromium with biological systems in yeasts and fungi: a review. J Basic Microbiol 50:21–36. https://doi.org/10.1002/jobm.200900170

Pradhan D, Sukla LB, Sawyer M, Rahman PKSM (2017) Recent bioreduction of hexavalent chromium in wastewater: A review. J Ind Eng Chem 55:1–20

Quintelas C, Fonseca B, Silva B, Figueiredo H, Tavares T (2009) Treatment of chromium(VI) solutions in a pilot-scale bioreactor through a biofilm of *Arthrobacter viscosus* supported on GAC. Biores Technol 100:220–226. https://doi.org/10.1016/j.biortech.2008.05.010

Sanea P, Chaudharia S, Nemadea P, Sontakke S (2018) Photocatalytic reduction of chromium (VI) using combustion synthesized TiO2. J Environ Chem Eng 6:68–73

Sathishkumar K, Murugan K, Benelli G, Higuchi A, Rajasekar A (2016) Bioreduction of hexavalent chromium by *Pseudomonas stutzeri* L1 and *Acinetobacter baumannii* L2. Ann Microbiol 67:91–98. https://doi.org/10.1007/s13213-016-1240-4

Seo H, Roh Y (2015) Biotransformation and its application: biogenic nano-catalyst and metal-reducing-bacteria for remediation of Cr (VI)-contaminated water. J Nanosci Nanotechnol 15:5649–5652. https://doi.org/10.1166/jnn.2015.10446

Sharma S, Adholeya A (2011) Detoxification and accumulation of chromium from tannery effluent and spent chrome effluent by *Paecilomyces lilacinus* fungi. Int Biodeterior Biodegrad 65:309–317. https://doi.org/10.1016/j.ibiod.2010.12.003

Sivakumar D (2016) Biosorption of hexavalent chromium in a tannery industry wastewater using fungi species. Global J Environ Sci Manag 2:105–124. https://doi.org/10.7508/GJESM.2016.02.002

Su H, Fang Z, Tsang PE, Zheng L, Cheng W, Fang J, Zhao D (2016) Remediation of hexavalent chromium contaminated soil by biochar-supported zero-valent iron nanoparticles. J Hazard Mater 318:533–540

Sultana M-Y, Akratos CS, Pavlou S, Vayenas DV (2014) Chromium removal in constructed wetlands: A review. Int Biodeterior Biodegradation 96:181–190. https://doi.org/10.1016/j.ibiod.2014.08.009

Thatoi H, Das S, Mishra J, Rath BP, Das N (2014) Bacterial chromate reductase, a potential enzyme for bioremediation of hexavalent chromium: a review. J Environ Manag 146:383–399. https://doi.org/10.1016/j.jenvman.2014.07.014

Tziotzios G, Dermou E, Politi D, Vayenas DV (2008) Simultaneous phenol removal and biological reduction of hexavalent chromium in a packed bed reactor. J Chem Technol Biotechnol 83:829–835. https://doi.org/10.1002/jctb.1876

Upadhyay S, Tarafdar A, Sinha A (2018) Assessment of *Serratia* sp. isolated from iron ore mine in hexavalent chromium reduction:. kinetics, fate and variation in cellular morphology. Environ Techn https://doi.org/10.1080/09593330.2018.1521875

US Environmental Protection Agency (1987) Office of Drinking Water. Health advisory chromium, Washington, DC

Varadharajan C, Han R, Beller HR, Yang L, Marcus MA, Michel M, Nico PS (2015) Characterization of Chromium Bioremediation Products in Flow-Through Column Sediments Using Micro–X-ray Fluorescence and X-ray Absorption Spectroscopy. J Environ Qual 44 (3):729–738. https://doi.org/10.2134/jeq2014.08.0329

Viti C, Marchi E, Decorosi F, Giovannetti L (2014) Molecular mechanisms of Cr(VI) resistance in bacteria and fungi. FEMS Microbiol Rev 38:633–659. https://doi.org/10.1111/1574-6976.12051

Wang W (2018) Chromium (VI) removal from aqueous solutions through powdered activated carbon countercurrent two-stage adsorption. Chemosphere 190:97–102

Wang L, Wang N, Zhu L, Yu H, Tang H (2008) Photocatalytic reduction of Cr(VI) over different TiO2 photocatalysts and the effects of dissolved organic species. J Hazard Mater 152:93–99. https://doi.org/10.1016/j.jhazmat.2007.06.063

Wu X, Zhu X, Song T, Zhang L, Jia H, Wei P (2015) Effect of acclimatization on hexavalent chromium reduction in a biocathode microbial fuel cell. Bioresour Technol 180:185–191. https://doi.org/10.1016/j.biortech.2014.12.105

Yadav AK, Kumar N, Sreekrishnan TR, Satya S, Bishnoi NR (2010) Removal of chromium and nickel from aqueous solution in constructed wetland: mass balance, adsorption-desorption and FTIR study. Chem Eng J 160:122–128. https://doi.org/10.1016/j.cej.2010.03.019

Zouboulis AI, Loukidou MX, Matis KA (2004) Biosorption of toxic metals from aqueous solutions by bacterial strain isolated from metal-polluted soils. Process Biochem 39:909–916. https://doi.org/10.1016/S0032-9592(03)00200-0

Chapter 5
Dissolved Arsenic in Groundwater Bodies: A Short Review of Remediation Technologies

M. S. V. Naga Jyothi, S. Gayathri, T. Pushparaj Gandhi, and Shihabudheen M. Maliyekkal

Abstract Arsenic is a crystalline metalloid present in the earth's crust in different minerals. Various natural processes and human interventions have mobilized As into the groundwater system over the years. The dissolved As presence in drinking water is now a significant health concern in India and many other countries. As is a proven carcinogen, and long-term ingestion of low concentrations of As can lead to serious health issues. As in drinking water was first regulated by the United States Public Health Service (USPHS), which set a maximum permissible concentration (MPC) of 50 μg/L. Later, based on lung and bladder cancer risk, the United States Environmental Protection Agency (USEPA) promulgated a maximum contaminant level (MCL) of 10 μg/L. Indian standard for drinking water also recommends a total As concentration of 10 μg/L as the acceptable limit. However, some of the developed countries have As standards lower than 10 μg/L as well. The removal of As in drinking water is inevitable, considering its widespread presence in groundwater, potential human toxicity, and stringent regulation in drinking water. Several point-of-use (PoU) and community-based treatment technologies have been developed to provide As-free water to the affected population. The processes adopted in these technologies include coagulation and filtration, sorptive filtration, ion-exchange, electrocoagulation, and membrane filtration. The present chapter reviews these technologies critically and their suitability in the Indian context. The chapter also provides insight into the As occurrence in the environment, its speciation, toxicity, and health effects.

Keywords Arsenic toxicity · Treatment technologies · Point-of-use treatment · Groundwater

---

M. S. V. Naga Jyothi · S. Gayathri · T. Pushparaj Gandhi · S. M. Maliyekkal (✉)
Department of Civil and Environmental Engineering, Indian Institute of Technology Tirupati, Tirupati 517 506, India
e-mail: shihab@iittp.ac.in

S. Singh et al. (eds.), *Pollution Control Technologies*, Energy, Environment, and Sustainability, https://doi.org/10.1007/978-981-16-0858-2_5

5.1 Introduction

Freshwater is crucial to sustaining life, and however, it is a limited resource. Among freshwater resources, groundwater is considered more dependable in terms of quality and quantity as compared to surface water. The water drawn from an underground reservoir is typically free from microbial contamination and requires less treatment. It is an indispensable source of drinking water supply in many countries around the world. However, natural causes and human interventions have significantly reduced the quality and quantity of groundwater over the years. The increasing presence of dissolved chemicals in groundwater is a primary concern now. Dissolved arsenic (As) is one such chemical contaminant widely reported in groundwater reserves. According to recent reports, 107 countries around the globe have reported cases of As contamination in groundwater. India, Bangladesh, Pakistan, Thailand, Taiwan, Africa, and Vietnam are some of the most affected nations (Shahid et al. 2018; Ghosh et al. 2019). Around 70 million people in India are exposed to As poisoning through drinking As-contaminated water (Chakrabarti et al. 2019). The magnitude of contamination in Bangladesh is much higher, as 50–80 million people are suspected of being at risk of drinking As-contaminated water (Gilbert 2004; Chakrabarti et al. 2019).

The scale of this disastrous effect in terms of mortality is more than the accidents in Bhopal, India (1984), and Chernobyl, Ukrain (1986) (Smith et al. 2000). Besides drinking water, As can enter the human body through other exposure mechanisms. Accumulation of As in crops irrigated with As laden water or As-containing pesticides can lead to its entry into the food chain (Ahsan and Del Valls 2011). Another way it reaches humans is through seafood, as arsenobetaine and arsenosugar compounds (Majumder and Banik 2019; Jinadasa and Fowler 2019). According to the International Agency for Research on Cancer (IARC), As is a Group I carcinogen, and chronic intake causes lung, kidney, and skin cancers (Laxman et al. 2016). Illness such as hyperpigmentation, keratosis, cardiovascular, developmental, hematologic, renal, hepatic, nervous, and respiratory disorders are also reported (Tchounwou et al. 2019). In these contexts, it is essential to control the excess As present in groundwater.

This chapter gives an overview of the various As treatment technologies such as coagulation and filtration, electrocoagulation, adsorption, ion-exchange, membrane filtration, capacitive deionization, and phytoremediation. Some of the well-known field methods on As remediation are also discussed. This chapter also discusses the physical and chemical properties, occurrence, drinking water standards, and origin of As in groundwater.

5.2 Properties, Occurrence, and Sources of Arsenic

According to the encyclopedia of microbiology, the name As is derived from the Persian word 'zarnikh,' meaning 'yellow orpiment (AS_2S_3)' (Mudhoo et al. 2011). Albertus Magnus from Regensburg possibly first discovered As in AD 1250.

However, a more reliable report says German pharmacist Johann Schroeder made the As discovery in 1649 (Mudhoo et al. 2011). It belongs to the 'nitrogen family' with an atomic number of 33 and an electronic configuration of [Ar] $3d^{10}\,4s^2\,4p^3$. It is a metalloid with both metallic and non-metallic properties. Chemically, it has a close resemblance to phosphorus. Three hundred and twenty species of As-containing minerals are present in nature (Fleischer 1983). However, minerals such as orpiment, arsenopyrite, and realgar are considered as principal As ores (Hossain 2006). As is also found in sedimentary environments being adsorbed by Fe(III) and Mn(IV) after the weathering of sulfide minerals (Nickson et al. 1998, 2000). It is reported that arsenopyrite is the prime source of As pollution in Bangladesh and West Bengal (Mudhoo et al. 2011).

5.3 Arsenic Speciation

As is a redox-sensitive element that can exist in various oxidation states (+5, +3, 0, −3). The two predominant As species found in subsurface water are arsenate [As (V)] and arsenite [As(III)] (Kartinen and Martin 1995). The distribution of these species depends on the Eh and pH conditions of groundwater (Fig. 5.1). Arsenate ($HAsO_4^{2-}$) generally predominates under oxidizing conditions in the pH range of 6–9, whereas arsenite (H_3AsO_3) predominates when conditions become sufficiently reducing (below pH 9.22). Arsine sulfides and arsine gas are the expected forms in a highly reducing environment, such as anoxic and anaerobic sediments. However, the presence of these species varies depending upon the solution pH and other environmental conditions.

5.4 Origin of Arsenic in Groundwater Resources

A large number of groundwater sources across the globe are contaminated with As. As is usually mobilized by natural weathering, volcanic eruption, and forest fires following biological and geochemical reactions (Panagiotaras and Nikolopoulos 2015). Anthropogenic activities like mining of metals, coal combustion, oil extraction, pesticide use, and disposal of waste from various industries can also mobilize As in the environment (Shrivastava et al. 2015). However, most environmental As problems are primarily due to natural reasons rather than anthropogenic or geothermal activities (Welch et al. 2000; Lackovic et al. 2000). The biogeochemical reactions in As-rich minerals such as the reductive dissolution of iron oxide, oxidation of sulfide minerals, adsorption/desorption, and geothermal processes can help to understand the origin of the contamination and take appropriate mitigation measures (Herath et al. 2016).

Though a few hypotheses are reported stating the origin of As in groundwater systems, the reductive dissolution of Fe minerals seems to be the significant cause

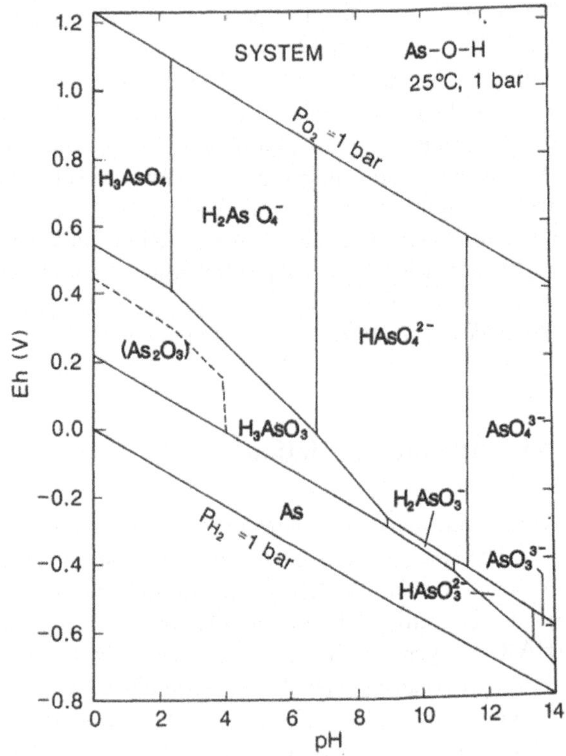

Fig. 5.1 Eh–pH diagram for As at 25 °C and 101.3 kPa. (Brookins 1988). Reprinted from 'Eh-pH diagrams for geochemistry' with permission of Springer

of As release into the groundwater. For instance, in the presence of organic matter in aquifer formation, biological reduction of mineral scorodite ($FeAsO_4{\cdot}2H_2O$) to Fe(II) by iron dissimilatory reducing bacterium *S. algae* releases As(V) into the aqueous medium (Welch et al. 2000). In another way, As(III) is mobilized from ferrihydrite [Fe(III)OAs$(OH)_2$] by reductive dissolution through the sulfidization process, as shown in Eq. (1) (O'Day et al. 2004). Subsequently, the released sulfide reacts and exchanges with the As present in the Fe crystal structure as per chemical Eqs. (2) and (3) (Herath et al. 2016).

$$As_2S_{3(s)} + 6H_2O \rightarrow 2H_3AsO_{3(aq)} + 3HS^-_{(aq)} + 3H^+_{(aq)} \quad (1)$$

$$Fe(III)OAs + HS^-_{(aq)} \xrightarrow{\text{Ligand exchange}} Fe(III)S^-_{(s)} + H_3AsO_{3(aq)} + S^-_{(aq)} \quad (2)$$

$$Fe(III)OAs(OH)_{2(s)} + HS^-_{(aq)} \xrightarrow{\text{Reductive dissolution}} Fe(II) + H_3AsO_{3(aq)} + S^-_{(aq)} \quad (3)$$

The mobility of As species in subsurface water varies with pH, organic matter, redox potential, metal oxides, and other ions present in groundwater. Carbonate ion in groundwater can replace the As in Fe mineral by ligand exchange, thereby mobilize As (Anawar et al. 2004). Adsorption/desorption process by oxides of aluminum, magnesium, and clay minerals also controls the dissolved As concentration in groundwater (Bhattacharya et al. 2009).

5.5 Arsenic Toxicity and Health Effects

Arsenic is a potent toxicant that may exist in the natural system in a variety of chemical forms. It includes inorganic As(III) and As(V), and mono-, di-, and tri-methylated arsenic compounds (Sarkar and Paul 2016). The toxicity and mobility of these compounds differ depending on their chemical forms and valence state (National Research Council 1999; Thomas et al. 2001). The decreasing order of toxicity of different organic and inorganic As species is arsine > arsenite > arsenate > alkyl arsenic acids > arsonium compounds and metallic arsenic (EPA 1988; National Research Council 1999). Among these chemical forms, As (III) and As(V) are commonly found in groundwater. As(III) is generally considered to be more toxic compared to As(V).

The primary human exposure to As is through the oral route, through food and drinking water. Short-term exposure (ingestion of a large amount of As in a short time) can cause severe abdominal cramps, internal bleeding, and even death. Chronic exposure can cause the condition of garlic breath with excessive salivation and sweating, stomatitis, oedema, hyperkeratosis, liver enlargement, jaundice, and renal damage. It may also lead to peripheral arteriosclerosis (black foot disease), peripheral neuritis, and encephalopathy (Hughes 2002; Ratnaike 2003; Mazumder 2008). Children have less tolerance to even low concentrations because of their low body weight, immature absorption, and excretion mechanism (Tippairote et al. 2020). Prolonged consumption of As can also cause bladder, lungs, kidney, prostate, and skin cancers (National Research Council 1999; Palma-Lara et al. 2020).

5.6 Arsenic Problems Across the World

In recent years, the high concentrations of inorganic As in groundwater have become an epidemiological problem. Groundwater abundant in As is found in many regions around the globe (Smedley and Kinniburgh 2002). Table 5.1 shows the list of affected nations. In India, the first case of As contamination in groundwater was recorded in 1978 in West Bengal (Acharyya et al. 2000). In West Bengal, As-enriched groundwater is detected in 111 blocks across twelve districts (Sanyal et al. 2015). Of these, severely affected districts are Murshidabad, Malda, Nadia, North and South 24 Parganas. According to a survey conducted in the Malda district, the concentration of As in groundwater varied about 4.2–884 μg/L (Golui et al. 2017). States, including Bihar, Uttar Pradesh, Jharkhand, Chhattisgarh,

Table 5.1 Major As-affected nations

Affected nation	Range of As concentrations (μg/L)	References
India	10–3200	Sarkar and Paul (2016)
Bangladesh	1–2500	
China	50–2000	
Vietnam	1–3050	
Thailand	1–5000	
Taiwan	10–1820	
Inner Mongolia	1–2400	
Argentina	1–9900	
Chile	100–1000	
Mexico	8–620	
Hungary; Romania	2–176	
Greece	1–1840	
UK	1–80	
Canada, USA	1–10,000	
Ghana	1–175	
Nepal (Terai)	1–2620	Shrestha et al. (2003)
Turkey (Simav plain)	0.5–561	Gunduz et al. (2010)
Southern France (Aude)	40	Khaska et al. (2015)
Germany (Bavaria)	10–150	Heinrichs and Udluft (1999)
Finland (Pirkanmaa region)	0.05–2230	Backman et al. (2006)
West of Iran	0.4–986.9	Keshavarzi et al. (2011)

Assam, and Manipur, also have reported dissolved As cases in groundwater (Rahman et al. 2009; Chakraborti et al. 2018). Approximately 6 million people are at high health risk (Rahaman et al. 2013; Dey et al. 2014).

5.7 Water Quality Criteria and Standards

Arsenic in water was first regulated by the United States Public Health Service (USPHS), which set a maximum permissible concentration (MPC) of 50 μg/L in 1942 (Pontius et al. 1994). Based on lung and bladder cancer risk, the United States Environmental Protection Agency (USEPA) promulgated a maximum contaminant level (MCL) of 10 μg/L and the maximum contaminant level goal (MCLG) of zero for As on January 22, 2001 (USEPA g2001). This new standard became effective on January 23, 2006. The WHO guideline value is 10 μg/L, which is derived from the studies based on the health effects of As (WHO 2011). The guideline value was suggested based on a healthy 70 kg person drinking an average of 2 L of water

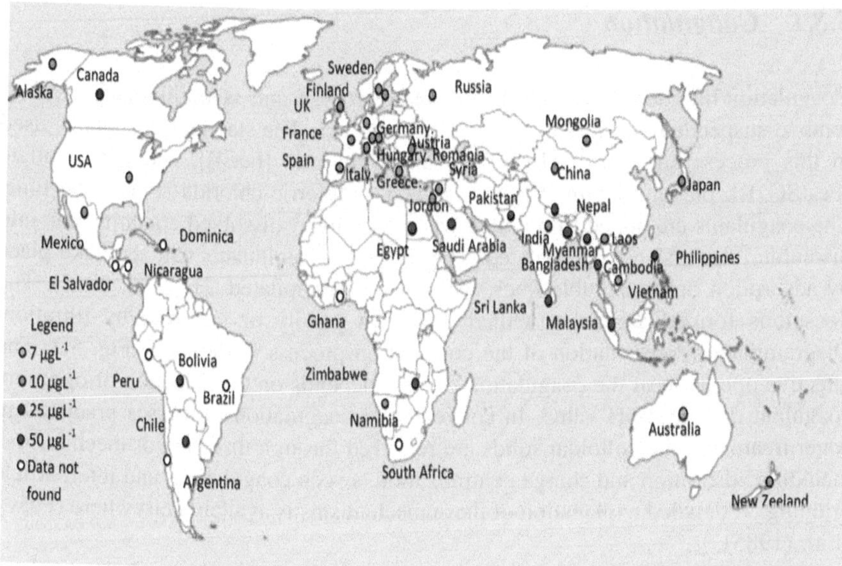

Fig. 5.2 Drinking water standards for As across the world. Reprinted from 'Remediation of inorganic As in groundwater for safe water supply: A critical assessment of technological solutions' with the permission of Elsevier

per day in his/her life span. Figure 5.2 shows the arsenic drinking water standards worldwide (Mondal et al. 2013). Bangladesh, one of the most affected regions, is still following the national standard of 50 μg/L. However, people from Bangladesh and West Bengal tend to drink more water than European or US residents because of hot climatic conditions. Another critical issue to be addressed is the poor dietary levels of the persons living in affected areas of West Bengal and Bangladesh. People with less nutritional levels are more affected by As compared to healthy persons (Mitra et al. 2004). Considering these facts, one can argue that the As standards in India and Bangladesh should be lowered than the international standards.

5.8 Arsenic Remediation Technologies

Many treatment methods are available for As removal from the drinking water system. A brief description of the processes involved in these treatment technologies is discussed below.

5.8.1 Coagulation

Coagulation/filtration is a well-documented process and is traditionally used to remove suspended and dissolved solids from water. The standard coagulants used in this process are alum [$Al_2(SO_4)_3$], ferric chloride [$FeCl_3$], and ferric sulfate [$Fe_2(SO_4)_3$], polyaluminum chlorides (PAC), polyferric chloride (PFC), and lime. The coagulants are added to polluted water to convert dissolved contaminants into insoluble form and get precipitated. The removal of pollutants can also take place by adsorption onto insoluble flocs, which is co-precipitated (Hu et al. 2012). The flocs thus formed are either settled down by gravity or removed by filtration. Diagrammatic representation of the coagulation process is shown in Fig. 5.3. The effective operation of the coagulation process depends on the selection of optimum coagulant dose and pH value. In conventional coagulation processes practiced in water treatment, the colloidal solids are removed through three major mechanisms, including adsorption and charge neutralization, sweep coagulation, and interparticle bridging. A detailed explanation of these mechanisms is available elsewhere (Peavy et al. (1985).

Coagulation with alum: Alum is one of the most commonly used chemicals in the coagulation/filtration process. Alum coagulation can effectively remove As, and suspended solids from water, resulting in significant improvement of water quality. However, many factors like pH, alum dosage, and As concentration, govern its efficiency in removing these pollutants. Alum dosage required for As removal is usually much higher than that needed for turbidity removal in conventional water treatment processes. The excess addition of alum may lead to an increase in dissolved aluminum content in treated water if proper flocculation does not occur. In alum coagulation, pH is maintained in the range of 5–7.5 for optimum removal of As(V) (Ahmed and Rahaman 2000). However, As(III) cannot be removed by this process since it is present in the neutral form (H_3AsO_3) at natural pH. Pre-oxidation by an oxidizing agent like chlorine enhances the total As removal capacity by converting uncharged As(III) to As(V) (Kartinen and Martin 1995). In the absence

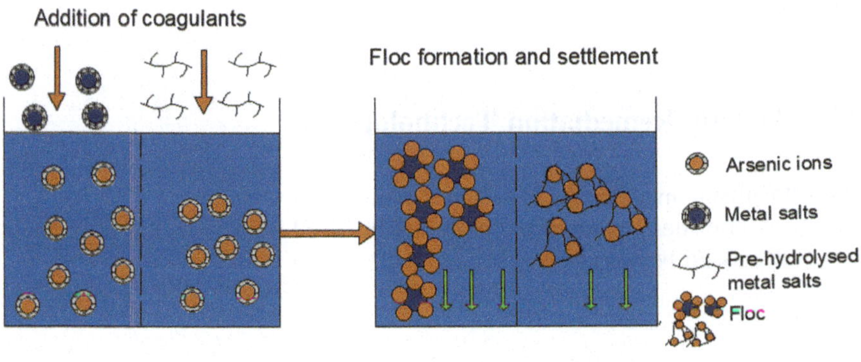

Fig. 5.3 Schematic diagram of the coagulation process using metal salts and polymers

of chlorine, 30 mg/L of alum could remove only 10% of the initial concentration of 300 μg/L As in water, whereas in the presence of chlorine, the As removal rate increased to 90% (Kartinen and Martin 1995).

Coagulation with ferric salts: Ferric chloride and ferric sulfate are the two commonly used iron-based coagulants for As removal in water. Fe-based coagulants are reported to be more effective than Al-based coagulants. However, they are less efficient in removing color and turbidity. Sorg and Logsdon (1978) stated that ferric coagulation is most effective in the pH range of 5–8 for As removal. Logsdon et al. (1974) reported that with ferric sulfate, at an initial concentration of 0.3 mg/L, the removal efficiency varied from 40 to 60%, which was almost four times higher than that of alum coagulation for the same set of conditions. Dosage of 10 mg/L ferric chloride and ferrous sulfate showed 100% removal of As(V) at an initial concentration of 0.1 mg/L (Donmez and Akbal 2011).

Coagulation with pre-hydrolyzed coagulants: The metal salts-based coagulants are chemical intensive and generate excess sludge. The use of pre-hydrolyzed coagulants like PAC and PFC can reduce the production of chemical sludge. Besides, the pre-hydrolyzed coagulants require only less pH adjustment during coagulation due to partially hydrolyzed products. Baskan and Pala (2010) reported that an alum dosage of 80 mg/L removed 96.63% of 0.5 mg/L of As(V) at pH 9. However, PAC with less dosage of 25 mg/L showed >80% removal of As(V) at an initial concentration of 0.3 mg/L at pH 9.7. PAC, along with Fe-Mn, improved removal efficiency to 96.23% at the same experimental condition (Eslami et al. 2019). The addition of Fe-Mn increased the adsorption sites for As(V). Eslami et al. (2019) also reported that PFC dosage of 25 mg/L could remove > 85% of As(V). Upon the addition of Fe-Mn, it increased to 98.26%. However, some of these coagulants are less stable at high pH due to the formation of neutral precipitates (Gao et al. 2005).

Lime softening: As removal by the addition of lime resembles coagulation with metal salts in the treatment mechanism. Here, the precipitated calcium hydroxide [$Ca(OH)_2$] acts as a sorbing flocculent for As. However, it has generally been observed that As removal efficiency by lime is relatively low, usually between 40% and 70%. The highest removal is achieved in the pH range 10.6–11.4 (Logsdon et al. 1974; Sorg and Logsdon 1978). Rosehart and Lee (1972) reported that the lime-softening process effectively removes 95% of soluble As at a Ca/As ratio of 9:8. However, the amount of lime required is much higher, and can be recommended for As removal only if hardness removal is also desired. High pH of the treated water and requirement of high magnesium hardness renders it unsuitable for practical purposes.

Electrocoagulation (EC): Unlike conventional coagulation processes, electrocoagulation uses an electrical charge to produce coagulant species in-situ. An electrolytic cell that consists of a pair of electrodes (cathode and anode) is employed to perform the reaction. The current applied through the electrodes triggers the electrochemical reaction, resulting in the oxidation of sacrificial anode, and releases metal ions into the water. The reduction at cathode generates hydroxyl ions and hydrogen gas. Finally, these reactions result in the generation of metal hydroxides,

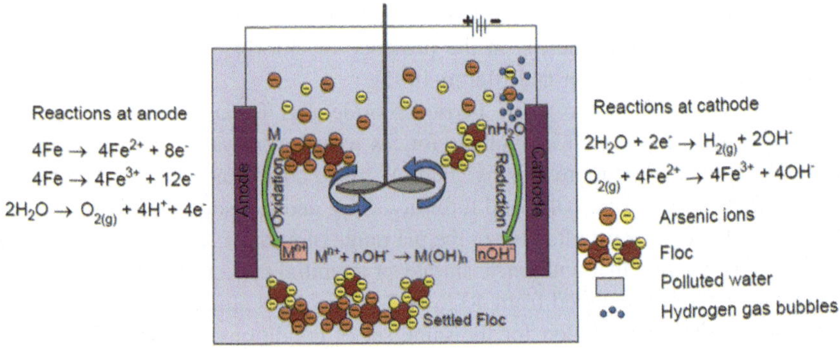

Fig. 5.4 Schematic diagram showing the electrocoagulation process

which are responsible for coagulation, as shown in Fig. 5.4 (Mollah et al. 2001; Binh et al. 2018). Commonly used anode electrodes in As remediation are iron and aluminum. However, the iron electrode is more suitable than the Al electrode because of its higher affinity to As.

In the electrolytic reaction, the anode electrode [Fe/Al] dissolutes as monomeric or polymeric hydroxyl metal species based on the solution pH (Daniel and Prabhakara Rao 2012). Besides, there is the formation of Fe/Al oxides [e.g., $Al(OH)_3$], hydrous metal oxide (HFO), amorphous $Fe(OH)_3$, goethite (α-FeOOH), and lepidocrocite (γ-FeOOH), which adsorbs As ions (Chen 2004). Thus, the mechanisms responsible for As removal include precipitation, adsorption, and co-precipitation. Factors that affect the EC of As are the type and shape of the electrode, current density, applied current, operating time, and pH (Kobya et al. 2020). Table 5.2 shows the performance of the EC reactor under various operating conditions. A study on As removal by EC reported that the presence of co-anions reduces the removal efficiency and their impact order is as follows: phosphates > chloride > silicate > bicarbonate. Some cations like calcium and magnesium showed a positive effect on As removal at an optimum pH of 8 (Zhao et al. 2010). Sludge production in electrocoagulation is less when compared to that of coagulation with alum or ferric salts (Kumar and Goel 2010).

One of the main drawbacks of the coagulation process is the formation of chemical sludge. Dewatered sludge containing dissolved solids and heavy metals are susceptible to leaching when disposed of in landfills or natural water sources (Mandal et al. 2016). The best approach for sustainable disposal of As contaminated sludge is to make it into a stabilized form by the solidification process (Sullivan et al. 2010). Replacing or addition of a certain percentage of sludge in the production of bricks and concrete is one such way of solidification process. The release of elements from solids, liquid, and multiphasic wastes can be quantified from leaching tests (Clancy et al. 2013). Roy and his co-workers reported that amount of As in the leachate collected from the leaching test on concrete cubes [sludge added-0.1–1% (w/w)] and bricks [sludge added-1–3% (w/w)] is within the

Table 5.2 As removal by using Fe or Al electrodes of different type (Kobya et al. 2020)

EC reactor	Water type	Initial pH	Electrode type	As species	Initial concentration of As (μg/L)	i or j (A, A/m^2)	Operating time (min)	R (%)
BR	GW	8.6	Fe anode	As(III) and As (V)	36.03–1020.5	0.025–0.1 A	2–16 min	97–99.9
BR	SW	4.0	Fe anode	As(III)	50,000	5.4 A/m^2	45	99.9
CR	GW	7.2	Fe anode	As(V)	133	15–45 A/m^2	10.5–0.75	99
BR	GW	7.6	Fe ball	As(III) and As (V)	285	0.3 A	20	99.3
BR	SW	7	Al plate	As(III)	150	2.5 A/m^2	4	93.5
BR	SW	7	Fe-Al hybrid plate	As(III)	500	47 A/m^2	2	99.9

Note BR—Batch Reactor, SW—Surface Water, GW—Groundwater, CR—Continuous Reactor, i—Applied current, j—Current density

permissible limit as per US EPA (5 mg/L) (Roy et al. 2018). As in the sludge can also be removed by volatilization using an anaerobic bioreactor. In India and Bangladesh, mixing of cow dung with As laden sludge is practiced. In this process, the soluble As species are reduced to arsine gas by the microorganisms present in cow dung (Mudgal 2001).

5.8.2 *Adsorption*

Adsorption by a solid surface is one of the appropriate technologies for removing As from drinking water. The process involves the transfer of adsorbate from a bulk liquid phase to the water film layer on the adsorbent followed by diffusion through the film, surface, and finally through pores (Patel 2019). This process is cost-effective, easy to handle, and versatile in operation. Another attraction is the minimum sludge production and reuse potential (Singh et al. 2015). However, the active surface area, solution pH, surface functional groups on the adsorbent, water chemistry, and temperature affect the process. Over the years, several As sorptive media have been tested. These include natural minerals, waste materials, metal oxides, nanomaterials, and their derivatives. A brief description of these sorbents is presented below.

Arsenic removal by low-cost adsorbents: Adsorbents derived from locally available natural materials are attractive due to their low cost. Materials such as natural soils, minerals, and food, industrial, and agricultural waste are studied for their As removal potential in contaminated drinking water (Mohan and Pittman 2007). Numerous attempts have also been made to enhance their performance through physical and chemical treatments. Pre-treatment with acid and base, or coating with metal oxides, enhances porosity, stability, and adsorption capacity

Table 5.3 Examples of adsorbents derived from industrial, food, and agricultural wastes

Waste-based adsorbents	Initial concentration of As (mg/L)	pH		Adsorption capacity (mg/g)		References
		As(III)	As(V)	As(III)	As(V)	
Red mud	0.002–0.03	7.25	3.5	0.66	0.51	Mohan and Pittman (2007)
Fe(III)/Cr(III) hydroxide waste	20–100	–	4	–	11.02	Namasivayam and Senthilkumar (1998)
Chrome sludge	5–500	–	8.04	–	21	Low and Lee (1995)
Coconut coir pith	5–100	–	7	–	12.51	Anirudhan and Unnithan (2007)

(Shakoor et al. 2016). The performance of a few waste materials as adsorbent toward As removal is summarized in Table 5.3.

Clay minerals are also used for As removal due to its low cost and abundance. These are hydrous aluminum silicates with trace amounts of iron, magnesium, and other cations with high surface area and exchange capacity (Adeyemo et al. 2017). The adsorption capacities of these materials are dependent on solution pH and competing substances like phosphates and organic matter present in water. Vithanage et al. (2006) reported that natural red earth (NRE) exhibited ~100% adsorption of both the As species when the total As was 28.84 μg/L at a pH range of 4–8. The maximum adsorption for kaolinite, montmorillonite, and illite ranged from 0.52 to 0.86 mg As(V)/g at an initial concentration of 10–200 mg/L at pH 5 (Mohapatra et al. 2007). Arsenic ions get adsorbed by inner-sphere complexation with hydroxyl groups on clay minerals (Asere et al. 2019). Clay minerals are chemically modified by Fe or surfactants for improving the performance. Montmorillonite doped with Fe showed adsorption capacity of 16.13 mg/g of As (III) and 15.15 mg/g of As(V) at pH range of 4–10 (Ren et al. 2014). Organo-vermiculites modified with hexadecyltrimethylammonium (HDTMA) showed maximum adsorption capacity of 15.65 mg/g of As(III) and 18.02 mg/g of As(V) at the initial As concentration of 1500 mg/L (Tuchowska et al. 2019). Kaolin clay prepared by intercalating hexadecyltrimethylammonium bromide (HDTMA-Br) cationic surfactant showed maximum adsorption capacities of 2.33 and 2.88 mg/g of As(III) and As(V), respectively, which is higher than unmodified kaolin clay mineral (Mudzielwana et al. 2019).

Arsenic removal by metal oxides/hydroxides: Various metal and metal oxides/hydroxides have been employed for As removal in drinking water. Iron, zirconium, and manganese-based composites are typically used for As removal. Among them, iron-based compounds like akaganeite (β-FeOOH), goethite (α-FeOOH), and ferrihydrites ($Fe_{10}O_{14}(OH)_2$) have a large surface area and high affinity toward As. Many studies reported that As is adsorbed on to the iron oxide surface through complexation or electrostatic interaction (Hao et al. 2018). Co-anions like chlorides and nitrates have no significant effect on As removal by metal oxides/hydroxides (Yang et al. 2015). However, the presence of PO_4^{3-} ions can reduce efficiency.

Another element like zirconium also has an affinity toward multivalent ions like As (III) and As(V). Commercially available hydrous Zr oxide (302-HZO) can remove As in a broad range of pH < 8 (Bortun et al. 2010). However, silicates and phosphates are the primary competing ions to As. Since Zr is expensive, it can be turned into binary oxides like Zr-Al, Zr-Mn, Zr-Al–Fe, etc. Studies have shown that manganese dioxide [α-MnO_2 and β-MnO_2] is effective in oxidizing As(III) to As (V) (Moore et al. 1990; Tournassat et al. 2002). Besides, it has a high As(V) adsorption capacity in the presence of divalent cations (Takamatsu et al. 1985). A few examples of metal oxides-based As removal are given in Table 5.4.

Arsenic removal by activated alumina-based adsorbents: Though numerous adsorbents have been reported with varying As adsorption capacities, activated alumina (AA) is perhaps the most promising adsorbent (Singh and Pant 2004b). AA is usually prepared through dehydration of aluminum hydroxide [$Al(OH)_3$] at high temperatures, resulting in porous Al_2O_3 with high surface area and active sites. It is less expensive than membrane separation, more versatile than the ion-exchange process, and is a potential process for small-scale water treatment plants (Chen et al. 1999; Frey et al. 2000). Lin and Wu (2001) stated that the adsorption of As(III) and As(V) was governed by both surface charge and the chemical form of AA. The optimum removal of As(III) by AA is 0.18 mg/g at pH 7.6 (Singh and Pant 2004b). Mesoporous alumina (MA) with the surface area of 307 m^2/g and pore size of 3.5 nm is found to be superior to conventional AA in removing As(III). It has an intake capacity of 47 mg/g due to the sponge-like interlinked pore system (Kim et al. 2004). In comparison, AA coated with manganese oxide (MOCA) showed As (III) adsorption capacity of 42.48 mg/g at an optimum pH range of 4–7.5. Removal mechanism of As(III) by MOCA involves a two-step process: oxidation of As(III) to As(V) and retention of As(V) on MOCA surface by inner-sphere complexation (Maliyekkal et al. 2009). The As adsorption capacity of AA is also varied with the presence of co-ions. Youngran et al. (2007) observed that chlorides (500 mg/L) and nitrates (20 mg/L) do not affect the adsorption of 200 μg/L of As(V) on AA. However, the presence of phosphates (1 mg/L) and silica (10 mg/L) showed a significant reduction in removal of As(V) due to their high affinity toward the oxide surface than As.

Table 5.4 Examples of metal oxides/hydroxides

Bimetal oxide	Initial concentration of As (mg/L)	pH	Adsorption capacity (mg/g)		References
			As(III)	As(V)	
Zr-Mn	0.11–0.33	7.8	1.07	2.22	Sorochkina et al. (2015)
Magnetite	2	7	1.79	3.38	Ajith and Swain (2019)
Goethite	2	7	3.16	3.71	Ajith and Swain (2019)
Fe-Zr	5–40	7	120	46.10	Ren et al. (2011)
Fe–Mn (Mn/Fe-1:6)	5–40	7	56.10	42	Zhang et al. (2012)

Nanoscale materials: Nanoscale materials are an attractive option for removing pollutants in water due to their unique properties like high specific surface area, reactivity, and selectivity. Nanoscale metal oxides of copper, iron, zirconium, cobalt, and aluminum have been well studied for As removal (Maiti et al. 2019). Adsorbents like nanoscale titanium dioxide (TiO_2) and zero-valent iron (nZVI) are also studied for the said purpose (Singh et al. 2015). Apart from metal and metal oxides, layer double hydroxide (LDH), graphene oxide (GO), carbon nanotubes, metal organic frameworks, and its combination with conventional adsorbents like metal oxides and activated carbon are studied for As removal (Liu et al. 2020). However, direct use of nanoparticles in the purification of water is limited due to leaching of particles along with treated water, higher solubility compared to their bulk counterpart, ion-induced agglomeration, and difficulty in solid–liquid separation. Material loss through leaching and reduction in binding sites due to agglomeration diminishes the adsorption capacity. Several efforts are made to address these limitations by immobilization on a suitable matrix and surface modification of the nanoparticles. Feng et al. (2012) reported that the coating of ascorbic acid on paramagnetic Fe_3O_4 nanoparticles increases its dispersibility and chemical stability. The nZVI, a commonly used adsorbent for As removal, is highly prone to agglomeration due to high surface reactivity. It is reported that alumina supported nZVI prevents agglomeration and allows easy separation (Jain and Agarwal 2017). LDHs, also known as synthetic clay, is a cationic layered structure with anions between the interlayer regions. Due to high surface area, ion-exchange capacity, and hydrophilic nature, LDHs have As adsorption capacities in the range of 5–615 mg/g (Goh et al. 2008). Fe-impregnated 1,3,5-benzenetricarboxylic metal-organic coordination polymer (Fe-BTC) shown 6.5 times more adsorption capacity than that of 50 nm pristine iron nanoparticles (Zhu et al. 2012). Adsorption capacities of some of the nanoscale materials are summarized in Table 5.5. Nanocomposites can be engineered to enhance selectivity toward As in the presence of cations and anions. Fan et al. (2012) used surface-ion imprinting technique to selectively remove As(V) in the presence of lead, cadmium, nickel, and copper ions. At an initial concentration of 25 mg/L, removal of As(V) is > 85% in all cases due to the arrangement of pores and ligands for the coordination of As ion. An addition of zwitterions like glycine increased removal of As(V) among Se(VI) and Cr(VI) ions (Asiabi et al. 2017). Typically, phosphate and silicate ions compete with As(V) due to their strong inner complexation ability with hydroxyl groups (Yao et al. 2014). Similarly, As(III) removal is influenced by phosphates, silicates, and sulfates (Yu et al. 2015). Also, phosphate, As(V), and As(III) have similar tetrahedral geometry which competes with adsorption sites (Yu et al. 2015).

Ion-exchange: It is a physicochemical process in which a specific ion on the solid phase is exchanged with an equivalent number of ions in the feedwater, as shown in Fig. 5.5. The exchange takes place between ions of the same electric charge. Cation exchangers exchange positively charged ions, whereas anion exchangers exchange negatively charged ions. Some amphoteric exchangers can exchange both anions and cations. The selection of resin is based on its capacity to adsorb a particular contaminant selectively. Water quality parameters (pH,

Table 5.5 Examples of nanoscale adsorbents

Nanoscale adsorbent	Initial concentration (mg/L)		pH		Adsorption capacity (mg/g)		References
	As(III)	As(V)	As (III)	As (V)	As(III)	As (V)	
Copper (II) oxide	0.5–1	–	7	–	1.08	–	Goswami et al. (2012)
Magnetite–maghemite	0.7–3		5		0.96	0.98	Chowdhury et al. (2011)
Ultrafine superparamagnetic iron(III) oxide	–	100	–	7	–	45	Kilianová et al. (2013)
m- TiO_2-αFe_2O_3	5	–	3		80%	99%	Zhou et al. (2008)
Reduced graphene oxide-Fe_2O_3 and TiO_2	3–80		7	6	77.76	99.05	Babu et al. (2016)
Nanoiron/oyster shell composite	1.8	–	6.8	–	100%		Fan et al. (2015)
Magnetic iron nanoparticles-modified microfibrillated cellulose	–	10.48–800	–	2	–	184.3	Hokkanen et al. (2015)
CNT/CuO	1		7	5	2.42	2.83	Singh et al. (2016)
Chitosan zero-valent iron nanoparticles	1–20	1–20	7	7	94	119	Gupta et al. (2012)
Modified GO using Fe–Mn binary oxide	0.2–7	0.2–7	7	7	10.16	11.49	Zhu et al. (2015)
Cellulose incorporated in Ca/Al layered double hydroxide (LDH)	0.2–200		5.5		10.91	41.18	Bessaies et al. (2020)
Cellulose incorporated in Zn/Al layered double hydroxide (LDH)	0.2–200		5.5		30.15	42	Bessaies et al. (2020)
Mg–Fe-Zr layered double hydroxide/Fe_3O_4 composite		40–360		3		188	Suvokhiaw et al. (2016)
Fe_2Co_1 MOF-74 nanoparticles	1–250		4.3	4.3	266.52	292.3	Sun et al. (2019)
Zn-MOF-74	800	800	12	7	204.9	320.5	Yu et al. (2019)

competing ions, alkalinity, and influent contaminant concentration) need to be considered in deciding the applicability of this method in the water treatment process (LeMire et al. 2010). Strong base anionic (SBA) resins are immensely ionized and can work over a wide range of pH in hydroxide form, whereas weak base anionic (WBA) resins work at pH < 7. SBA resins have a high affinity to As (V) (Clifford and Ghurye 2002). However, the removal efficiency of As(III) is less because it is uncharged (Ficklin 1983; Wang et al. 2000). Hence, pre-oxidation of As(III) to As(V) is required for its effective removal (Clifford and Lin 1986; Fox 1989). Kartinen and Martin (1995) described that ion-exchange could result in an As reduction of >95%, bringing the concentration in treated water to <2 μg/L. In a batch study, N, N-dimethylaminoethyl methacrylate grafted with polyethylene/polypropylene fibers showed a removal efficiency of 94.67% for initial concentration of 120 μg/L (Kavaklı et al. 2014). Hybrid anion exchange impregnated with zirconium oxide nanoparticles (HAIX-Zr) showed As(V) adsorption capacity of 17 mg/g with an initial concentration of 1 mg/L (Padungthon et al. 2015). Competing ions like SO_4^{2-}, Cl^- and HCO_3^- decrease the As removal efficiency due to their high affinity toward resins. Chelating ion-exchange resin loaded with Cu^{2+} ions can overcome interference by competing ions (An et al. 2010). It is reported that polymer ligand exchanger DOW-3 N-Cu achieved a breakthrough at 6000-bed volumes for the influent containing 94 μg/L of As(V) in the presence of chlorides, sulfates, and bicarbonates (An et al. 2005).

Regeneration: Regeneration allows the use of adsorbent/ion-exchange resins to its full potential. It is also a necessary process to reduce the treatment cost and solid waste generation. The process involves replenishment and activation of the adsorbent sites for further cycles. The selection of regenerant is based on the chemical nature of adsorbent, adsorbate species, and adsorption mechanism. Copper loaded coconut husk regenerated with 30% H_2O_2 in 0.5 M HNO_3 showed >80% As(III) recovery by the end of three cycles (Manju et al. 1998). It is attributed to the formation of neutral As(III) species that desorb from the positively charged

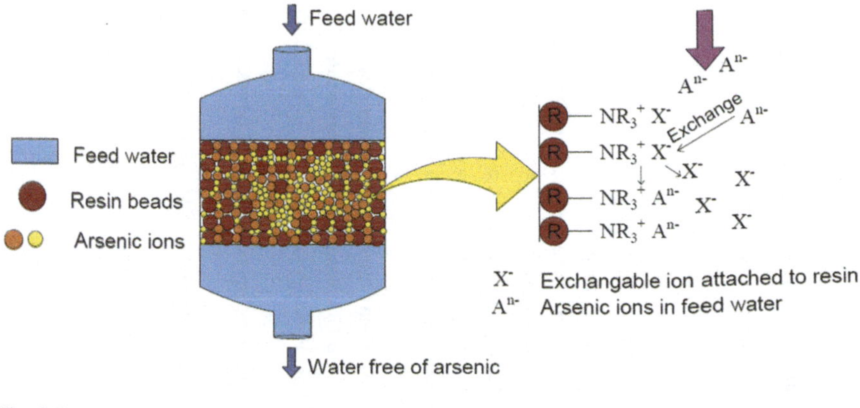

Fig. 5.5 Schematic diagram showing ion-exchange process

activated carbon. (Manju et al. 1998). For oxyanions like As, typically, alkaline pH is favorable for regeneration. Under alkaline pH, the adsorbed As gets repelled from negatively charged adsorbent sites. Sodium hydroxide (NaOH) is well reported as a desorbing agent for As. The adsorption capacity of the AA drops to only 11% by the end of five times regeneration with NaOH (Singh and Pant 2004a). However, the desorption of As(V) from red mud shows only 40% at pH 11.6 adjusted by NaOH due to chemisorption (Genç-Fuhrman et al. 2004). Besides desorption efficiency, high alkalinity leads to the dissolution of metal ions from the adsorbent (Ghosh and Gupta 2012). In a study, at pH 12, As(V) is desorbed completely along with adsorbent loss of 16.5% (Inchaurrondo et al. 2019). The nanoscale ZVI and Fe-Cu binary oxide do not show a significant reduction in removal efficiency of As (III) and As(V) up to 4 cycles of regeneration (Zhu et al. 2009; Zhang et al. 2013).

Regeneration of DOW-3 N-Cu ligand exchanger with 4% sodium chloride at pH 9.1 did not show much variation in As(V) intake capacity up to 8 cycles of regeneration. The same study reported that the treated regenerant was reused successfully in further regeneration cycles after removing As by the addition of $FeCl_3$ (An et al. 2005). Padungthon et al. (2015) reported nearly 90% desorption of As(V) from Zr-loaded hybrid anion exchange resin by a mixed solution of 3% NaCl and 3% NaOH, and the resin beads showed no sign of chemical instability.

5.8.3 Membrane Filtration

The membrane filtration process consists of a selective membrane with pores and serves as a selective barrier for the passage of ions in water (Cheryan 1998). The process is employed widely in removing contaminants like suspended solids, viruses, bacteria, mono and multivalent ions. The driving force for the process is based on the pressure difference and chemical potential across the membrane. There are two types of pressure-driven membranes, i.e., high pressure-driven (reverse osmosis, RO, and nanofiltration, NF) and low pressure-driven (ultrafiltration, UF, and microfiltration, MF) membranes. RO and NF membranes require high pressure of 50–1000 psi, whereas MF and UF require 5–100 psi pressure (Shih 2005). The filtering of pollutants in the low-pressure processes occurs by physical sieving due to a large pore size. In contrast, in high-pressure processes, it is caused by chemical diffusion through dense pore size. The selection of the membrane is based on the size of the contaminants to be removed, and the sample matrix, which is shown in Fig. 5.6.

RO membranes are more suitable for As removal in water (Kartinen and Martin 1995). The factors such as pH, ionic strength, pressure applied, pore size, and surface charge can affect the process. Schneiter and Middlebrooks (1983) studied As removal from groundwater by RO. For feedwater with an As concentration of 100 μg/L, rejection of As(V) and As(III) was found to be 90% and 58%, respectively. In another study, Kartinen and Martin (1995) observed As(V) removal of 97% and As(III) removal ranging from 40 to 80% by the RO process. Thus, the

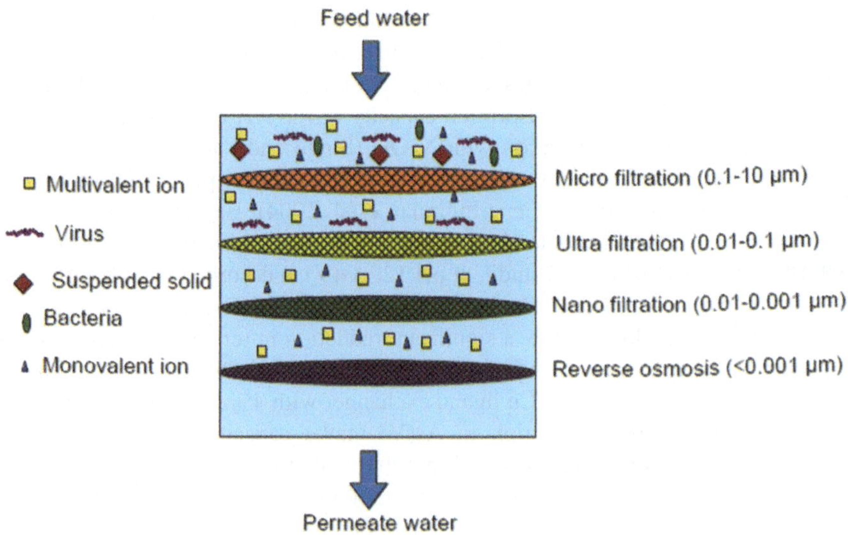

Fig. 5.6 Membrane filtration (Ghosh et al. 2019)

feedwater must be treated with an oxidizing agent to convert As(III) to As(V) to achieve a substantial As reduction in the permeate. The life of the membranes largely depends on the quality of feedwater. The presence of iron, manganese, suspended solids, microorganism, and organic matter can trigger RO membrane fouling (Mondal et al. 2006). Hence, RO requires high-quality feed streams and often requires a pre-treatment to remove fouling agents. Besides that, re-mineralization of treated water is required as essential minerals get removed during the treatment. Other concerns for the RO membrane include high capital, water rejection, and maintenance costs. The cost of treated water depends on capital cost as well as the operating cost of the system. The cost of the energy necessary to run the RO process shares almost 35% of the total cost (Abejón et al. 2015).

NF membranes are developed to overcome some of the above-listed demerits of RO membranes (Hou 2017; Tanne et al. 2019). NF membrane consists of a porous inert layer and a negatively charged hydrophobic rejection layer (Pal et al. 2009). Steric hindrance by large hydrated radius and Donnan exclusion by the negatively charged membrane are the removal mechanisms of As(V). The mechanism of As (III) removal is through steric hindrance (Nguyen et al. 2009). According to a study, As(III) rejection can be enhanced by prior oxidation to As(V) (Criscuoli and Figoli 2019). Besides this, As rejection is also based on the preferential passage of more mobile ions. Nguyen et al. (2009) stated that filtration of As(V) is enhanced in the presence of Cl^- and HCO_3^- and reduced in the presence of SO_4^{2-}. Rejection of As (V) is higher by 1% in the presence of divalent cations compared to the presence of monovalent sodium. However, there was no specific trend in As(III) removal in salts' presence (Xia et al. 2007). Some of the NF membranes reported in the literature are given in Table 5.6.

Table 5.6 Examples of NF membranes for As removal

Type of membrane	Flow rate (L/min)	Initial concentration (μg/L)		Pressure (bar)	Ionic concentration (mg/L)	Removal efficiency (%)		References
		As (III)	As(V)			As(III)	As (V)	
HL (Polyamide-TFC)	9	–	435	11	242.1	–	97	Figoli et al. (2020)
Polyamide blended membranes, with chitosan–claynanoparticles (FA-CSBF)	0.0006–0.005	100	–	0.5	–	99	–	Zeeshan et al. (2020)
Thin film composite (TFC) NF-300	10	–	7492.1	50	485	–	99.99	Harisha et al. (2010)
Polyamide thin-film composite (TFC)	60	100	100	5	–	90	93	Harfoush et al. (2018)
NF270	–	100	–	10	2000 NaCl and $MgSO_4$	57.96 (With the dose of 10 mg/L of hypochlorite, raised to 86)	–	Elcik et al. (2016)

5.8.4 Capacitive Deionization

It is an advanced electrochemical method to remove ions present from aqueous systems by electrosorption. It consists of a stack of porous carbon electrodes connected to the current collector. When an electric potential is applied across the electrode, the ions of the opposite charge from the spacer channel through which water flows are adsorbed onto the porous electrode. This process is based on the electrical double layer (EDL) mechanism. According to this, charged ions will be collected onto the stern layer of carbon interstices. Ions will be collected to micro and mesopores of carbon material based upon the charge of species and surface functional groups. After the complete filling of the pores, the ions can be desorbed by reversing voltage. While regeneration, some of the desorbed ions will get attached back to the electrode, which reduces the efficiency in further cycles by filling pores with the contaminant. The limitation can be overcome by the membrane CDI technique, where the ion-exchange membrane lets the desorbed ions completely leached from the system, which is shown in Fig. 5.7 (Choi et al. 2019). Compared to conventional techniques for As removal, it has many advantages like easy operation, high selectivity, and no usage of chemicals; hence secondary pollution can be eliminated. These carbon electrodes can be further modified with doping of metals or organic compounds to increase its ion selectivity induced removal (Chen et al. 2020; Syam Babu and Nidheesh 2020).

Dai et al. (2017) stated that electrode made of activated carbon has As(V) adsorption capacity 6.57 mg/g whereas electrode made of reduced graphene-iron composite gave 13 mg/g at an initial concentration of 10 mg/L at 1.2 V. Fan et al.

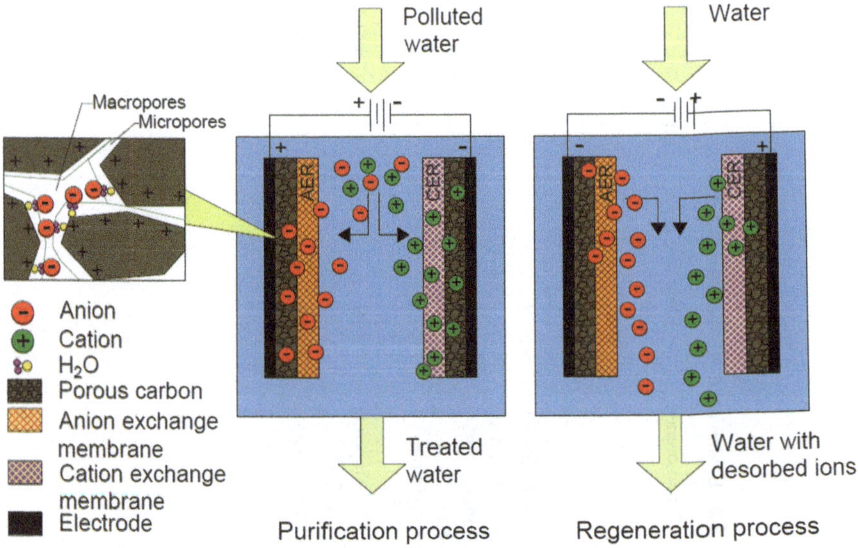

Fig. 5.7 Diagrammatic representation of membrane capacitive deionization technique

(2016) used electrode made of activated carbon and polyvinylidene fluoride using N,N-dimethylacetamide as the solvent, and applied on titanium plate. It was found that the intake of As is high at a higher voltage and initial As concentration. As(V) is adsorbed on carbon electrode electrostatically, whereas As(III) undergoes electro-oxidation and electrosorption.

The presence of organic matter and other ions can decrease the efficiency of As removal (Laxman et al. 2016; Choi et al. 2019). Zhang et al. (2016) conducted a statistical experimental study using a solar-powered CDI unit. It was found that higher pH and low salinity favors arsenate removal, and efficiency of above 80% is obtained.

5.8.5 Phytoremediation

Phytoremediation is a low-cost and eco-friendly process that uses green plants to treat and control the pollutants from soil, water, and air. The process is typically useful for removing low concentrations of the pollutants. The process efficacy varies with the type of plants used, characteristics of contaminated water, concentration and toxicity of the contaminants, soil condition, hydrology, nutrient sustainability, meteorology, contaminant characteristics, and type of ecosystem. Though the process is simple, affordable, and eco-friendly, the phytotoxicity and bioavailability of the target pollutants limit its application. Large land area requirements and longer treatment time are also the limitations of this method. Figure 5.8 shows the mechanisms of As removal by plants through the phytoremediation process. Kumar et al. (2015) have reported that more than 450 plant species of 45 different families can hyper-accumulate As. A few examples of the plants that accumulate and tolerate As are listed in Table 5.7. Studies showed that plant growth-promoting bacteria (PGPB) improve the efficiency of phytoremediation (Alka et al. 2020). A few recent studies proved that plants develop dense root systems and high biomass production in the presence of As (Pardo et al. 2017; Kaur et al. 2018). Phytovolatilization uses plants to volatilize As into the atmosphere with or without the help of rhizospheric microbes. In this process, plants uptake As from the environment and release them into the atmosphere by using the transpiration process (Jakob et al. 2010). The inorganic form of As converted into volatile mono-, di-, and tri-methylated species through a series of methylation reactions by using microbes (Cullen and Reimer 1989; Mukhopadhyay and Rosen 2002). The vapors released into the atmosphere by the *Arbundo donax L* contain volatile trimethylarsine compound (Mahmood 2010). Even some macrophytes like rigid hornwort, waterweed, *Myriophyllum*, *Petries starwort*, can uptake As in water (Robinson et al. 2005). A tolerable limit of 15 μg/kg of body weight per week is suggested by WHO/Food agricultural organization (Kohlmeyer et al. 2003). A summary of various As treatment technologies is given in Table 5.8.

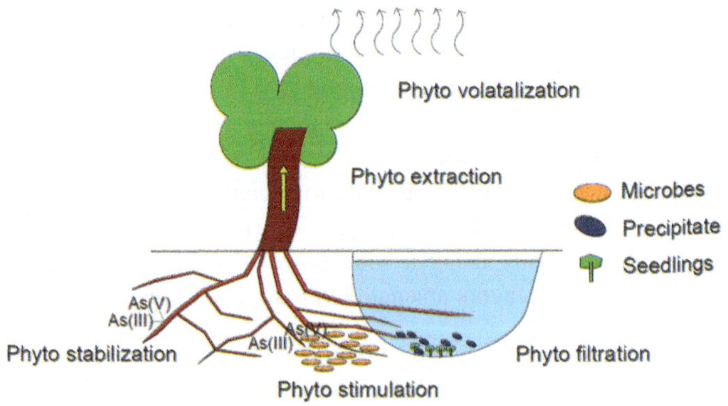

Fig. 5.8 Diagram showing the phytoremediation process

5.9 Methods of Arsenic Removal Practiced in India and Bangladesh

Arsenic is a silent killer affecting a large population worldwide. As toxicity is one of the significant environmental health problems of the modern era. Drinking As laden water renders not only health problems but also has a considerable impact on the socio-economic development of an affected individual. This problem is severe for the weaker section of the society. Unfortunately, most of the affected area is densely populated with high levels of rural poverty. Though the major economic problem is due to medical costs related to As toxicity, the economic burden of prejudice and superstitions of arsenicosis is also a serious concern. Women and children are most affected, and in many affected areas, they have to walk miles to get safe drinking water. The majority of fatal deaths against As toxicity in rural and remote areas are due to the lack of free access to safe drinking water. Providing a sustainable supply of As-free water is required to address the problem effectively.

In association with some of the national and international agencies, the government has been making sincere efforts to provide As-free water to the affected population. In this context, several communities or households As removal units (ARUs) are developed. For a successful installation, these units are to be designed suiting the local conditions. Factors such as groundwater quality, population distribution, socio-economic status of people residing in the area, availability of electricity, and sludge disposal are considered for the selection of treatment methods (Alaerts et al. 2001). The provision of separate land space is allotted to dispose of spent filter media. In most of the cases, these units are run and maintained by the local community. Existing ARUs remove As by coagulation, adsorption, precipitation, and filtration mechanisms. Some of the ARUs installed in the field are summarized in Table 5.9.

Table 5.7 Examples of plant species reported for As removal using phytoremediation

Plant type	Initial amount of As in soil (mg/kg)	As concentration in plant organs		As removed by volatilization (mg)	Removal efficiency (%)		References
		As(III)	As(V)		As (III)	As(V)	
Arundo donax L combined with *Stenotrophomonas maltophilia* sp and *Agrobacterium* sp bacteria	13.33	14 (mg/kg) in root and rhizome		11.37	56.9		Guarino et al. (2020)
Pteris vittata	200	6358 (mg/kg) in fronds	6329 (mg/kg) in fronds	–	86.9	81.8	Zhao et al. (2015)
Pteris cretica	200	–	4875 (mg/kg)	–	–	94.3	Eze and Harvey (2018)

Table 5.8 Comparison of various As removal technologies

Treatment methods / Parameters	Coagulation	Electrocoagulation	Adsorption	Membrane filtration	CDI	Phytoremediation
Maintenance	Low	Medium	Medium	High	High	Low
Sludge generation	High	Medium	Low	Low	Low	Low
Cost of treated water (USD/m^3)	0.01–0.17	0.89–1.11	0.11–0.59	0.3	0.087	22.6
As(III) removal efficiency (%)	30–90	80–99.9	40–96	50–90	16–77	–
As(V) removal efficiency (%)	>85	80–99.9	55–99	70–99	50–99	–
Operator skill required	Low	High	Low	Medium	High	Low
Power consumption (kWh/m^3)	The minimum requirement for flocculation	0.06–1.23	The minimum requirement for pumping of water	3.4	0.5–1.5	The minimum requirement for pumping of water
References	US EPA (2003), Sarkar and Paul (2016), and Shan et al. (2019)	Can et al. (2014), Amrose et al. (2014), and Kobya et al. (2020)	US EPA (2003), Shan et al. (2019), and Liu et al. (2020)	US EPA (2003), Anjum et al. (2009), Schmidt et al. (2016), and Piacente (2017)	Bundschuh et al. (2008) and Chen et al. (2020)	Wan et al. (2016)

Table 5.9 Examples of ARU and centralized treatment plants practiced in India and Bangladesh

Name of the technology	Removal mechanism	Place of installation	Treatment capacity (L/d)	Cost/m^3 of water treated	Performance	References
SAFI filter	Adsorption	Bangladesh	–	–	86–100% efficiency. No labor and electricity required; washing for 15 days	Rahman et al. (2005)
Arsiron Nilogon ($NaHCO_3$, $KMnO_4$, and $FeCl_3$)	Oxidation; coagulation; adsorption	Assam (India)	10–200	0.13 USD	95% efficiency; low cost; easy installation; and does not require electricity; partially removes bacteria; skilled labor is not required.	Bordoloi et al. (2015)
ARUZIF	Adsorption	Assam, Bihar, Uttar Pradesh, West Bengal (India)	–	0.0013–0.013 USD	Simple to operate and requires less maintenance; no electricity required, cleaning required in every 3 months; less sludge generation	Banerji and Chaudhari (2017)
Low-cost laterite-based As filter	Adsorption	West Bengal (India)	120–500	–	98% removal efficiency	Bhardwaj et al. (2019)
Tablets-filter system—(Oxidizing agent, Fe(III) salt and AC)	Co-precipitation	West Bengal (India)	20L	–	93–100% removal efficiency; Tablets can be stored for 15 months. Training is required for usage	Das et al. (2001)
All India Institute of Hygiene and Public Health (AIIH&PH) Technology (NaOCl and Alum)	Oxidation—Co-precipitation	West Bengal (India)	Household—30 L/d community-12,000 L/d	0.067 USD per month	90% removal efficiency. Bacteriological problem if not cleaned for 7 days. Cleaning required every 15 days. The skilled person is required for operation	Das et al. (2001) and Ahmed (2005)

5.10 Conclusion

Dissolved As in groundwater is an invisible threat affecting more than 200 million people worldwide. It is a proven carcinogen, and chronic intake of As can cause cancerous and non-cancerous diseases. Many treatment technologies have developed over the years for the removal of As in water. The processes adopted in these technologies mainly include adsorption, coagulation and precipitation, membrane filtration, electrocoagulation, and ion-exchange. Though many of these methods are proven successful in removing As in water, each method has its limitation and may not be suitable in all the field conditions. So, several factors must be considered while selecting the right treatment technique. These include As concentration, feed-flow rate, water chemistry, As speciation, sludge generation, ease of operation, electricity availability, and treatment cost. The socio-economic conditions of the affected population and type of water supply source are also an essential factor in selecting the technology. Another factor to be considered is whether the affected area requires a centralized, semicentralized, or independent household treatment system.

Coagulation/precipitation is a traditionally used method for purification of water. However, the process requires a high amount of chemicals, and sludge handling will be difficult and costly. Compared to the conventional coagulation process, electrocoagulation produces less sludge and can be used in small community-level treatment. However, it requires high energy throughput and less efficient compared to non-conventional treatment methods. Ion-exchange and membrane filtration processes are expensive and may not be affordable to most affected populations in India and Bangladesh. The competition from co-ions and high-water rejection is significant concerns for ion-exchange and RO membrane processes, respectively. Capacitive deionization is an emerging technology and requires the development of a selective electrode for the effective removal of As. Adsorption on a solid media is a mature technology and is widely employed in the field. It is a proven technique for the treatment of water at the community and household levels. However, the efficacy of the adsorbent and treatment cost decides the success of the process. The handling of spent adsorbent is also a major challenge.

For the past two decades, with the help of non-governmental organizations and multilateral funding agencies, a lot has done to alleviate the problem in the As-affected areas. However, those efforts have yielded only little success, and the problem persists. Lack of awareness, equipment availability, and accessibility to remote locations also brings challenges for supplying As-free water in developing countries. A technology that is affordable, easily constructed, operated at a community or household level is needed to address the problem.

Acknowledgements Authors gratefully acknowledge the Department of Science and Technology (DST), Technology Mission Division, Government of India [Grant no: DST/TM/WTI/WIC/2K17/82(C)]. The authors are thankful to Springer publisher for rendering license to reprint the Eh-pH diagram of As from the book chapter titled 'Eh-pH diagrams for geochemistry.' We are grateful for

Elsevier publishing for permission to use a figure named Affected countries and their respective MCL from 'Remediation of inorganic As in groundwater for safe water supply: A critical assessment of technological solutions'. The authors also thank IIT Tirupati for support.

References

Abejón A, Garea A, Irabien A (2015) Arsenic removal from drinking water by reverse osmosis: Minimization of costs and energy consumption. Sep Purif Technol 144:46–53. https://doi.org/10.1016/j.seppur.2015.02.017

Acharyya SK, Lahiri S, Raymahashay BC, Bhowmik A (2000) Arsenic toxicity of groundwater in parts of the Bengal basin in India and Bangladesh: the role of Quaternary stratigraphy and Holocene sea-level fluctuation. Environ Geol 39:1127–1137. https://doi.org/10.1007/s002540000107

Adeyemo AA, Adeoye IO, Bello OS (2017) Adsorption of dyes using different types of clay: a review. Appl Water Sci 7:543–568. https://doi.org/10.1007/s13201-015-0322-y

Ahmed MF, Rahaman MM (2000) Water Supply and Sanitation-Low Income Urban Communities. International Training Network (ITN) Centre, BUET

Ahsan DA, Del Valls TA (2011) Impact of Arsenic Contaminated Irrigation Water in Food Chain: An Overview From Bangladesh. Int J Environ Res 5:627–638. https://doi.org/10.22059/ijer.2011.370

Ajith N, Swain KK (2019) Study on the performance and interaction of different synthetic iron oxides for arsenic uptake using 76As radiotracer. Appl Radiat Isot 153: https://doi.org/10.1016/j.apradiso.2019.108807

Alaerts G, Khouri N, Kabir B (2001) Strategies to mitigate arsenic contamination of water supply. Arsenic in drinking water United Nations synthesis report on arsenic in drinking water Available: https://www.whoint/water_sanitation_health/dwq/arsenicun8pdf. Accessed 4 June 2005

Alka S, Shahir S, Ibrahim N, Chai TT, Bahari ZM, Abd Manan F (2020) The role of plant growth promoting bacteria on arsenic removal: A review of existing perspectives. Environ Technol & Innov 17: https://doi.org/10.1016/j.eti.2020.100602

Amrose SE, Bandaru SRS, Delaire C, van Genuchten CM, Dutta A, DebSarkar A, Orr C, Roy J, Das A, Gadgil AJ (2014) Electro-chemical arsenic remediation: Field trials in West Bengal. Sci Total Environ 488–489:539–546. https://doi.org/10.1016/j.scitotenv.2013.11.074

An B, Fu Z, Xiong Z, Zhao D, SenGupta AK (2010) Synthesis and characterization of a new class of polymeric ligand exchangers for selective removal of arsenate from drinking water. React Funct Polym 70:497–507. https://doi.org/10.1016/j.reactfunctpolym.2010.01.006

An B, Steinwinder TR, Zhao D (2005) Selective removal of arsenate from drinking water using a polymeric ligand exchanger. Water Res 39:4993–5004. https://doi.org/10.1016/j.watres.2005.10.014

Anawar HM, Akai J, Sakugawa H (2004) Mobilization of arsenic from subsurface sediments by effect of bicarbonate ions in groundwater. Chemosphere 54:753–762. https://doi.org/10.1016/j.chemosphere.2003.08.030

Anirudhan TS, Unnithan MR (2007) Arsenic(V) removal from aqueous solutions using an anion exchanger derived from coconut coir pith and its recovery. Chemosphere 66:60–66. https://doi.org/10.1016/j.chemosphere.2006.05.031

Anjum S, Gautam D, Gupta B, Ikram S (2009) Arsenic removal from water: An overview of recent technologies. J Chem 48

Asere TG, Stevens CV, Du Laing G (2019) Use of (modified) natural adsorbents for arsenic remediation: A review. Sci Total Environ 676:706–720. https://doi.org/10.1016/j.scitotenv.2019.04.237

Asiabi H, Yamini Y, Shamsayei M (2017) Highly selective and efficient removal of arsenic(V), chromium(VI) and selenium(VI) oxyanions by layered double hydroxide intercalated with zwitterionic glycine. J Hazard Mater 339:239–247. https://doi.org/10.1016/j.jhazmat.2017.06.042

Babu CM, Vinodh R, Sundaravel B, Abidov A, Peng MM, Cha WS, Jang H-T (2016) Characterization of reduced graphene oxide supported mesoporous Fe2O3/TiO2 nanoparticles and adsorption of As(III) and As(V) from potable water. J Taiwan Inst Chem Eng 62:199–208. https://doi.org/10.1016/j.jtice.2016.02.005

Backman B, Luoma S, Ruskeeniemi T, Karttunen V, Talikka M, Kaija J (2006) Sundaram Geological Survey of Finland. Miscellaneous Publications, Espoo

Banerji T, Chaudhari S (2017) A Cost-Effective Technology for Arsenic Removal: Case Study of Zerovalent Iron-Based IIT Bombay Arsenic Filter in West Bengal. In: Nath KJ, Sharma VP (eds) Water and Sanitation in the New Millennium. Springer, New Delhi, pp 127–137

Bessaies H, Iftekhar S, Doshi B, Kheriji J, Ncibi MC, Srivastava V, Sillanpää M, Hamrouni B (2020) Synthesis of novel adsorbent by intercalation of biopolymer in LDH for the removal of arsenic from synthetic and natural water. J Environ Sci 91:246–261. https://doi.org/10.1016/j.jes.2020.01.028

Bhardwaj A, Rajput R, Misra K (2019) Status of arsenic remediation in India. In: Advances in Water Purification Techniques. Elsevier, pp 219–258

Bhattacharya P, Hasan MA, Sracek O, Smith E, Ahmed KM, von Brömssen M, Huq SMI, Naidu R (2009) Groundwater chemistry and arsenic mobilization in the Holocene flood plains in south-central Bangladesh. Environ Geochem Health 31:23–43. https://doi.org/10.1007/s10653-008-9230-5

Bilici Baskan M, Pala A (2010) A statistical experiment design approach for arsenic removal by coagulation process using aluminum sulfate. Desalination 254:42–48. https://doi.org/10.1016/j.desal.2009.12.016

Binh PT, Van Anh NT, Thuy MTT, Xuan MT, The LC (2018) Effect of potential on arsenic treatment using technology of electrocoagulation. Vietnam J Chem 56:478–482. https://doi.org/10.1002/vjch.201800033

Bordoloi S, Gogoi S, Dutta RK (2015) A Low-Cost Arsenic Removal Method for Application in the Brahmaputra-Ganga Plains: Arsiron Nilogon. In: Ramanathan A, Johnston S, Mukherjee A, Nath B (eds) Safe and Sustainable Use of Arsenic-Contaminated Aquifers in the Gangetic Plain: A Multidisciplinary Approach. Springer, Cham, pp 289–298

Bortun A, Bortun M, Pardini J, Khainakov SA, García JR (2010) Effect of competitive ions on the arsenic removal by mesoporous hydrous zirconium oxide from drinking water. Mater Res Bull 45:1628–1634. https://doi.org/10.1016/j.materresbull.2010.07.011

Brookins DG (1988) Arsenic. In: Eh-pH Diagrams for Geochemistry. Springer, Berlin, Heidelberg

Bundschuh J, Armienta M, Birkle P, Bhattacharya P, Matschullat J, Mukherjee A (eds) (2008). In: Natural arsenic in groundwaters of Latin America. CRC Press

Can BZ, Boncukcuoglu R, Yilmaz AE, Fil BA (2014) Effect of some operational parameters on the arsenic removal by electrocoagulation using iron electrodes. J Environ Health Sci Engineer 12:95. https://doi.org/10.1186/2052-336X-12-95

Chakrabarti D, Singh SK, Rashid MH, Rahman MM (2019) Arsenic: Occurrence in Groundwater. In: Encyclopedia of Environmental Health. Elsevier, pp 153–168

Chakraborti D, Singh SK, Rahman MM, Dutta RN, Mukherjee SC, Pati S, Kar PB (2018) Groundwater arsenic contamination in the Ganga River Basin: a future health danger. Int J Environ Res Public Health 15:180. https://doi.org/10.3390/ijerph15020180

Chen G (2004) Electrochemical technologies in wastewater treatment. Sep Purif Technol 38:11–41. https://doi.org/10.1016/j.seppur.2003.10.006

Chen H, Frey MM, Clifford D, McNeill LS, Edwards M (1999) Arsenic treatment considerations. J-Am Water Work Assoc 91:74–85

Chen R, Sheehan T, Ng JL, Brucks M, Su X (2020) Capacitive deionization and electrosorption for heavy metal removal. Environ Sci: Water Res Technol 6:258–282. https://doi.org/10.1039/C9EW00945K

Cheryan M (1998) Ultrafiltration and microfiltration handbook. Technomic Pub.Co., Lancaster, PA

Choi J, Dorji P, Shon HK, Hong S (2019) Applications of capacitive deionization: Desalination, softening, selective removal, and energy efficiency. Desalination 449:118–130. https://doi.org/10.1016/j.desal.2018.10.013

Chowdhury SR, Yanful EK, Pratt AR (2011) Arsenic removal from aqueous solutions by mixed magnetite–maghemite nanoparticles. Environ Earth Sci 64:411–423. https://doi.org/10.1007/s12665-010-0865-z

Clancy TM, Hayes KF, Raskin L (2013) Arsenic Waste Management: A Critical Review of Testing and Disposal of Arsenic-Bearing Solid Wastes Generated during Arsenic Removal from Drinking Water. Environ Sci Technol 47:10799–10812. https://doi.org/10.1021/es401749b

Clifford D, Lin CC (1986) Arsenic Removal From Groundwater in Hanford, California—A Preliminary Report. University of Houston, Department of Civil and Envir Engineering

Clifford DA, Ghurye GL (2002) Metal-oxide adsorption, ion exchange, and coagulation-microfiltration for arsenic removal from water. Environmental chemistry of arsenic 217–245

Criscuoli A, Figoli A (2019) Pressure-driven and thermally-driven membrane operations for the treatment of arsenic-contaminated waters: A comparison. J Hazard Mater 370:147–155. https://doi.org/10.1016/j.jhazmat.2018.07.047

Cullen WR, Reimer KJ (1989) Arsenic speciation in the environment. Chem Rev 89:713–764

Dai M, Zhang M, Xia L, Li Y, Liu Y, Song S (2017) Combined Electrosorption and Chemisorption of As(V) in Water by Using Fe-rGO@AC Electrode. ACS Sustainable Chem Eng 5:6532–6538. https://doi.org/10.1021/acssuschemeng.7b00633

Daniel R, Prabhakara Rao AVS (2012) An efficient removal of arsenic from industrial effluents using electro-coagulation as clean technology option. Int J Environ Res 6:711–718

Das D, Chatterjee A, Samanta G, Chowdhury TR, Mandal BK, Dhar R, Chanda CR, Lodh D, Chowdhury PP, Basu GK (2001) A simple household device to remove arsenic from groundwater and two years performance report of arsenic removal plant for treating groundwater with community participation. In: BUETUNU International Workshop on Technologies for Arsenic Removal from Drinking Water. Citeseer, pp 5–7

Dey TK, Banerjee P, Bakshi M, Kar A, Ghosh S (2014) Groundwater arsenic contamination in West Bengal: current scenario, effects and probable ways of mitigation. International Letters of Natural Sciences 8

Donmez M, Akbal F (2011) The removal of As(V) from drinking waters by coagulation process using iron salts. Int J Environ Ecol Eng 5:340–342

Elcik H, Celik SO, Cakmakci M, Özkaya B (2016) Performance of nanofiltration and reverse osmosis membranes for arsenic removal from drinking water. Desalination and Water Treatment 57:20422–20429. https://doi.org/10.1080/19443994.2015.1111812

EPA S (1988) Special report on ingested inorganic arsenic, skin cancer: nutritional essentiality. EPA/625/3-87/013. Washington, DC: US Environmental Protection Agency

Eslami H, Ehrampoush MH, Esmaeili A, Salmani MH, Ebrahimi AA, Ghaneian MT, Falahzadeh H, Fard RF (2019) Enhanced coagulation process by Fe-Mn bimetal nano-oxides in combination with inorganic polymer coagulants for improving As(V) removal from contaminated water. J Clean Prod 208:384–392. https://doi.org/10.1016/j.jclepro.2018.10.142

Eze VC, Harvey AP (2018) Extractive recovery and valorisation of arsenic from contaminated soil through phytoremediation using Pteris cretica. Chemosphere 208:484–492. https://doi.org/10.1016/j.chemosphere.2018.06.027

F. Ahmed (2005) F. Ahmed, K. Minnatullah, A. Talbi, Arsenic Mitigation Technologies in South and East Asia. Toward a More Effective Operational Response: Arsenic Contamination in Ground Water in South and East Asian Countries, World Bank, Washington, DC, 2005, pp 166–207. Vol. II, Technical Report No. 31303, Water and Sanitation Program—South Asia and the World Bank. Accessed 28 Apr 2020

Fan C-S, Tseng S-C, Li K-C, Hou C-H (2016) Electro-removal of arsenic(III) and arsenic(V) from aqueous solutions by capacitive deionization. J Hazard Mater 312:208–215. https://doi.org/10.1016/j.jhazmat.2016.03.055

Fan H-T, Fan X, Li J, Guo M, Zhang D, Yan F, Sun T (2012) Selective Removal of Arsenic(V) from Aqueous Solution Using A Surface-Ion-Imprinted Amine-Functionalized Silica Gel Sorbent. Ind Eng Chem Res 51:5216–5223. https://doi.org/10.1021/ie202655x

Fan L, Zhang S, Zhang X, Zhou H, Lu Z, Wang S (2015) Removal of arsenic from simulation wastewater using nano-iron/oyster shell composites. J Environ Manage 156:109–114. https://doi.org/10.1016/j.jenvman.2015.03.044

Feng L, Cao M, Ma X, Zhu Y, Hu C (2012) Superparamagnetic high-surface-area Fe3O4 nanoparticles as adsorbents for arsenic removal. J Hazard Mater 217–218:439–446. https://doi.org/10.1016/j.jhazmat.2012.03.073

Ficklin WH (1983) Separation of arsenic (III) and arsenic (V) in ground waters by ion-exchange. Talanta 30:371–373. https://doi.org/10.1016/0039-9140(83)80084-8

Figoli A, Fuoco I, Apollaro C, Chabane M, Mancuso R, Gabriele B, Rosa RD, Vespasiano G, Barca D, Criscuoli A (2020) Arsenic-contaminated groundwaters remediation by nanofiltration. Sep Purif Technol 238: https://doi.org/10.1016/j.seppur.2019.116461

Fleischer M (1983) Glossary of Mineral Species, Mineral. Record, Tucson, AZ

Fox KR (1989) Field Experience With Point-of-Use Treatment Systems for Arsenic Removal. J Am Water Work Assoc 81:94–101. https://doi.org/10.1002/j.1551-8833.1989.tb03169.x

Frey MM, Chwirka J, Narasimhan R (2000) Cost implications of a lower arsenic MCL. AWWA Research Foundation

Gao B-Y, Chu Y-B, Yue Q-Y, Wang B-J, Wang S-G (2005) Characterization and coagulation of a polyaluminum chloride (PAC) coagulant with high Al13 content. J Environ Manage 76:143–147. https://doi.org/10.1016/j.jenvman.2004.12.006

Genç-Fuhrman H, Tjell JC, McConchie D (2004) Adsorption of arsenic from water using activated neutralized red mud. Environ Sci Technol 38:2428–2434

Ghosh D, Gupta A (2012) Economic justification and eco-friendly approach for regeneration of spent activated alumina for arsenic contaminated groundwater treatment. Resour Conserv Recycl 61:118–124. https://doi.org/10.1016/j.resconrec.2012.01.005

Ghosh S, Debsarkar A, Dutta A (2019) Technology alternatives for decontamination of arsenic-rich groundwater—A critical review. Environ Technol & Innov 13:277–303. https://doi.org/10.1016/j.eti.2018.12.003

Gilbert SG (2004) A small dose of toxicology: the health effects of common chemicals. CRC Press, Boca Raton

Goh K-H, Lim T-T, Dong Z (2008) Application of layered double hydroxides for removal of oxyanions: A review. Water Res 42:1343–1368. https://doi.org/10.1016/j.watres.2007.10.043

Golui D, Guha Mazumder DN, Sanyal SK, Datta SP, Ray P, Patra PK, Sarkar S, Bhattacharya K (2017) Safe limit of arsenic in soil in relation to dietary exposure of arsenicosis patients from Malda district, West Bengal—A case study. Ecotoxicol Environ Saf 144:227–235. https://doi.org/10.1016/j.ecoenv.2017.06.027

Goswami A, Raul PK, Purkait MK (2012) Arsenic adsorption using copper (II) oxide nanoparticles. Chem Eng Res Des 90:1387–1396. https://doi.org/10.1016/j.cherd.2011.12.006

Guarino F, Miranda A, Castiglione S, Cicatelli A (2020) Arsenic phytovolatilization and epigenetic modifications in Arundo donax L. assisted by a PGPR consortium. Chemosphere 251:126310. https://doi.org/10.1016/j.chemosphere.2020.126310

Gunduz O, Simsek C, Hasozbek A (2010) Arsenic pollution in the groundwater of Simav Plain, Turkey: its impact on water quality and human health. Water Air Soil Pollut 205:43. https://doi.org/10.1007/s11270-009-0055-3

Gupta A, Yunus M, Sankararamakrishnan N (2012) Zerovalent iron encapsulated chitosan nanospheres—A novel adsorbent for the removal of total inorganic Arsenic from aqueous systems. Chemosphere 86:150–155. https://doi.org/10.1016/j.chemosphere.2011.10.003

Hao L, Liu M, Wang N, Li G (2018) A critical review on arsenic removal from water using iron-based adsorbents. RSC Adv 8:39545–39560. https://doi.org/10.1039/C8RA08512A

Harfoush M, Mirbagheri SA, Ehteshami M, Nejati S (2018) Arsenic removal from drinking water using low-pressure nanofiltration under various operating conditions. Water Pract Technol 13:295–302. https://doi.org/10.2166/wpt.2018.042

Harisha RS, Hosamani KM, Keri RS, Nataraj SK, Aminabhavi TM (2010) Arsenic removal from drinking water using thin film composite nanofiltration membrane. Desalination 252:75–80. https://doi.org/10.1016/j.desal.2009.10.022

Heinrichs G, Udluft P (1999) Natural arsenic in Triassic rocks: A source of drinking-water contamination in Bavaria, Germany. Hydrogeol J 7:468–476. https://doi.org/10.1007/s100400050219

Herath I, Vithanage M, Bundschuh J, Maity JP, Bhattacharya P (2016) Natural Arsenic in Global Groundwaters: Distribution and Geochemical Triggers for Mobilization. Curr Pollution Rep 2:68–89. https://doi.org/10.1007/s40726-016-0028-2

Hokkanen S, Repo E, Lou S, Sillanpää M (2015) Removal of arsenic (V) by magnetic nanoparticle activated microfibrillated cellulose. Chem Eng J 260:886–894. https://doi.org/10.1016/j.cej.2014.08.093

Hossain MF (2006) Arsenic contamination in Bangladesh—An overview. Agr Ecosyst Environ 113:1–16. https://doi.org/10.1016/j.agee.2005.08.034

Hou Y (2017) An Improved Method of Arsenic (III) Removal By Reverse Osmosis Membrane, 71

Hu C, Liu H, Chen G, Qu J (2012) Effect of aluminum speciation on arsenic removal during coagulation process. Sep Purif Technol 86:35–40. https://doi.org/10.1016/j.seppur.2011.10.017

Hughes MF (2002) Arsenic toxicity and potential mechanisms of action. Toxicol Lett 133:1–16. https://doi.org/10.1016/S0378-4274(02)00084-X

Inchaurrondo N, di Luca C, Mori F, Pintar A, Žerjav G, Valiente M, Palet C (2019) Synthesis and adsorption behavior of mesoporous alumina and Fe-doped alumina for the removal of dominant arsenic species in contaminated waters. J Environ Chem Eng 7: https://doi.org/10.1016/j.jece.2019.102901

Jain A, Agarwal M (2017) Kinetic equilibrium and thermodynamic study of arsenic removal from water using alumina supported iron nano particles. J Water Process Eng 19:51–59. https://doi.org/10.1016/j.jwpe.2017.07.001

Jakob R, Roth A, Haas K, Krupp EM, Raab A, Smichowski P, Gómcz D, Feldmann J (2010) Atmospheric stability of arsines and the determination of their oxidative products in atmospheric aerosols (PM 10): evidence of the widespread phenomena of biovolatilization of arsenic. J Environ Monit 12:409–416

Jinadasa BKKK, Fowler SW (2019) A Critical Review of Arsenic Contamination in Sri Lankan Foods. J Food Qual Hazards Control. https://doi.org/10.18502/jfqhc.6.4.1991

Kartinen EO, Martin CJ (1995) An overview of arsenic removal processes. Desalination 103:79–88. https://doi.org/10.1016/0011-9164(95)00089-5

Kaur P, Singh S, Kumar V, Singh N, Singh J (2018) Effect of rhizobacteria on arsenic uptake by macrophyte Eichhornia crassipes (Mart.) Solms. Int J Phytorem 20:114–120

Kavaklı C, Akkaş Kavaklı P, Turan BD, Hamurcu A, Güven O (2014) Quaternized dimethylaminoethyl methacrylate strong base anion exchange fibers for As(V) adsorption. Radiat Phys Chem 102:84–95. https://doi.org/10.1016/j.radphyschem.2014.04.011

Keshavarzi B, Moore F, Mosaferi M, Rahmani F (2011) The source of natural arsenic contamination in groundwater, west of Iran. Water Qual, Expo Health 3:135–147. https://doi.org/10.1007/s12403-011-0051-x

Khaska M, La Salle CLG, Verdoux P, Boutin R (2015) Tracking natural and anthropogenic origins of dissolved arsenic during surface and groundwater interaction in a post-closure mining context: Isotopic constraints. J Contam Hydrol 177:122–135. https://doi.org/10.1016/j.jconhyd.2015.03.008

Kilianová M, Prucek R, Filip J, Kolařík J, Kvítek L, Panáček A, Tuček J, Zbořil R (2013) Remarkable efficiency of ultrafine superparamagnetic iron (III) oxide nanoparticles toward arsenate removal from aqueous environment. Chemosphere 93:2690–2697. https://doi.org/10.1016/j.chemosphere.2013.08.071

Kim Y, Kim C, Choi I, Rengaraj S, Yi J (2004) Arsenic Removal Using Mesoporous Alumina Prepared via a Templating Method. Environ Sci Technol 38:924–931. https://doi.org/10.1021/es0346431

Kobya M, Soltani RDC, Omwene PI, Khataee A (2020) A review on decontamination of arsenic-contained water by electrocoagulation: Reactor configurations and operating cost along with removal mechanisms. Water Qual, Expo Health 17: https://doi.org/10.1016/j.eti.2019.100519

Kohlmeyer U, Jantzen E, Kuballa J, Jakubik S (2003) Benefits of high resolution IC–ICP–MS for the routine analysis of inorganic and organic arsenic species in food products of marine and terrestrial origin. Anal Bioanal Chem 377:6–13. https://doi.org/10.1007/s00216-003-2064-1

Kumar NS, Goel S (2010) Factors influencing arsenic and nitrate removal from drinking water in a continuous flow electrocoagulation (EC) process. J Hazard Mater 173:528–533. https://doi.org/10.1016/j.jhazmat.2009.08.117

Kumar S, Dubey RS, Tripathi RD, Chakrabarty D, Trivedi PK (2015) Omics and biotechnology of arsenic stress and detoxification in plants: current updates and prospective. Environ Int 74:221–230

Lackovic JA, Nikolaidis NP, Dobbs GM (2000) Inorganic Arsenic Removal by Zero-Valent Iron. Environ Eng Sci 17:29–39. https://doi.org/10.1089/ees.2000.17.29

Laxman K, Bhattacharya P, Ahmed T, Uheida AS, Dutta J (2016) Capacitive deionization for removal of arsenic from water. In: 6th International Congress on Arsenic in the Environment, AS 2016, 19 June 2016 through 23 June 2016. CRC Press/Balkema, pp 521–522

LeMire LE, Teixeira MA, Reed BE (2010) Removal of As(V) using an iron-impregnated ion exchange bead. Sep Sci Technol 45:2051–2063. https://doi.org/10.1080/01496395.2010.504433

Lin TF, Wu JK (2001) Adsorption of arsenite and arsenate within activated alumina grains: equilibrium and kinetics. Water Res 35:2049–2057. https://doi.org/10.1016/s0043-1354(00)00467-x

Liu B, Kim K-H, Kumar V, Kim S (2020) A review of functional sorbents for adsorptive removal of arsenic ions in aqueous systems. J Hazard Mater 388: https://doi.org/10.1016/j.jhazmat.2019.121815

Logsdon GS, Sorg TJ, Symons JM (1974) Removal of heavy metals by conventional treatment. In: Proc. 16th Water Quality Conference-Trace Metals in Water Supplies: Occurrence, Significance, and Control, University Bulletin

Low KS, Lee CK (1995) Chrome Waste as Sorbent for the Removal of Arsenic(V) from Aqueous Solution. Environ Technol 16:65–71. https://doi.org/10.1080/09593331608616246

Mahmood NMQ (2010) Phytoremediation of arsenic (As) and mercury (Hg) contaminated soil. World Applied Sciences Journal 8:113–118

Maiti A, Mishra S, Chaudhary M (2019) Nanoscale materials for arsenic removal from water. In: Nanoscale Materials in Water Purification. Elsevier, pp 707–733

Majumder S, Banik P (2019) Geographical variation of arsenic distribution in paddy soil, rice and rice-based products: A meta-analytic approach and implications to human health. J Environ Manage 233:184–199. https://doi.org/10.1016/j.jenvman.2018.12.034

Maliyekkal SM, Philip L, Pradeep T (2009) As(III) removal from drinking water using manganese oxide-coated-alumina: Performance evaluation and mechanistic details of surface binding. Chem Eng J 153:101–107. https://doi.org/10.1016/j.cej.2009.06.026

Mandal P, Debbarma SR, Saha A, Ruj B (2016) Disposal Problem of Arsenic Sludge Generated During Arsenic Removal from Drinking Water. Procedia Environ Sci 35:943–949. https://doi.org/10.1016/j.proenv.2016.07.084

Manju GN, Raji C, Anirudhan TS (1998) Evaluation of coconut husk carbon for the removal of arsenic from water. Water Res 32:3062–3070. https://doi.org/10.1016/S0043-1354(98)00068-2

Mazumder DNG (2008) Chronic arsenic toxicity & human health. INDIAN J MED RES 12

Mitra SR, Mazumder DNG, Basu A, Block G, Haque R, Samanta S, Ghosh N, Hira Smith MM, von Ehrenstein OS, Smith AH (2004) Nutritional factors and susceptibility to arsenic-caused skin lesions in West Bengal, India. Environ Health Perspect 112:1104–1109. https://doi.org/10.1289/ehp.6841

Mohan D, Pittman CU (2007) Arsenic removal from water/wastewater using adsorbents—A critical review. J Hazard Mater 142:1–53. https://doi.org/10.1016/j.jhazmat.2007.01.006

Mohapatra D, Mishra D, Chaudhury GR, Das RP (2007) Arsenic adsorption mechanism on clay minerals and its dependence on temperature. Korean J Chem Eng 24:426–430. https://doi.org/10.1007/s11814-007-0073-z

Mollah MYA, Schennach R, Parga JR, Cocke DL (2001) Electrocoagulation (EC)—science and applications. J Hazard Mater 84:29–41. https://doi.org/10.1016/S0304-3894(01)00176-5

Mondal P, Bhowmick S, Chatterjee D, Figoli A, Van der Bruggen B (2013) Remediation of inorganic arsenic in groundwater for safe water supply: a critical assessment of technological solutions. Chemosphere 92:157–170. https://doi.org/10.1016/j.chemosphere.2013.01.097

Mondal P, Majumder CB, Mohanty B (2006) Laboratory based approaches for arsenic remediation from contaminated water: recent developments. J Hazard Mater 137:464–479. https://doi.org/10.1016/j.jhazmat.2006.02.023

Moore JN, Walker JR, Hayes TH (1990) Reaction scheme for the oxidation of As (III) to As (V) by birnessite. Clays Clay Miner 38:549–555. https://doi.org/10.1346/CCMN.1990.0380512

Mudgal AK (2001) Draft review of the household arsenic removal technology options. Rural Water Supply Network (http://www.htnweb.com)

Mudhoo A, Sharma SK, Garg VK, Tseng C-H (2011) Arsenic: An Overview of Applications, Health, and Environmental Concerns and Removal Processes. CritAl Rev Environ Sci Technol 41:435–519. https://doi.org/10.1080/10643380902945771

Mudzielwana R, Gitari MW, Ndungu P (2019) Performance evaluation of surfactant modified kaolin clay in As(III) and As(V) adsorption from groundwater: adsorption kinetics, isotherms and thermodynamics. Heliyon 5: https://doi.org/10.1016/j.heliyon.2019.e02756

Mukhopadhyay R, Rosen BP (2002) Arsenate reductases in prokaryotes and eukaryotes. Environ Health Perspect 110:745–748

Namasivayam C, Senthilkumar S (1998) Removal of Arsenic(V) from Aqueous Solution Using Industrial Solid Waste: Adsorption Rates and Equilibrium Studies. Ind Eng Chem Res 37:4816–4822. https://doi.org/10.1021/ie970774x

National Research Council (1999) Arsenic in drinking water. National Academies Press

Nguyen CM, Bang S, Cho J, Kim K-W (2009) Performance and mechanism of arsenic removal from water by a nanofiltration membrane. Desalination 245:82–94. https://doi.org/10.1016/j.desal.2008.04.047

Nickson R, McArthur J, Burgess W, Ahmed KM, Ravenscroft P, Rahmanñ M (1998) Arsenic poisoning of Bangladesh groundwater. Nature 395:338. https://doi.org/10.1038/26387

Nickson RT, McArthur JM, Ravenscroft P, Burgess WG, Ahmed KM (2000) Mechanism of arsenic release to groundwater, Bangladesh and West Bengal. Appl Geochem 15:403–413. https://doi.org/10.1016/S0883-2927(99)00086-4

O'Day PA, Vlassopoulos D, Root R, Rivera N (2004) The influence of sulfur and iron on dissolved arsenic concentrations in the shallow subsurface under changing redox conditions. Proc Natl Acad Sci 101:13703–13708. https://doi.org/10.1073/pnas.0402775101

Padungthon S, German M, Wiriyathamcharoen S, SenGupta AK (2015) Polymeric anion exchanger supported hydrated Zr(IV) oxide nanoparticles: A reusable hybrid sorbent for selective trace arsenic removal. React Funct Polym 93:84–94. https://doi.org/10.1016/j.reactfunctpolym.2015.06.002

Pal P, Sen M, Manna A, Pal J, Pal P, Roy S, Roy P (2009) Contamination of groundwater by arsenic: a review of occurrence, causes, impacts, remedies and membrane-based purification. J Integr Environ Sci 6:295–316. https://doi.org/10.1080/19438150903185077

Palma-Lara I, Martínez-Castillo M, Quintana-Pérez JC, Arellano-Mendoza MG, Tamay-Cach F, Valenzuela-Limón OL, García-Montalvo EA, Hernández-Zavala A (2020) Arsenic exposure:

A public health problem leading to several cancers. Regul Toxicol Pharmacol 110: https://doi.org/10.1016/j.yrtph.2019.104539

Panagiotaras D, Nikolopoulos D (2015) Arsenic occurrence and fate in the environment; A geochemical perspective. J Earth Sci Clim Change 06:1. https://doi.org/10.4172/2157-7617.1000269

Pardo T, Bernal MP, Clemente R (2017) Phytostabilisation of severely contaminated mine tailings using halophytes and field addition of organic and inorganic amendments. Chemosphere 178:556–564

Patel H (2019) Fixed-bed column adsorption study: a comprehensive review. Appl Water Sci 9:45. https://doi.org/10.1007/s13201-019-0927-7

Peavy HS, Rowe DR, Tchobanoglous G (1985) Environmental Engineering. McGraw-Hill, New York

Piacente E (2017) Disposición de rechazos de osmosis en plantas de la Provincia de Santa Fe. Aguas Santafesinas SA Comunicación personal

Pontius FW, Brown KG, Chen C-J (1994) Health implications of arsenic in drinking water. J-Am Water Work Assoc 86:52–63

Rahaman S, Sinha AC, Pati R, Mukhopadhyay D (2013) Arsenic contamination: a potential hazard to the affected areas of West Bengal, India. Environ Geochem Health 35:119–132. https://doi.org/10.1007/s10653-012-9460-4

Rahman IMM, Hossain MM, Uddin MH, Nazimuddin M, Majid MA (2005) Appraisal of Two Indigenous Household Groundwater Arsenic Removal Technologies for Bangladesh under Field Conditions. J Agric Soc Sci 1:5

Rahman MM, Naidu R, Bhattacharya P (2009) Arsenic contamination in groundwater in the Southeast Asia region. Environ Geochem Health 31:9–21. https://doi.org/10.1007/s10653-008-9233-2

Ratnaike RN (2003) Acute and chronic arsenic toxicity. Postgrad Med J 79:391–396. https://doi.org/10.1136/pmj.79.933.391

Ren X, Zhang Z, Luo H, Hu B, Dang Z, Yang C, Li L (2014) Adsorption of arsenic on modified montmorillonite. Appl Clay Sci 97–98:17–23. https://doi.org/10.1016/j.clay.2014.05.028

Ren Z, Zhang G, Paul Chen J (2011) Adsorptive removal of arsenic from water by an iron–zirconium binary oxide adsorbent. J Colloid Interface Sci 358:230–237. https://doi.org/10.1016/j.jcis.2011.01.013

Robinson B, Marchetti M, Moni C, Schroeter L, Van Den Dijssel C, Milne G, Bolan N, Mahimairaja S (2005) Arsenic accumulation by aquatic and terrestrial plants. Managing Arsenic in the Environment: From Soil to Human Health. CSIRO Publishing, Collingwood, VIC, pp 235–247

Rosehart RG, Lee JY (1972) Effective methods of arsenic removal from gold mine wastes. Can Min J 6:53–57

Roy PK, Majumder A, Pal S, Banerjee G, Roy MB, Debbarma J, Mazumdar A (2018) Development of an environmentally sustainable approach for safe disposal of arsenic-rich sludge. In: Hussain CM (ed) Handbook of environmental materials management. Springer, Cham, pp 1–16

Sanyal SK, Gupta SK, Kukal SS, Jeevan Rao K (2015) Soil degradation, pollution and amelioration. State of Indian Agriculture-Soil, National Academy of Agricultural Sciences, New Delhi 234–266

Sarkar A, Paul B (2016) The global menace of arsenic and its conventional remediation—A critical review. Chemosphere 158:37–49. https://doi.org/10.1016/j.chemosphere.2016.05.043

Schmidt S-A, Gukelberger E, Hermann M, Fiedler F, Großmann B, Hoinkis J, Ghosh A, Chatterjee D, Bundschuh J (2016) Pilot study on arsenic removal from groundwater using a small-scale reverse osmosis system—Towards sustainable drinking water production. J Hazard Mater 318:671–678. https://doi.org/10.1016/j.jhazmat.2016.06.005

Schneiter RW, Middlebrooks EJ (1983) Arsenic and fluoride removal from groundwater by reverse osmosis. Environ Int 9:289–291. https://doi.org/10.1016/0160-4120(83)90087-9

Shahid M, Dumat C, Khan Niazi N, Khalid S, Natasha (2018) Global scale arsenic pollution: increase the scientific knowledge to reduce human exposure. VertigO—la revue électronique en sciences de l'environnement. https://doi.org/10.4000/vertigo.21331

Shakoor MB, Niazi NK, Bibi I, Murtaza G, Kunhikrishnan A, Seshadri B, Shahid M, Ali S, Bolan NS, Ok YS, Abid M, Ali F (2016) Remediation of arsenic-contaminated water using agricultural wastes as biosorbents. Crital Rev Environ Sci Technol 46:467–499. https://doi.org/10.1080/10643389.2015.1109910

Shan Y, Mehta P, Perera D, Varela Y (2019) Cost and efficiency of arsenic removal from groundwater: A review. United Nations University-Institute for Water, Environment and Health, Hamilton

Shih M-C (2005) An overview of arsenic removal by pressure-drivenmembrane processes. Desalination 172:85–97. https://doi.org/10.1016/j.desal.2004.07.031

Shrestha RR, Shrestha MP, Upadhyay NP, Pradhan R, Khadka R, Maskey A, Maharjan M, Tuladhar S, Dahal BM, Shrestha K (2003) Groundwater arsenic contamination, its health impact and mitigation program in Nepal. J Environ Sci Health, Part A 38:185–200. https://doi.org/10.1081/ESE-120016888

Shrivastava A, Ghosh D, Dash A, Bose S (2015) Arsenic Contamination in Soil and Sediment in India: Sources, Effects, and Remediation. Curr Pollution Rep 1:35–46. https://doi.org/10.1007/s40726-015-0004-2

Singh DK, Mohan S, Kumar V, Hasan SH (2016) Kinetic, isotherm and thermodynamic studies of adsorption behaviour of CNT/CuO nanocomposite for the removal of As (III) and As (V) from water. RSC Adv 6:1218–1230. https://doi.org/10.1039/C5RA20601D

Singh R, Singh S, Parihar P, Singh VP, Prasad SM (2015) Arsenic contamination, consequences and remediation techniques: A review. Ecotoxicol Environ Saf 112:247–270. https://doi.org/10.1016/j.ecoenv.2014.10.009

Singh TS, Pant KK (2004a) A comparative study for the removal of As(III) and As(V) by activated alumina. In: Natural arsenic in groundwater-Proceedings of the pre-congress workshop "Natural arsenic in groundwater", 32nd International geological congress. CRC Press, Florence, Italy, pp 211–221

Singh TS, Pant KK (2004b) Equilibrium, kinetics and thermodynamic studies for adsorption of As (III) on activated alumina. Separation and Purification Technology 36:139–147. https://doi.org/10.1016/S1383-5866(03)00209-0

Smedley PL, Kinniburgh DG (2002) A review of the source, behaviour and distribution of arsenic in natural waters. Appl Geochem 17:517–568. https://doi.org/10.1016/S0883-2927(02)00018-5

Smith AH, Lingas EO, Rahman M (2000) Contamination of drinking-water by arsenic in Bangladesh: A public health emergency. Bulletin WHO 78:1093–1101

Sorg TJ, Logsdon GS (1978) Treatment technology to meet the interim primary drinking water regulations for inorganics: Part 2. AWWA 70:379–393. https://doi.org/10.1002/j.1551-8833.1978.tb04198.x

Sorochkina K, Smotraiev R, Ratnaweera H (2015) Arsenic removal by zirconium oxyhydroxide adsorbents. In: Selected publications from the water harmony project: Water research and technology. Water Harmony, p 11

Sullivan C, Tyrer M, Cheeseman CR, Graham NJD (2010) Disposal of water treatment wastes containing arsenic — A review. Sci Total Environ 408:1770–1778. https://doi.org/10.1016/j.scitotenv.2010.01.010

Sun J, Zhang X, Zhang A, Liao C (2019) Preparation of Fe–Co based MOF-74 and its effective adsorption of arsenic from aqueous solution. J Environ Sci 80:197–207. https://doi.org/10.1016/j.jes.2018.12.013

Suvokhiaw S, Imyim A, Sukpirom N (2016) As (V) removal using a magnetic layered double hydroxide composite. Sep Sci Technol 51:2948–2957. https://doi.org/10.1080/01496395.2016.1231206

Syam Babu D, Nidheesh PV (2020) A review on electrochemical treatment of arsenic from aqueous medium. Chem Eng Commun 208:1–22. https://doi.org/10.1080/00986445.2020.1715956

Takamatsu T, Kawashima M, Koyama M (1985) The role of Mn2 + -rich hydrous manganese oxide in the accumulation of arsenic in lake sediments. Water Res 19:1029–1032. https://doi.org/10.1016/0043-1354(85)90372-0

Tanne N, Xu R, Zhou M, Zhang P, Wang X, Wen X (2019) Influence of pore size and membrane surface properties on arsenic removal by nanofiltration membranes. Front Environ Sci Eng 13:19. https://doi.org/10.1007/s11783-019-1105-8

Tchounwou PB, Yedjou CG, Udensi UK, Pacurari M, Stevens JJ, Patlolla AK, Noubissi F, Kumar S (2019) State of the science review of the health effects of inorganic arsenic: Perspectives for future research. Environ Toxicol 34:188–202. https://doi.org/10.1002/tox.22673

Thomas DJ, Styblo M, Lin S (2001) The cellular metabolism and systemic toxicity of arsenic. Toxicol Appl Pharmacol 176:127–144. https://doi.org/10.1006/taap.2001.9258

Tippairote T, Karnpanit W, Trachootham D (2020) Sources of arsenic exposure in well-nourished children. In: Fares A, Singh SK (eds) Arsenic Water Resources Contamination. Springer, Cham, pp 73–101

Tournassat C, Charlet L, Bosbach D, Manceau A (2002) Arsenic(III) Oxidation by Birnessite and Precipitation of Manganese(II) Arsenate. Environ Sci Technol 36:493–500. https://doi.org/10.1021/es0109500

Tuchowska M, Wołowiec M, Solińska A, Kościelniak A, Bajda T (2019) Organo-Modified Vermiculite: Preparation, Characterization, and Sorption of Arsenic Compounds. Minerals 9:483. https://doi.org/10.3390/min9080483

US EPA (2003) Arsenic treatment technology evaluation handbook for small systems

USEPA (2001) National primary drinking water regulations; Arsenic and clarifications to compliance and new source contaminants monitoring. In: Federal Register. https://www.federalregister.gov/documents/2001/01/22/01-1668/national-primary-drinking-water-regulations-arsenic-and-clarifications-to-compliance-and-new-source. Accessed 18 April 2020

Vithanage M, Chandrajith R, Bandara A, Weerasooriya R (2006) Mechanistic modeling of arsenic retention on natural red earth in simulated environmental systems. J Colloid Interface Sci 294:265–272. https://doi.org/10.1016/j.jcis.2005.07.026

Wan X, Lei M, Chen T (2016) Cost–benefit calculation of phytoremediation technology for heavy-metal-contaminated soil. Sci Total Environ 563–564:796–802. https://doi.org/10.1016/j.scitotenv.2015.12.080

Wang L, Chen A, Fields K (2000) Arsenic removal from drinking water by ion exchange and activated alumina plants (EPA/600/R-00/088). Office of Research and Development, USEPA, Cincinnati, OH, USA

Welch AH, Westjohn DB, Helsel DR, Wanty RB (2000) Arsenic in Ground Water of the United States: Occurrence and Geochemistry. Groundwater 38:589–604. https://doi.org/10.1111/j.1745-6584.2000.tb00251.x

WHO (2011) Guidelines for drinking-water quality, 4th edn. World Health Organization, Geneva

Xia S, Dong B, Zhang Q, Xu B, Gao N, Causseranda C (2007) Study of arsenic removal by nanofiltration and its application in China. Desalination 204:374–379. https://doi.org/10.1016/j.desal.2006.04.035

Yang G, Liu Y, Song S (2015) Competitive adsorption of As(V) with co-existing ions on porous hematite in aqueous solutions. J Environ Chem Eng 3:1497–1503. https://doi.org/10.1016/j.jece.2015.05.011

Yao S, Liu Z, Shi Z (2014) Arsenic removal from aqueous solutions by adsorption onto iron oxide/activated carbon magnetic composite. J Environ Health Sci Eng 12:58

Youngran J, Fan M, Van Leeuwen J, Belczyk JF (2007) Effect of competing solutes on arsenic(V) adsorption using iron and aluminum oxides. J Environ Sci 19:910–919. https://doi.org/10.1016/S1001-0742(07)60151-X

Yu F, Sun S, Ma J, Han S (2015) Enhanced removal performance of arsenate and arsenite by magnetic graphene oxide with high iron oxide loading. Phys Chem Chem Phys 17:4388–4397. https://doi.org/10.1039/C4CP04835K

Yu W, Luo M, Yang Y, Wu H, Huang W, Zeng K, Luo F (2019) Metal-organic framework (MOF) showing both ultrahigh As(V) and As(III) removal from aqueous solution. J Solid State Chem 269:264–270. https://doi.org/10.1016/j.jssc.2018.09.042

Zeeshan MH, Khan RU, Shafiq M, Sabir A (2020) Polyamide intercalated nanofiltration membrane modified with biofunctionalized core shell composite for efficient removal of arsenic and selenium from wastewater. J Water Process Eng 34: https://doi.org/10.1016/j.jwpe.2020.101175

Zhang G, Liu H, Qu J, Jefferson W (2012) Arsenate uptake and arsenite simultaneous sorption and oxidation by Fe-Mn binary oxides: Influence of Mn/Fe ratio, pH, Ca2 + , and humic acid. J Colloid Interface Sci 366:141–146. https://doi.org/10.1016/j.jcis.2011.09.058

Zhang G, Ren Z, Zhang X, Chen J (2013) Nanostructured iron(III)-copper(II) binary oxide: A novel adsorbent for enhanced arsenic removal from aqueous solutions. Water Res 47:4022–4031. https://doi.org/10.1016/j.watres.2012.11.059

Zhang W, Mossad M, Yazdi JS, Zou L (2016) A statistical experimental investigation on arsenic removal using capacitive deionization. Desalination Water Treat 57:3254–3260. https://doi.org/10.1080/19443994.2014.981761

Zhao D, Li H-B, Xu J-Y, Luo J, Ma LQ (2015) Arsenic extraction and speciation in plants: Method comparison and development. Sci Total Environ 523:138–145. https://doi.org/10.1016/j.scitotenv.2015.03.051

Zhao X, Zhang B, Liu H, Qu J (2010) Removal of arsenite by simultaneous electro-oxidation and electro-coagulation process. J Hazard Mater 184:472–476. https://doi.org/10.1016/j.jhazmat.2010.08.058

Zhou W, Fu H, Pan K, Tian C, Qu Y, Lu P, Sun C-C (2008) Mesoporous TiO2/α-Fe2O3: Bifunctional Composites for Effective Elimination of Arsenite Contamination through Simultaneous Photocatalytic Oxidation and Adsorption. J Phys Chem C 112:19584–19589. https://doi.org/10.1021/jp806594m

Zhu B-J, Yu X-Y, Jia Y, Peng F-M, Sun B, Zhang M-Y, Luo T, Liu J-H, Huang X-J (2012) Iron and 1,3,5-Benzenetricarboxylic metal–organic coordination polymers prepared by solvothermal method and their application in efficient As(V) removal from aqueous solutions. J Phys Chem C 116:8601–8607. https://doi.org/10.1021/jp212514a

Zhu H, Jia Y, Wu X, Wang H (2009) Removal of arsenic from water by supported nano zero-valent iron on activated carbon. J Hazard Mater 172:1591–1596. https://doi.org/10.1016/j.jhazmat.2009.08.031

Zhu J, Lou Z, Liu Y, Fu R, Baig SA, Xu X (2015) Adsorption behavior and removal mechanism of arsenic on graphene modified by iron–manganese binary oxide (FeMnO x/RGO) from aqueous solutions. RSC Adv 5:67951–67961. https://doi.org/10.1039/C5RA11601E

Chapter 6
Occurrence, Fate, and Removal of Microplastics in Sewage Treatment Plants (STPs)

Neha Parashar and Subrata Hait

Abstract Microplastics (MPs), usually plastic fragments smaller than 5 mm in size, are ubiquitous in the environment and pose an increasing threat to the entire ecosystem as an emerging micropollutant. MPs not only contribute to the accumulation of the plastics in the environment but also get transported into the food chain due to absorption and ingestion by aquatic species. They originate from either the direct environmental discharge of purposefully manufactured microscopic fragments as primary MPs or the fragmentation of large plastic debris by environmental factors as secondary MPs. It is widely speculated that the sewage treatment plants (STPs) are one of the prime conduits for releasing of MPs into the aquatic environment despite some degree of removal in the plants. Therefore, it is imperative to understand the occurrence and fate of MPs in STPs for their effective control. In this context, the current status of the identification, abundance, and removal of MPs in STPs is comprehensively explored. Various protocols for collection, pre-treatment, determination, and characterization of MPs in sewage are outlined and compared. The occurrence of MPs in the STPs in terms of their materialistic composition, shape and sizes, and concentration is summarized. The fate of MPs in various treatment stages in the STPs including entrainment in sewage sludge is also assessed. Further, future prospects on the development of advanced removal techniques are presented for effective MPs control.

Keywords Plastic pollution · Primary and secondary microplastics · Sewage · Treatment stages · Removal techniques

N. Parashar · S. Hait (✉)
Department of Civil and Environmental Engineering, Indian Institute of Technology Patna, Patna 801106, Bihar, India
e-mail: shait@iitp.ac.in

S. P. Singh et al. (eds.), *Pollution Control Technologies*, Energy, Environment, and Sustainability, https://doi.org/10.1007/978-981-16-0858-2_6

6.1 Introduction

Over the last decade, plastic pollution has gained its importance as a pressing problem around the globe (Derraik 2002; Jambeck et al. 2015; Lebreton et al. 2012; Murphy et al. 2016; Pruter 1987). Anthropogenic behavior has resulted in the deposition of plastic debris in different habitats encompassing a complex combination of materials, including synthetic polymers, which differ greatly from the Holocene period and therefore regarded as one of the vital indicators of the Anthropocene (Waters et al. 2016). Plastic particles will gradually break down into smaller pieces and are spread by oceanic or wind currents making their ways to the coastal environment due to their lightweight composition. Such tiny particles have the ability to adsorb persistent organic materials and toxic metals, and accumulate in the natural environment. Microplastics (MPs) are smaller plastic particles (<5 mm) in thickness, which are further categorized as primary and secondary MPs. The former one is purposely synthesized in factories with microscopic sizes and applied in abrasive products such as microbeads in several cosmetic products which enter the environment via sewage water. However, secondary MPs originate due to the fragmentation of bigger plastic particles under its exposure to daylight, air currents, water, and other microcontaminants (Murphy et al. 2016; Song et al. 2017). As the production along with the utility of plastics has steadily increased over the years, MPs occurrence and consumption have likewise escalated and are known to be the emerging micropollutants contaminating the water bodies (McCormick et al. 2014; Yonkos et al. 2014). Increased production of plastics as well as their persistent accumulation in the environment has received much attention from the scientific world and within the general community witnessing its adverse environmental impacts (Auta et al. 2017; Hernandez et al. 2019; Mason et al. 2018; Schirinzi et al. 2017; Yang et al. 2015). Over the years, an ample number of studies have been carried out about the sources, dispersal, and circulation of MPs exhibiting its toxicity within the water system (Carbery et al. 2018; Fossi et al. 2014; Guilhermino et al. 2018; Li et al. 2016; Rochman et al. 2013, 2014). Despite this, only a few researches have sought to establish the connecting link between the land-based sources and dispersal of plastic pieces in a variety of freshwater habitats (Lechner and Ramler 2015). From this perspective, the detection of possible land-based and aquatic sources of plastic debris remains one challenging issue. Among several important sources of MPs, sewage treatment plants (STPs) are regarded as the most prominent potential point source which releases MPs in the form of sewage sludge and ultimately reaches to the marine ecosystems via river flows (Carr et al. 2016). On the other hand, surface runoff is regarded as a potential transport pathway of MPs to the freshwater habitats, especially where sewage sludge containing MPs is used in the agricultural fields (Horton et al. 2017). Therefore, it is obligatory to develop reliable cum efficient experimental and quantification techniques to detect MPs in sewage influent, effluent, and sludge samples (Lassen et al. 2015). In these contexts, the present status of the occurrence, fate, and removal of MPs in STPs is extensively reviewed.

Various techniques for sample collection, processing, measurement, and characterization of MPs in sewage are comparatively summarized. The abundance of MPs in the STPs in terms of their composition, morphology, and concentration is explored. The fate of MPs at different stages of treatment in the STPs, viz. primary, secondary, and tertiary stages along with the dispersal in sewage sludge, is also examined. Further, the performance of advanced removal techniques for efficient removal of MPs is assessed. Moreover, the existing gaps in the literature are highlighted to suggest future research prospects in the area.

6.2 Overview of Plastic Pollution

Plastics—also known as the 'mermaid tears'—have infamously gained its importance for various properties like durability, stability, and affordability and were first reported in the 1630s as a material that is likely to undergo molding (Toxics Link 2018). The persuasive nature of plastic has made its cheap entry in the environment posing one major threat in recent years (Boots et al. 2019; Hernandez et al. 2019). Plastic production has surged exponentially during the last 15 years, accounting with over 400 million tonnes of annual production globally to date (Fig. 6.1). During the 1950s, annual global plastic production accounted to 1.5 million tonnes and has increased substantially to 250 million tonnes in the year 2009 with an annual rate of 9% which further increased significantly up to 311 million tonnes in the year 2014 with an annual rate of 25% (Hirai et al. 2011). Focusing on the current trend of plastic use and manufacturing, the global production is anticipated to ascend around 10.85 billion by 2100, and these figures are anticipated to range between 850 and 1124 million tonnes by the year 2050 (Rochman et al. 2013).

However, to predict the future this far without any degree of uncertainty is quite difficult. Indeed, it may sound unrealistic nowadays, but looking at the trend of

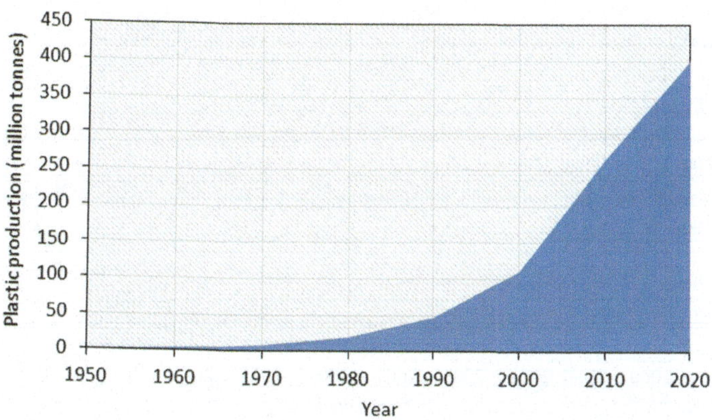

Fig. 6.1 Worldwide plastic production from 1950 onward

plastic production rate reaching up to 51% from 2010 (264 million tonnes) to 2020 (400 million tonnes) and historic output, the estimates are likely to surge exponentially in the near future.

MPs are essentially composed of polymers with 'micro' referring to the size category of plastic particles less than 5 mm (Hidalgo-Ruz et al. 2012). They are classified as synthetic (man-made) or semisynthetic based on the polymer content (Toxics Link 2018). Depending on the origin, MPs are further classified as primary (those MPs which are manufactured purposefully to be used as an ingredient in any product) and secondary (those MPs which are not manufactured but are the breakdown products of larger plastic debris). The earliest report on plastic pollution dates back to the 1960s, but significant studies on MPs have been conducted in the last ten years (Fig. 6.2). Nevertheless, the sudden increase in research on MPs pollution could be attributed to the increased rate of plastic production that has been doubled from 2011 to 2020 (Fig. 6.1).

While plastic pollution remains a visible problem, MPs originating via physical, biological, and chemical reactions are emerging as an important threat to human survival (Carbery et al. 2018; Magni et al. 2019; Schirinzi et al. 2017). The presence of MPs in a number of environmental samples with MPs reaching their sources of origin following different pathways closing the full circle has been reported. Recent studies have found the presence of MPs in bottled water, human stool, personal care cosmetic products, sea salts, and other living biota representing the scale of its perseverance (Table 6.1). Further, several studies showed the adverse effects of MPs pollution on marine and terrestrial ecosystems with the long-term forecasts showing escalating figure and likely to pose harmful effects on a much wider platform (Desforges et al. 2015; Fossi et al. 2014; Frias et al. 2014). Due to their invisible nature, MPs are constantly being used by the industries and manufacturers of different products such as cosmetics, food packaging, utensils, and textiles, a bigger portion of which either gets accumulated in water bodies or as plastic waste ends up in dump yards (WHO-UNICEF 2018).

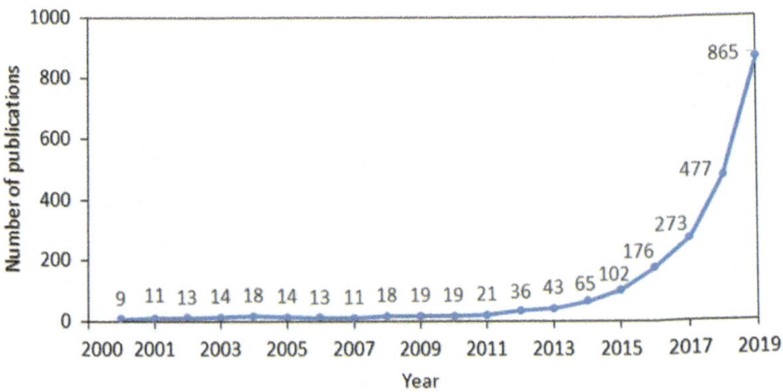

Fig. 6.2 Number of studies on MPs pollution from 2000 onward (Data sourced from the Scopus)

Table 6.1 Detection of MPs in a variety of environmental samples

Study area	Environmental sample	MPs content and size cut-off	Prominent MPs	Observations	References
India, Indonesia, USA, Lebanon, Thailand, Mexico, China, and Brazil	259 bottled water from 11 different brands	10.4 particles/L; 95% MPs in the range of 6.5–100 μm	Fragments, fibers (polypropylene)	93% of bottled water showed contaminated with MPs	Mason et al. (2018)
UK	Ryegrass (*L. perenne*), earthworm (*A. rosea*)	>2000 μm (macro-aggregated MPs), 63–250 μm (micro-aggregated MPs)	Polylactic acid (PLA), high-density polyethylene (HDPE), fibers	3.1% weight reduction observed in earthworm, 5.1% weight increase detected in earthworm without MPs exposure; shoot reduction in ryegrass and fewer seed germination due to exposure of fibers (PLA)	Boots et al. (2019)
UK, Japan, Finland, Poland, Austria, Italy, Russia, and Netherlands	Human stool	20 particles/10 g of stool; 50–500 μm	Polyethylene terephthalate (PET), polystyrene (PS)	Study established entry of MPs in human gut and human stool reporting 9 different types of MPs	Schwabl et al. (2019)
India	Personal care products	<1 mm	Microbeads	MPs detected in facial scrubs (67%) and face wash products (50%)	Toxics Link (2018)
China	Sea salts, rock/well salts, and lake salts	550–681 particles/kg (sea salts), 7–204 particles/kg (rock/well salts), 43–364 particles/kg (lake salts); 55% particles of size 200 μm	Fragments, fibers, pellets	Sea salts reported to contain higher MPs revealing greater MPs concentration in sea products and marine organisms	Yang et al. (2015)
Canada	Plastic tea bags	16 μg particles/cup of tea; 1–100 μm (small), 10–500 μm (large)	Nylon, polyethylene (PE)	3.1 billion released from a single plastic bag	Hernandez et al. (2019)
UK	Mosquito (*Culex*) larvae	2–15 μm	Microbeads, polystyrene (PS)	First study to show that MPs are getting into mosquitoes and contaminate the new food chain	Al-Jaibachi et al. (2019)
India	Human blood sample	13–60 nm	Polystyrene (PS)	Nanoplastics aggregate in blood and obstruct the body fluid flows rendering blood protein non-functional	Gopinath et al. (2019)

6.3 Determination and Characterization of MPs in Sewage

6.3.1 *Sample Collection Techniques*

Environmental samples are widely chosen for MP analysis. However, till date, no standard protocol has been approved for its sampling and analytical procedures. Although numerous studies have been conducted to compare the protocols adopted globally (Gatidou et al. 2019; Hidalgo-Ruz et al. 2012; Kang et al. 2018; Li et al. 2018), the determination of MPs, however, necessitates the following procedural steps: (i) sample collection, (ii) purification and recovery, and (iii) pretreatment, classification, and characterization. Various samples like surface water, sewage sludge, influent or effluent in an STP, and other biotic samples decide the steps required for sampling and preparation (Carr et al. 2016; Murphy et al. 2016; Talvitie et al. 2015, 2017). Methods for collection of STPs influent, final discharge, and sewage sludge are retrieved based on those used in environmental water bodies, but only some limited research has reported that there is no settlement on a standard method (Park et al. 2020; Simon et al. 2018). Collection of MPs is usually done by sieving or filtration using trawl nets, trapping nets, sieves, and filters of different shapes and sizes (mesh/pore sizes) to separate the MPs of different size categories (Carr et al. 2016; Corradini et al. 2019; Magnusson and Norén 2014; Mason et al. 2016). Grab or composite sample volume varies from some milliliters or liters for the raw wastewater to several liters to m^3 for the effluents (Lares et al. 2018; Magnusson and Norén 2014; Murphy et al. 2016; Talvitie et al. 2015; Ziajahromi et al. 2016). The reason behind the occurrence of a varying range of plastic particles between the incoming sewage water and final discharges relates to the larger concentration of the organic matter in the influents, and as a result, clogging of sampling devices such as filters or sieves can be triggered when a large volume of wastewater is collected (Dyachenko et al. 2017; Gies et al. 2018; Leslie et al. 2017). On the contrary, the final stage–discharge has significantly decreased the concentration of organic solids; hence, filtration of larger volumes is achievable. Furthermore, fewer MPs are predicted to be found in treated wastewater, so when larger particles are targeted for identification, sampling of higher volumes is therefore crucial (Magnusson and Norén 2014; Simon et al. 2018). For sampling the larger volume of STP influents (0.1–30 L), glass bottles and steel buckets are used, whereas custom-made pump filter systems, plankton nets, and commercial metal sieves are used to handle STP effluents (2–232,000 L approximately). Such differences in size cut-offs inevitably lead to greater variations in the concentration of MPs collected (Leslie et al. 2017; Magnusson and Norén 2014).

For instance, several authors have measured various categories of MPs from the influents and effluents employing filters representing different pore sizes. The results showed a contrasting concentration of MPs availability with a high discrepancy of the sample volumes possibly because of random sampling protocols and analytical procedures to quantify the MPs detected (Mason et al. 2016; Michielssen et al. 2016; Talvitie et al. 2015, 2017; WHO-UNICEF 2018).

6.3.2 MPs Extraction and Purification Methods

It is very imperative to select an appropriate sampling volume for different sampling sources within an STP as it may result in obtaining a reliable concentration of MPs, especially for the analysis of raw wastewater as the MPs count often exceeds 100 particles/L (Dris et al. 2017; Talvitie et al. 2017). Carrying out preliminary experiments would be useful in choosing an appropriate sample volume for a particular size category. Pre-treatment of collected samples before the isolation of MPs proved to be helpful in removing the organic and inorganic matter. Once separated, degradation of organic matter may be performed by the use of certain oxidizing agents like hydrogen peroxide or enzymatic reaction, while inorganic contents like sand particles or dust can be removed simply by density separation (Carr et al. 2016; Gies et al. 2018; Mason et al. 2016; Murphy et al. 2016). For the isolation of plastic debris from organic-rich samples (i.e., STP influents), wet peroxide oxidation digestion method is proven to be reliable but is time-consuming (Dyachenko et al. 2017). Wet peroxide oxidation (WPO) method is predominantly used to remove organic matter from samples under variable reaction time and temperature depending upon the concentration of the samples, while in the case of effluents, MPs are directly isolated from the samples by using filtration without employing any reaction treatment (Dris et al. 2017; Mintenig et al. 2017; Sutton et al. 2016; Vermaire et al. 2017).

Fluorescent dyes for the staining and detection of the smaller MPs are suggested in most of the studies (Hu et al. 2019). Nevertheless, it is also important to avoid contamination during the MPs isolation and pre-treatment. Several studies have suggested to investigate possible contamination simultaneously in order to prevent errors during the counting of MPs. For example, more emphasis should be given while cleaning the equipment such as containers, Petri dishes, and filters, avoiding the use of materials which are made up of synthetic fibers and if possible, use of glass or metal fitted equipment must be used instead of employing plastic containers (Gies et al. 2018; Simon et al. 2018; Talvitie et al. 2015). Treatment process may have an effect on the abundances and size distributions of the MPs in the collected sample. Any kind of alteration in MPs during the treatment process may have the potential to influence its counting, which could further add unknown risks associated with their accumulation in aquatic habitats resulting from field applications of effluents or sewage biosolids (Mahon et al. 2017). The use of different MPs isolation techniques showed variation between the samples collected. Therefore, to carry out interstudy comparisons for MPs extraction/purification, one harmonized method for different genres of samples is required (Xu et al. 2020).

6.3.3 MPs Detection and Analysis Techniques

The isolated MPs can be optically identified before any microscopic characterization encompassing three different methods determining the materialistic composition and size cut-offs, such as visual and biochemical identification, spectroscopic techniques, and thermoanalytical techniques (Braun et al. 2018). Chemical structure showing different polymeric compositions may be determined by comparing and contrasting absorption and emission spectrum against the reference spectrum. One such technique is the use of Fourier transform infrared spectroscopy (FT-IR), which is a well-known and comparatively fast technique having the ability to identify plastic particles between the ranges of 10–20 μm. Coupled with optical microscopy, FT-IR proved to be able to detect polymers even of lower size distribution. Since 70% of plastic particles that are confirmed as MPs by visual microscopes are not even proved as plastics by FT-IR, for reliable identification of polymeric chemical structure, spectroscopic techniques are strongly recommended over visual microscopy (Mintenig et al. 2017). Complexity in differentiating between natural and synthetic polymeric materials may lead to either overestimation or underestimation undergoing identification by the use of visual microscopes. Coupling of microscopy with other spectroscopic techniques such as Raman spectroscopy is likely to identify plastic particles within the ranges of 1–20 μm, but again this may subject to disturbances with certain barriers like 'biofilms' leading to error in detection, relatively slow analysis, and also requires expensive instrumentation (Carr et al. 2016; Li et al. 2018; Mahon et al. 2017; Mason et al. 2016).

On the other hand, larger plastic masses are the requirement to run thermoanalytical methods as compared to that of spectroscopic techniques. Mass spectroscopy can be used under inert conditions, decomposing specific components of the individual polymers before analysis. This provides information on the additives used during polymer synthesis and can be reliable in the identification of nanoplastic particles. The use of conventional methods is vital. For example, inductively coupled plasma mass spectrometry can be employed to identify different types of polymer fragments upon degradation (Braun et al. 2018). In both tagging and spectroscopic studies, software packages are often used to identify and count MPs, which could be helpful in their classification by comparing them with library spectra. It is also important to mention that spectroscopic methods are quite expensive, require highly trained staff, and often present uncertainties due to the limited sample scanned during analysis (Vollertsen and Hansen 2017).

6.4 Occurrence and Morphology of MPs in STPs

Shape, size, and materialistic composition of any polymer are physical characteristics of MPs which trigger their transportation and may contribute to the removal efficacy of different STPs (Kang et al. 2018). Several researches were undertaken during the last decade from countries like the USA, North America, and Australia highlighting the crucial role of raw or treated wastewater from STPs as a prominent route for a diverse group of plastic debris entering the ecosystem (Dris et al. 2017; Gallagher et al. 2016; Gies et al. 2018; Lares et al. 2018; Mason et al. 2018). A brief

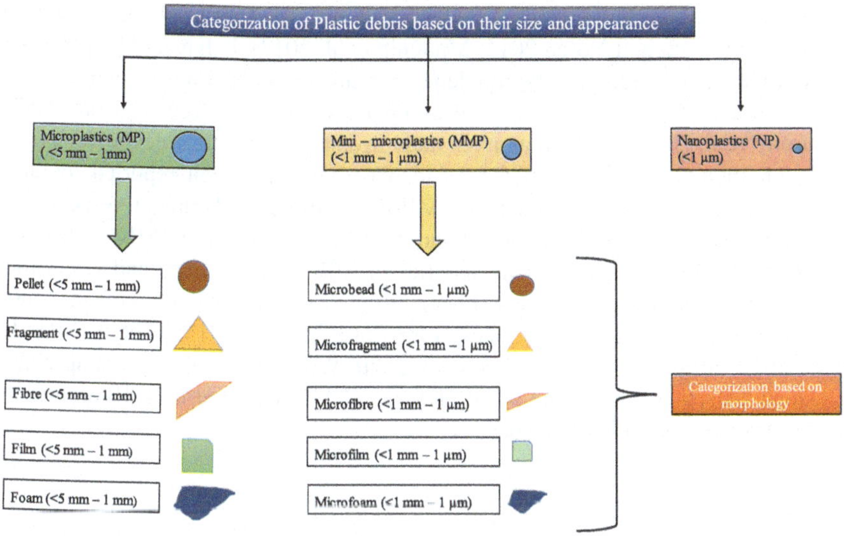

Fig. 6.3 Categorization of plastic based on size and morphology

summary of the categorization of MPs based on the size and morphology has been summarized and depicted in Fig. 6.3. MPs which are reported to be in abundance include polyethylene (PE), polypropylene (PP), polyester, polyamide (PA), polystyrene (PS), and nylon fibers, and the number count ranges between 1 and 3160 particles/L and 0.0007–125 particles/L for influent and effluent, respectively, in an STP (Gatidou et al. 2019). Polyethylene (PE) is identified as the largest contributor to plastic production in the world having a lower density than water (Hu et al. 2019). It has been reported that polyester is the characteristic plastic detected in the majority of STPs worldwide, the reason being its large-scale utilization as synthetic fibers in garment processing, and sewage water contains a greater proportion of such synthetic fibers from laundry effluents (Falco et al. 2018; Hernandez et al. 2019; Napper and Thompson 2016).

For example, a typical household-based washing machine has the potential to release as much MPs as 6 million fibers from 5 kg of fabrics (Falco et al. 2018). In a German STP, higher concentrations of MPs (approximately 24–33 particles per liter) were detected in the effluent of three municipal wastewater treatment plants based in the Netherlands (Dubaish and Liebezeit 2013; Leslie et al. 2017). In a more detailed analysis, final discharged effluent was analyzed from 12 German STPs, and plastic particles (<500 μm) were detected. As for the type of MPs, polyethylene, polystyrene, polyester, polyamide, and polyvinyl chloride were the most common. This study also reported that MPs with sizes greater than 500 μm reached up to 0.05 MPs/L of the effluent (Mintenig et al. 2017). MPs numbers have been found ranging in between 10^3 and 170.9×10^3 particles/kg by dry weight in sewage sludge samples indicating the potential of sewage treatment plants in the retention of plastic debris from the discharged effluent (Carr et al. 2016; Li et al. 2018; Magnusson and Norén 2014; Mintenig et al. 2017). Polymers like polyester, polyethylene, polyamide, polypropylene, and alkyds are the most common MPs types detected from the inlet as well as final stage effluent from a Scottish STP (Murphy et al. 2016). Further, another study investigated the treatment efficiency of 7 Dutch municipal STPs, and the mean plastic concentration reported to range between 68 and 910 MPs/L and 51–81 MPs/L in influent and effluent, respectively, of particle sizes ranging between 10 and 5000 μm (Leslie et al. 2017). This suggests that in STPs, physical characteristics of MPs such as shape and size varied frequently, with flakes/fragments, fibers, film, foam, and pellets being the most common shapes (Dubaish and Liebezeit 2013; Leslie et al. 2017; Li et al. 2018; Lv et al. 2019; Mason et al. 2016; Park et al. 2020; Wolff et al. 2019; Xu et al. 2020). Detailed analysis of the varied shape and size of MPs detected in STPs around the world has been provided in Table 6.2.

6.5 Fate and Removal of MPs at Different Treatment Stages in STPs

Raw wastewater enters a sewage treatment plant through different sources such as domestic washing machines, commercial laundries, textiles, stormwater, surface runoff, and sometimes also from factories (Mahon et al. 2017). The fate and transport of MPs at different treatment stages in an STP have been qualitatively depicted in Fig. 6.4. Municipal wastewater treatment plants receive MPs resulting from products like toothpaste, personal care products, plastic tea bags, fabrics generated from household washing machines, etc., and ultimately end up at the STPs (Browne et al. 2011). Recently, several studies have shown a varying range of MPs removal at different treatment stages of STPs as summarized in Table 6.2. Conventional STPs typically include two stages of wastewater treatment, viz. primary and secondary, and at times advanced treatment stages to decontaminate the raw wastewater by treating them with methods like physical, biological, and chemical techniques (Gatidou et al. 2019; Hu et al. 2019). A large amount of suspended organic solids entered STP during the primary treatment, which undergoes processes like screening/grit chamber, settling, filtration, floatation, skimming, etc. However, the effluent coming out after the primary treatment still contains a higher concentration of MPs, even though the primary treatment showed 25% removal efficiency (Poerio et al. 2019). The secondary treatment techniques have reported reducing MPs count with reduction of around 75% and include treatments like aeration, coagulation, flocculation, gravity filter, anaerobic membrane filtration, activated sludge treatment methods, etc. (Carr et al. 2016; Corradini et al. 2019; Mason et al. 2016; Mintenig et al. 2017). The final or tertiary treatment is deployed wherever required depending on the pollutant removal rate following secondary treatment possessing a removal rate of up to 99% (Mintenig et al. 2017; Sun et al. 2019). During the last decade, with

Table 6.2 Removal of MPs at treatment stages in STPs employing various technologies

Country	Sample	Treatment	Type of MPs	MPs content	Removal efficiency (%)	Identification method	References
Korea	Influent and effluent	Advanced treatment using phosphorus	Fragments, fibers	Influent (10–470 MPs/L), effluent (0.004–0.51 MPs/L)	98.7–99.99	Dissection Microscope and FT-IR spectroscopy	Park et al. (2020)
China	Sludge	Sludge was treated by using an isolation device with fitted glass funnel	Pellets, fibers, fragments, films	NM	100	FT-IR spectroscopy	Xu et al. (2020)
China	Influent, effluent, and sludge	Oxidation ditch (OD), membrane bioreactor (MBR)	Fibers, fragments, films, flakes	0–4 MPs/L	OD (97), MBR (99.5)	FT-IR spectroscopy	Lv et al. (2019)
Germany	Secondary clarifier effluent	Chemical and biological treatment	Polyester, polypropylene, polyethylene, polystyrene	3000–5900 MPs/m^3	NM	Raman spectroscopy	Wolff et al. (2019)
Chile	Influent and secondary effluent	Primary and secondary treatment	Fibers	34 particles/g	97	Visual observation and micro-FT-IR	Corradini et al. (2019)
Turkey	Influent and secondary effluent	Primary and secondary treatment	Polyester	Influent (1–6.5 million particles/day), effluent (220,000–1.5 million particles/day)	73–79	Visual observation and micro-FT-IR	Gündoğdu et al. (2018)
China	Sludge	Secondary treatment	Polyolefin, acrylic fibers, polyethylene, polyamide	Approximately 22.7 × 103 MPs/kg dry wt.	NM	Micro-FT-IR	Li et al. (2018)

(continued)

Table 6.2 (continued)

Country	Sample	Treatment	Type of MPs	MPs content	Removal efficiency (%)	Identification method	References
Canada	Primary influent, secondary effluent, primary sludge and secondary sludge	Secondary treatment	Fibers, fragment, foam granule, sheet, pellets	Primary influent (1.76 trillion MPs), secondary effluent (0.3 trillion MPs), primary sludge (1.28 trillion MPs), secondary sludge (0.36 trillion MPs)	97–99	Visual observation and FT-IR	Gies et al. (2018)
Denmark	Influent and tertiary effluent	Biological, activated sludge and rapid sand filter	Polyethylene, polypropylene, polyamide, polystyrene, etc.	0.56 g MPs/capita/year), influent (27%), effluent (25–27%)	98.3	Focal plane array (FPA)-based FT-IR	Simon et al. (2018)
Korea	Influent, effluent, and sludge	Anaerobic anoxic—aerobic (A2O), sequence batch reactor (SBR), media process	Fibers, fragments	Influent (29.85 MPs/g), effluent (0.435 MPs/g), sludge (14.895 MPs/g)	98		Lee and Kim (2018)
Finland	Influent, primary clarifier effluent, effluent after disinfection	Conventional activate sludge (CAS) and membrane bioreactor (MBR)	Polyester, polyamide, polyethylene, various polymers	MBR (0.4 MPs/L), CAS (1.0 MPs/L)	NM	Visual observation and FT-IR/ Raman spectroscopy	Lares et al. (2018)
Germany	Tertiary effluent	Post-filtration	Polyethylene, polypropylene, polyamide, polyvinyl chloride, polyester, etc.	0.01 particles/L, 0.02 fibers/L	97	Micro-FT-IR	Mintenig et al. (2017)

(continued)

Table 6.2 (continued)

Country	Sample	Treatment	Type of MPs	MPs content	Removal efficiency (%)	Identification method	References
Netherlands	Influent, effluent, and sludge	Primary, secondary, and tertiary treatment	Fibers, foils, spheres	Influent (68–910 MPs/L), effluent (51–81 MPs/L), sludge (510–760 MPs/kg)	NM	Visual observation and FT-IR spectroscopy	Leslie et al. (2017)
Australia	Tertiary effluent	Reverse osmosis	Polyvinyl chloride, polystyrene, polypropylene, polyethylene terephthalate, polyethylene, nylon	0.21–0.28 MPs/L	NM	Visual observation and FT-IR	Ziajahromi et al. (2016)
Slovenia	Secondary effluent	Activated sludge, gravity filter, and anaerobic membrane filtration		NM	52		Kalčíková et al. (2017)
Finland	Tertiary effluent	Rapid sand filter (RSF), dissolved air floatation (DAF), disk filter (DF), membrane bioreactor (MBR)	Fibers, fragments, flakes, films, spherical	RSF (0.02 MPs/L), DAF (0.1 MPs/L), DF (0.03 MPs/L), MBR (0.005 MPs/L)	RSF (97), DAF (95), DF (98.5), MBR (99.9)	ATR-FT-IR and focal plane array (FPA)-based FT-IR	Talvitie et al. (2017)
Scotland	Secondary effluent	Biological and chemical treatment	Flakes, fibers, film, beads	0.25 MPs/L	98.4	Visual observation and FT-IR	Murphy et al. (2016)
USA		Activated sludge process (ASP),		ASP (5.9 MPs/L), GF (2.6 MPs/L), AMF (0.5 MPs/L)	ASP (95.9), GF	Visual observation	Michielssen et al. (2016)

(continued)

Table 6.2 (continued)

Country	Sample	Treatment	Type of MPs	MPs content	Removal efficiency (%)	Identification method	References
	Secondary and tertiary effluent	gravity filter (GF), anaerobic membrane filtration (AMF)	Fiber, fragment, microbeads, nurdle, paint chip, others		(97.2), AMF (99.4)		
USA	Tertiary effluent	Gravity filter	Irregularly shaped polyethylene	0.0008 MPs/L	99.9	Visual observation and FT-IR spectroscopy	Carr et al. (2016)
USA	Secondary and tertiary effluent	Primary, secondary, and tertiary treatment	Fragments, fibers, films, foam, pellets	0.05 particles/L, 0.03 fibers/L	NM	Dissection microscope	Mason et al. (2016)
Ireland	Sludge	Anaerobic digestion, thermal drying, lime stabilization	Fibers, fragments, films	4196–15,385 MPs/kg dry wt.	NM	ATR-FT-IR spectroscopy and scanning electron microscopy (SEM)	Mahon et al. (2017)
Finland	Tertiary effluent	Biological filtration	Synthetic particle, textile fibers	8.6 particles/L, 4.9 fibers/L	97–98	Stereo microscope	Talvitie et al. (2015)
Sweden	Influent and effluent	Chemical, mechanical, and biological treatment	Fibers, fragments, flakes	Influent (15,000 MPs/m^3), effluent (1.1–1.8 MPs/m^3)	NM	Stereo microscope and FT-IR	Magnusson and Norén (2014)
St. Petersburg	Influent, secondary effluent, and final effluent	Mechanical purification	Synthetic fibers, black particles, textile fibers	Influent (3787 MPs/L), after secondary treatment effluent (356 MPs/L), final effluent (148 MPs/L)	NM	Visual observation	HELCOM (2014)

NM: Not mentioned

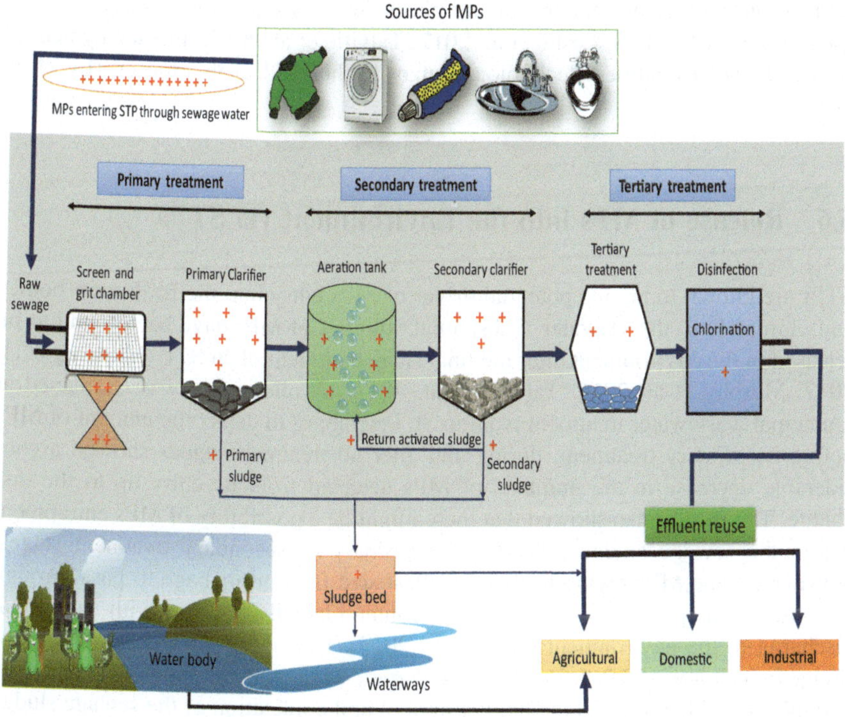

Fig. 6.4 Fate and transport of MPs at different treatment stages in an STP

the increase in plastic production worldwide studies on the treatment of influent have been continuously carried out to enhance the quality of final discharge. Furthermore, technologies deployed in various treatment stages within an STP are not expressly designed to reduce the MPs count. However, such techniques proved to be helpful in enhancing the final water quality (Mason et al. 2016; Talvitie et al. 2017). Conventional STPs with its different treatment methods may reduce MPs with a removal rate up to 99%, eliminating most of the plastic debris during the initial phases with the average removal efficacy ranging between 72 and 99.4% (Leslie et al. 2017; Mason et al. 2016). Nevertheless, due to large quantities being handled, even a small amount of MP released per liter will cause substantial MP load entering the environment (Murphy et al. 2016). Similar studies stated that the removal of MPs during the preliminary treatment through solid skimming, sludge settling processes, and the final discharge from later treatment stages may encompass some amount of MPs in them. Therefore, regardless the high recovery rate, traditional STPs can actually act as a substantial source of MPs, provided the larger volume of final effluent is discharged into the natural environment (Mason et al. 2016; Mintenig et al. 2017; Murphy et al. 2016; Talvitie et al. 2017). Although some studies have suggested that various treatment techniques in a general STP like screening, sedimentation, flotation, coagulation–flocculation, and filtration processes are able to remove an appreciable amount of the plastic waste from wastewater (Carr et al. 2016; Lares et al. 2018; Leslie et al. 2017; Talvitie et al. 2017), the deployment of some advanced treatment technologies may further add to its efficacy (Lv et al. 2019).

6.6 Release of MPs into the Environment via STPs

STPs are known to be one potential driver of MPs entry into the freshwater bodies and ultimately to the seawater. Substantial volumes of MPs have been found to be released in the environment after the final stage treatment of an STP (Mintenig et al. 2017; Murphy et al. 2016; Talvitie et al. 2017). A pilot study was conducted at municipal wastewater treatment plant in St. Petersburg to detect the amount of MPs before its primary treatment, during and after all treatment stages showed a considerable decrease in the numbers of MPs detected from its entry up to the discharge. The results also showed that there might be a possibility of MPs entrapment within the sewage sludge released after primary or secondary treatment stages. However, some MPs owing to their smallest size fraction manage to pass through such treatment processes and likely to contaminate the system with their final discharge in the environment (HELCOM 2014). Depending on the production of sludge in a Chinese STP, 1.56–1014 MP particles/year was estimated to be released into the natural habitats via the field or agricultural application of the sewage sludge (Li et al. 2018). Similar results from other studies have identified sewage sludge as one potential transport medium allowing tiny plastic debris to reach the surrounding habitats (Mahon et al. 2017). Researchers have also suggested such treatment processes may have an effect on the degradation pattern of the MPs in wastewater. MPs get altered passing through different treatment stages, and their potential mobility may also be influenced accordingly (Mahon et al. 2017). This could further exaggerate any unidentified threat associated with the persistence of MPs in the incoming wastewater and the effluents. Despite the treatment processes, there are chances of MPs accumulation in the discharged sewage sludge over time which automatically ends up contaminating the freshwater, seawater, or terrestrial ecosystems (da Costa et al. 2016; Michielssen et al. 2016; Mintenig et al. 2017; Talvitie et al. 2015, 2017). Sewage sludge has been reported to be contaminated with increased MPs counts that eventually used as fertilizer on agricultural fields, thus creating a pathway to enter agricultural soils (Corradini et al. 2019; Zubris and Richards 2005). Sources and the transfer following the fate of MPs within a traditional sewage treatment plant involving different treatment stages have been highlighted in Fig. 6.4.

Therefore, better knowledge on MPs origin and their transportation pathways across different treatment stages in an STP will enable researchers to revise, improve, and modify the current existing technologies, thereby facilitating their efficient removal from the system (Carr et al. 2016). Further, more quantitative studies are required from different regions across the globe so as to identify the behavior and fate of MPs entering and leaving any treatment system.

6.7 Advanced Techniques for MPs Removal

Wastewater treatment techniques have been continuously needed to enhance the final effluent quality. Operational performances of different treatment plants have failed to avoid the retention of MPs at their one or other stages attributing to the fact that such plants are not specifically designed to precisely filter these tiny plastic particles (Li et al. 2018; Kang et al. 2018; Talvitie et al. 2015, 2017). During the last decade, several studies using advanced wastewater treatment regimens have shown that MPs could be efficiently removed (Carr et al. 2016; Mintenig et al. 2017; Talvitie et al. 2017). Talvitie et al. (2017) examined the efficacy of different advanced treatment techniques for the reduction of MPs concentration in final effluent at four different municipal wastewater treatment plants located in Finland. Advanced treatment approaches like rapid sand filter, disk filter, and dissolved air floatation were used during the tertiary treatment stage, while secondary treatment includes membrane bioreactor. Results from the study showed MPs removal of around 97% (from 0.7 to 0.02 MPs/L) by the rapid sand filter, 40–98.5% (from 0.5–2.0 to 0.03–0.3 MPs/L) by the disk filter, and 95% (from 2.0 to 0.1 MPs/L) by the dissolved air flotation, whereas membrane bioreactor was able to remove around 99.9% (from 6.9 to 0.005 MPs/L) of MPs. Similar studies from different countries have also reported that advanced wastewater treatment technologies may substantially reduce MPs discharged from sewage treatment plants (Carr et al. 2016; Lee and Kim 2018; Lv et al. 2019; Michielssen et al. 2016; Park et al. 2020). In another study for the removal of MPs from sewage sludge, a unique isolation device using a fritted glass funnel was developed with nearly 100% recovery efficiency (Xu et al. 2020).

Recently, dynamic membrane technology including ultrafiltration has emerged as a plausible alternative for the treatment of sewage and sewage sludge (Poerio et al. 2019). Dynamic membrane technology has attracted great attention due to its low cost than the frequently used membranes, exhibited large permeation flux saving the membrane module quantity requirement and low-energy input owing to operation under gravity diving mode (Saleem et al. 2017; Zhang et al. 2019). Ultrafiltration is another feasible alternative used in advanced wastewater treatment technologies as it permits to attain a high quality of drinking water, contributes to low energy consumption, and possesses greater removal efficiency of MPs due to its compact plant size (Molinari et al. 2006, 2009; Molsehyani et al. 2019). A more detailed summary of the recently developed advanced treatment technologies for sewage wastewater and sludge has been presented in Table 6.2. The implication of these newly designed and advanced techniques is reported to be beneficial in reducing the MPs pollution from STPs.

6.8 Conclusions

MPs are omnipresent in the natural ecosystem. Important contributors of MPs into the natural ecosystem are land-based runoffs and drainage effluent. Still, given the information available at this time, a detailed assessment of relative MPs concentration emerging from different sources is quite a tedious process. To date, there is no mutual agreement to avail the MPs sampling and processing protocols at different treatment stages in an STP, reflecting the necessity to improvise its collection and determination deploying methodologies with proper quality control and assurance. Shape, size, and polymer material are characteristic features of MPs that may affect the way they are handled in STPs and can have an effect on the toxicity and effectiveness of treatment processes. A brief summary of MPs based on their morphological appearances along with types detected in STPs is discussed. Treated wastewater effluent is known to be the potential drivers of MPs pollution to the freshwater and agricultural soils. However, well-equipped and advanced treatment techniques in the sewage treatment plants may significantly eliminate maximum MPs count from the final discharge. Under typical operating conditions, traditional STPs are proved to be effectively removing MPs from influents. However, some plastic debris (<20 μm) manages to retain within the final effluents. All advanced wastewater treatment technologies developed so far along with their MPs removal efficacy are briefly presented. Furthermore, the increasing prevalence of MPs in sludge samples indicates the urgent requirement of a more robust technical approach, with a specific emphasis on estimating the sources of MPs, its fate, and prevention from returning back into the natural environment.

References

Al-Jaibachi R, Cuthbert RN, Callaghan A (2019) Examining effects of ontogenic microplastic transference on Culex mosquito mortality and adult weight. Sci Total Environ 651:871–876. https://doi.org/10.1016/j.scitotenv.2018.09.236

Auta HS, Emenike CU, Fauziah SH (2017) Distribution and importance of microplastics in the marine environment: A review of the sources, fate, effects, and potential solutions. Environ Int 102:165–176. https://doi.org/10.1016/j.envint.2017.02.013

Boots B, Russell CW, Green DS (2019) Effects of Microplastics in Soil Ecosystems: Above and below Ground. Environ Sci Technol 53:11496–11506. https://doi.org/10.1021/acs.est.9b03304

Browne MA, Crump P, Niven SJ, Teuten E, Tonkin A, Galloway T, Thompson R (2011) Accumulation of microplastic on shorelines woldwide: Sources and sinks. Environ Sci Technol 45:9175–9179. https://doi.org/10.1021/es201811s

Carbery M, O'Connor W, Palanisami T (2018) Trophic transfer of microplastics and mixed contaminants in the marine food web and implications for human health. Environ Int 115:400–409. https://doi.org/10.1016/j.envint.2018.03.007

Carr SA, Liu J, Tesoro AG (2016) Transport and fate of microplastic particles in wastewater treatment plants. Water Res 91:174–182. https://doi.org/10.1016/j.watres.2016.01.002

Corradini F, Meza P, Eguiluz R, Casado F, Huerta-Lwanga E, Geissen V (2019) Evidence of microplastic accumulation in agricultural soils from sewage sludge disposal. Sci Total Environ 671:411–420. https://doi.org/10.1016/j.scitotenv.2019.03.368

da Costa JP, Santos PSM, Duarte AC, Rocha-Santos T (2016) (Nano)plastics in the environment—Sources, fates and effects. Sci Total Environ 566–567:15–26. https://doi.org/10.1016/j.scitotenv.2016.05.041

De Falco F, Gullo MP, Gentile G, Di Pace E, Cocca M, Gelabert L, Brouta-Agnésa M, Rovira A, Escudero R, Villalba R, Mossotti R, Montarsolo A, Gavignano S, Tonin C, Avella M (2018) Evaluation of microplastic release caused by textile washing processes of synthetic fabrics. Environ Pollut 236:916–925. https://doi.org/10.1016/j.envpol.2017.10.057

Derraik JGB (2002) The pollution of the marine environment by plastic debris: A review. Mar Pollut Bull 44:842–852. https://doi.org/10.1016/S0025-326X(02)00220-5

Desforges JPW, Galbraith M, Ross PS (2015) Ingestion of Microplastics by Zooplankton in the Northeast Pacific Ocean. Arch Environ Contam Toxicol 69:320–330. https://doi.org/10.1007/s00244-015-0172-5

Dris R, Gasperi J, Bounoua B, Rocher V, Tassin B (2017) Microplastics in different compartments of the urban water cycle: From the sources to the rivers. In: Jorgensen B, Pahl S, Thompson RC, Vanderlinden J-P (eds) Fate and Impact of Microplastics in Marine Ecosystems: From the Coastline to the Open Sea. Elsevier, Amsterdam

Dubaish F, Liebezeit G (2013) Suspended microplastics and black carbon particles in the Jade system, southern North Sea. Water Air Soil Pollut 224:1352. https://doi.org/10.1007/s11270-012-1352-9

Dyachenko A, Mitchell J, Arsem N (2017) Extraction and identification of microplastic particles from secondary wastewater treatment plant (WWTP) effluent. Anal Methods 9:1412–1418. https://doi.org/10.1039/c6ay02397e

Fossi MC, Coppola D, Baini M, Giannetti M, Guerranti C, Marsili L, Panti C, de Sabata E, Clò S (2014) Large filter feeding marine organisms as indicators of microplastic in the pelagic environment: The case studies of the Mediterranean basking shark (*Cetorhinus maximus*) and fin whale (*Balaenoptera physalus*). Mar Environ Res 100:17–24. https://doi.org/10.1016/j.marenvres.2014.02.002

Frias JPGL, Otero V, Sobral P (2014) Evidence of microplastics in samples of zooplankton from Portuguese coastal waters. Mar Environ Res 95:89–95. https://doi.org/10.1016/j.marenvres.2014.01.001

Gallagher A, Rees A, Rowe R, Stevens J, Wright P (2016) Microplastics in the Solent estuarine complex, UK: An initial assessment. Mar Pollut Bull 102:243–249. https://doi.org/10.1016/j.marpolbul.2015.04.002

Gatidou G, Arvaniti OS, Stasinakis AS (2019) Review on the occurrence and fate of microplastics in sewage treatment plants. J Hazard Mater 367:504–512. https://doi.org/10.1016/j.jhazmat.2018.12.081

Gies EA, LeNoble JL, Noël M, Etemadifar A, Bishay F, Hall ER, Ross PS (2018) Retention of microplastics in a major secondary wastewater treatment plant in Vancouver, Canada. Mar Pollut Bull 133:553–561. https://doi.org/10.1016/j.marpolbul.2018.06.006

Guilhermino L, Vieira LR, Ribeiro D, Tavares AS, Cardoso V, Alves A, Almeida JM (2018) Uptake and effects of the antimicrobial florfenicol, microplastics and their mixtures on freshwater exotic invasive bivalve *Corbicula fluminea*. Sci Total Environ 622–623:1131–1142. https://doi.org/10.1016/j.scitotenv.2017.12.020

Gündoğdu S, Çevik C, Güzel E, Kilercioğlu S (2018) Microplastics in municipal wastewater treatment plants in Turkey: a comparison of the influent and secondary effluent concentrations. Environ Monit Assess 190:626. https://doi.org/10.1007/s10661-018-7010-y

HELCOM (2014) Preliminary study on synthetic microfibers and particles at a municipal waste water treatment plant. Balt Mar Environ Prot Comm HELCOM. https://helcom.fi/media/publications/Microplastics-at-a-municipal-waste-water-treatment-plant.pdf

Hernandez LM, Xu EG, Larsson HCE, Tahara R, Maisuria VB, Tufenkji N (2019) Plastic Teabags Release Billions of Microparticles and Nanoparticles into Tea. Environ Sci Technol 53:12300–12310. https://doi.org/10.1021/acs.est.9b02540

Hidalgo-Ruz V, Gutow L, Thompson RC, Thiel M (2012) Microplastics in the marine environment: A review of the methods used for identification and quantification. Environ Sci Technol 46:3060–3075. https://doi.org/10.1021/es2031505

Hirai H, Takada H, Ogata Y, Yamashita R, Mizukawa K, Saha M, Kwan C, Moore C, Gray H, Laursen D, Zettler ER, Farrington JW, Reddy CM, Peacock EE, Ward MW (2011) Organic micropollutants in marine plastics debris from the open ocean and remote and urban beaches. Mar Pollut Bull 62:1683–1692. https://doi.org/10.1016/j.marpolbul.2011.06.004

Horton AA, Walton A, Spurgeon DJ, Lahive E, Svendsen C (2017) Microplastics in freshwater and terrestrial environments: Evaluating the current understanding to identify the knowledge gaps and future research priorities. Sci Total Environ 586:127–141. https://doi.org/10.1016/j.scitotenv.2017.01.190

Hu Y, Gong M, Wang J, Bassi A (2019) Current research trends on microplastic pollution from wastewater systems: a critical review. Rev Environ Sci Biotechnol 18:207–230. https://doi.org/10.1007/s11157-019-09498-w

Jambeck J, Geyer R, Wilcox C, Siegler TR, Perryman M, Andrady A, Narayan R, Law KL (2015) Plastic waste inputs from land into the ocean. Science 347:768–771. https://doi.org/10.1126/science.1260352

Kalčíková G, Alič B, Skalar T, Bundschuh M, Gotvajn AŽ (2017) Wastewater treatment plant effluents as source of cosmetic polyethylene microbeads to freshwater. Chemosphere 188:25–31. https://doi.org/10.1016/j.chemosphere.2017.08.131

Kang H-J, Park H-J, Kwon O-K, Lee W-S, Jeong D-H, Ju B-K, Kwon J-H (2018) Occurrence of microplastics in municipal sewage treatment plants: a review. Environ Health Toxicol 33: https://doi.org/10.5620/eht.e2018013

Lares M, Ncibi MC, Sillanpää M, Sillanpää M (2018) Occurrence, identification and removal of microplastic particles and fibers in conventional activated sludge process and advanced MBR technology. Water Res 133:236–246. https://doi.org/10.1016/j.watres.2018.01.049

Lassen C, Hansen SF, Magnusson K, Hartmann NB, Rehne Jensen P, Nielsen TG, Brinch A (2015) Microplastics occurrence, effects and sources of releases. https://backend.orbit.dtu.dk/ws/files/118180844/Lassen_et_al._2015.pdf

Lebreton LCM, Greer SD, Borrero JC (2012) Numerical modelling of floating debris in the world's oceans. Mar Pollut Bull 64:653–661. https://doi.org/10.1016/j.marpolbul.2011.10.027

Lechner A, Ramler D (2015) The discharge of certain amounts of industrial microplastic from a production plant into the River Danube is permitted by the Austrian legislation. Environ Pollut 200:159–160. https://doi.org/10.1016/j.envpol.2015.02.019

Lee H, Kim Y (2018) Treatment characteristics of microplastics at biological sewage treatment facilities in Korea. Mar Pollut Bull 137:1–8. https://doi.org/10.1016/j.marpolbul.2018.09.050

Leslie HA, Brandsma SH, van Velzen MJM, Vethaak AD (2017) Microplastics en route: Field measurements in the Dutch river delta and Amsterdam canals, wastewater treatment plants, North Sea sediments and biota. Environ Int 101:133–142. https://doi.org/10.1016/j.envint.2017.01.018

Li J, Qu X, Su L, Zhang W, Yang D, Kolandhasamy P, Li D, Shi H (2016) Microplastics in mussels along the coastal waters of China. Environ Pollut 214:177–184. https://doi.org/10.1016/j.envpol.2016.04.012

Li X, Chen L, Mei Q, Dong B, Dai X, Ding G, Zeng EY (2018) Microplastics in sewage sludge from the wastewater treatment plants in China. Water Res 142:75–85. https://doi.org/10.1016/j.watres.2018.05.034

Lv X, Dong Q, Zuo Z, Liu Y, Huang X, Wu WM (2019) Microplastics in a municipal wastewater treatment plant: Fate, dynamic distribution, removal efficiencies, and control strategies. J Clean Prod 225:579–586. https://doi.org/10.1016/j.jclepro.2019.03.321

Magni S, Binelli A, Pittura L, Avio CG, Della Torre C, Parenti CC, Gorbi S, Regoli F (2019) The fate of microplastics in an Italian Wastewater Treatment Plant. Sci Total Environ 652:602–610. https://doi.org/10.1016/j.scitotenv.2018.10.269

Magnusson K, Norén F (2014) Screening of microplastic particles in and down-stream a wastewater treatment plant. IVL Swedish Environ Res Inst C 55:22. https://doi.org/naturvardsverket-2226

Toxics Link (2018) Personal Eco Care Product: Microplastics in Cosmetics, Toxics Link, New Delhi, India. https://www.toxicslink.org/docs/Microplastic.pdf

Mahon AM, O'Connell B, Healy MG, O'Connor I, Officer R, Nash R, Morrison L (2017) Microplastics in sewage sludge: Effects of treatment. Environ Sci Technol 51:810–818. https://doi.org/10.1021/acs.est.6b04048

Mason SA, Garneau D, Sutton R, Chu Y, Ehmann K, Barnes J, Fink P, Papazissimos D, Rogers DL (2016) Microplastic pollution is widely detected in US municipal wastewater treatment plant effluent. Environ Pollut 218:1045–1054. https://doi.org/10.1016/j.envpol.2016.08.056

Mason SA, Welch VG, Neratko J (2018) Synthetic polymer contamination in bottled water. Front Chem 6:407. https://doi.org/10.3389/fchem.2018.00407

McCormick A, Hoellein TJ, Mason SA, Schluep J, Kelly JJ (2014) Microplastic is an abundant and distinct microbial habitat in an urban river. Environ Sci Technol 48:11863–11871. https://doi.org/10.1021/es503610r

Michielssen MR, Michielssen ER, Ni J, Duhaime MB (2016) Fate of microplastics and other small anthropogenic litter (SAL) in wastewater treatment plants depends on unit processes employed. Environ Sci Water Res Technol 2:1064–1073. https://doi.org/10.1039/c6ew00207b

Mintenig SM, Int-Veen I, Löder MGJ, Primpke S, Gerdts G (2017) Identification of microplastic in effluents of waste water treatment plants using focal plane array-based micro-Fourier-transform infrared imaging. Water Res 108:365–372. https://doi.org/10.1016/j.watres.2016.11.015

Molinari R, Argurio P, Poerio T (2006) Ultrafiltration of polymer-metal complexes for metal ion removal from wastewaters. Macromol Symp 235:206–214. https://doi.org/10.1002/masy.200650325

Molinari R, Argurio P, Poerio T (2009) Membrane processes based on complexation reactions of pollutants as sustainable wastewater treatments. Sustainability 1:978–993. https://doi.org/10.3390/su1040978

Moslehyani A, Ismail AF, Matsuura T, Rahman MA, Goh PS (2019) Recent Progresses of Ultrafiltration (UF) Membranes and Processes in Water Treatment. Elsevier Inc

Murphy F, Ewins C, Carbonnier F, Quinn B (2016) Wastewater Treatment Works (WwTW) as a Source of Microplastics in the Aquatic Environment. Environ Sci Technol 50:5800–5808. https://doi.org/10.1021/acs.est.5b05416

Napper IE, Thompson RC (2016) Release of synthetic microplastic plastic fibers from domestic washing machines: Effects of fabric type and washing conditions. Mar Pollut Bull 112:39–45. https://doi.org/10.1016/j.marpolbul.2016.09.025

Park HJ, Oh MJ, Kim PG, Kim G, Jeong DH, Ju BK, Lee WS, Chung HM, Kang HJ, Kwon JH (2020) National reconnaissance survey of microplastics in municipal wastewater treatment plants in Korea. Environ Sci Technol 54:1503–1512. https://doi.org/10.1021/acs.est.9b04929

Poerio T, Piacentini E, Mazzei R (2019) Membrane processes for microplastic removal. Molecules 24(22):4148. https://doi.org/10.3390/molecules24224148

Pruter AT (1987) Sources, quantities and distribution of persistent plastics in the marine environment. Mar Pollut Bull 18:305–310. https://doi.org/10.1016/S0025-326X(87)80016-4

Rochman CM, Hoh E, Kurobe T, Teh SJ (2013) Ingested plastic transfers hazardous chemicals to fish and induces hepatic stress. Sci Rep 3:1–7. https://doi.org/10.1038/srep03263

Rochman CM, Kurobe T, Flores I, Teh SJ (2014) Early warning signs of endocrine disruption in adult fish from the ingestion of polyethylene with and without sorbed chemical pollutants from the marine environment. Sci Total Environ 493:656–661. https://doi.org/10.1016/j.scitotenv.2014.06.051

Saleem M, Alibardi L, Cossu R, Lavagnolo MC, Spagni A (2017) Analysis of fouling development under dynamic membrane filtration operation. Chem Eng J 312:136–143. https://doi.org/10.1016/j.cej.2016.11.123
Schirinzi GF, Pérez-Pomeda I, Sanchís J, Rossini C, Farré M, Barceló D (2017) Cytotoxic effects of commonly used nanomaterials and microplastics on cerebral and epithelial human cells. Environ Res 159:579–587. https://doi.org/10.1016/j.envres.2017.08.043
Schwabl P, Koppel S, Konigshofer P, Bucsics T, Trauner M, Reiberger T, Liebmann B (2019) Detection of various microplastics in human stool: A prospective case series. Ann Intern Med 171:453–457. https://doi.org/10.7326/M19-0618
Simon M, van Alst N, Vollertsen J (2018) Quantification of microplastic mass and removal rates at wastewater treatment plants applying Focal Plane Array (FPA)-based Fourier Transform Infrared (FT-IR) imaging. Water Res 142:1–9. https://doi.org/10.1016/j.watres.2018.05.019
Song YK, Hong SH, Jang M, Han GM, Jung SW, Shim WJ (2017) Combined effects of UV exposure duration and mechanical abrasion on microplastic fragmentation by polymer type. Environ Sci Technol 51:4368–4376. https://doi.org/10.1021/acs.est.6b06155
Sun J, Dai X, Wang Q, van Loosdrecht MCM, Ni BJ (2019) Microplastics in wastewater treatment plants: Detection, occurrence and removal. Water Res 152:21–37. https://doi.org/10.1016/j.watres.2018.12.050
Sutton R, Mason SA, Stanek SK, Willis-Norton E, Wren IF, Box C (2016) Microplastic contamination in the San Francisco Bay, California, USA. Mar Pollut Bull 109:230–235. https://doi.org/10.1016/j.marpolbul.2016.05.077
Talvitie J, Heinonen M, Pääkkönen JP, Vahtera E, Mikola A, Setälä O, Vahala R (2015) Do wastewater treatment plants act as a potential point source of microplastics? Preliminary study in the coastal Gulf of Finland, Baltic Sea. Water Sci Technol 72:1495–1504. https://doi.org/10.2166/wst.2015.360
Talvitie J, Mikola A, Koistinen A, Setälä O (2017) Solutions to microplastic pollution—Removal of microplastics from wastewater effluent with advanced wastewater treatment technologies. Water Res 123:401–407. https://doi.org/10.1016/j.watres.2017.07.005
Vermaire JC, Pomeroy C, Herczegh SM, Haggart O, Murphy M (2017) Microplastic abundance and distribution in the open water and sediment of the Ottawa River, Canada, and its tributaries. Facets 2:301–314. https://doi.org/10.1139/facets-2016-0070
Vollertsen, J, Hansen, AA (2017) Microplastic in Danish wastewater: Sources, occurrences and fate. Environmental Project No. 1906, The Danish Environmental Protection Agency. https://www2.mst.dk/Udgiv/publications/2017/03/978-87-93529-44-1.pdf
Waters CN, Zalasiewicz J, Summerhayes C, Barnosky AD, Poirier C, Gałuszka A, Cearreta A, Edgeworth M, Ellis EC, Ellis M, Jeandel C, Leinfelder R, McNeill JR, Richter DDB, Steffen W, Syvitski J, Vidas D, Wagreich M, Williams M, Zhisheng A, Grinevald J, Odada E, Oreskes N, Wolfe AP (2016) The Anthropocene is functionally and stratigraphically distinct from the Holocene. Science 80:351. https://doi.org/10.1126/science.aad2622
WHO-UNICEF (2018) Drinking water, sanitation and hygiene in schools. Global baseline report 2018, WHO-UNICEF Joint Monitoring Programme for Water Supply, Sanitation and Hygiene. https://www.unicef.org/media/47671/file/JMP-WASH-in-Schools-ENG.pdf
Wolff S, Kerpen J, Prediger J, Barkmann L, Müller L (2019) Determination of the microplastics emission in the effluent of a municipal waste water treatment plant using Raman microspectroscopy. Water Res X 2: https://doi.org/10.1016/j.wroa.2018.100014
Xu Q, Gao Y, Xu L, Shi W, Wang F, LeBlanc GA, Cui S, An L, Lei K (2020) Investigation of the microplastics profile in sludge from China's largest Water reclamation plant using a feasible isolation device. J Hazard Mater 388: https://doi.org/10.1016/j.jhazmat.2020.122067
Yang D, Shi H, Li L, Li J, Jabeen K, Kolandhasamy P (2015) Microplastic Pollution in Table Salts from China. Environ Sci Technol 49:13622–13627. https://doi.org/10.1021/acs.est.5b03163
Yonkos LT, Friedel EA, Perez-Reyes AC, Ghosal S, Arthur CD (2014) Microplastics in four estuarine rivers in the chesapeake bay, U.S.A. Environ Sci Technol 48:14195–14202. https://doi.org/10.1021/es5036317

Zhang S, Wang J, Liu X, Qu F, Wang X, Wang X, Li Y, Sun Y (2019) Microplastics in the environment: A review of analytical methods, distribution, and biological effects. TrAC—Trends Anal Chem 111:62–72. https://doi.org/10.1016/j.trac.2018.12.002

Ziajahromi S, Neale PA, Leusch FDL (2016) Wastewater treatment plant effluent as a source of microplastics: review of the fate, chemical interactions and potential risks to aquatic organisms. Water Sci Technol 74:2253–2269. https://doi.org/10.2166/wst.2016.414

Zubris KAV, Richards BK (2005) Synthetic fibers as an indicator of land application of sludge. Environ Pollut 138:201–211. https://doi.org/10.1016/j.envpol.2005.04.013

Braun U, Jekel M, Gerdts G, Ivleva N, Reiber J (2018) Microplastics analytics: sampling, preparation and detection methods. Discussion paper. Ecologic Institute, Berlin.

Gopinath PM, Saranya V, Vijayakumar S (2019) Assessment on interactive prospectives of nanoplastics with plasma proteins and the toxicological impacts of virgin, coronated and environmentally released-nanoplastics. Sci Rep 9:1–15. https://doi.org/10.1038/s41598-019-45139-6

Chapter 7
Membrane Technology for Desalination and Wastewater Recycling

Najmul Haque Barbhuiya and Swatantra P. Singh

Abstract Over the last few decades, the effectiveness of the conventional water treatment processes has become limited due to its failure to meet stringent water quality regulations, the emergence of non-conventional contaminants, and space limitations. Membrane technology can serve as a viable option to meet the limitations of the conventional treatment processes due to low chemical use, higher efficiency, easy operation, smaller footprint, and better quality treatment. But its effective usage is hindered due to fouling, shorter life span, and selectivity–permeability trade-off. Thus, there is a need for new-generation membranes that can surpass the limitations of the conventional membrane technologies, with less or no decrement in the efficiency of the treatment. With the advancements in material sciences and membrane fabrication processes, it could be possible to overcome the problems of conventional membrane treatment processes. In this chapter, a brief introduction of the membrane technology, its advantages over conventional treatment processes, commonly used pressure-driven membranes and membrane fabrication processes, the problems of conventional membranes, and some of the promising new-generation membrane materials and membrane systems, and water and wastewater treatment have been discussed.

Keywords Membrane treatment · Catalytic membrane · Phase inversion · New-generation membrane · Graphene · Fouling · Selectivity–permeability trade-off

---

N. H. Barbhuiya · S. P. Singh
Environmental Science and Engineering Department (ESED), Indian Institute of Technology Bombay, Mumbai, India

S. P. Singh (✉)
Centre for Research in Nanotechnology and Science (CRNTS), Indian Institute of Technology Bombay, Mumbai, India
e-mail: swatantra@iitb.ac.in

S. P. Singh et al. (eds.), *Pollution Control Technologies*, Energy, Environment, and Sustainability, https://doi.org/10.1007/978-981-16-0858-2_7

7.1 Introduction

According to the United Nations, by 2025, around 2.7 billion people will face problems accessing the potable water (UN Water 2012) and, by 2050, at least one-fourth of the world population will be living in a country having chronic or periodic shortages of freshwater. There will be a surge in demand for clean water in domestic, agricultural, and industrial sectors (Santhosh et al. 2016). Many industries are releasing untreated wastewater with a high concentration amount of pollutants into water (Nawab et al. 2016, 2017). Environmental disasters and various anthropogenic activities have contributed to the increase of toxic pollutants in the environment (Geise et al. 2010). Around 14,000 people die every day due to serious health problems caused by $\sim$2 million tonnes of the wastes discharged into water (UNESCO-WWAP 2003). These statistics are likely to worsen further due to the rapid increase of water contamination by pathogens and various pollutants such as heavy metals, pharmaceutical compounds, pesticides, and endocrine disrupting chemicals (Bohn et al. 2009). Moreover, the existing geopolitical pressures and lack of access to freshwater can result in regional conflicts, thus threatening global peace and stability (Eliasson 2015).

The effectiveness of conventional water treatment processes has become limited over the last few decades. The growing population, industrialization and climate change are increasing the water scarcity. Desalination and reuse of industrial and municipal wastewater are the solutions for generating new freshwater resources. However, industrial and municipal wastewater contains toxic substances that cannot be removed by conventional treatment processes. Reports have suggested that conventional treatment plants are not much effective against the emerging contaminants (Heberer 2002). Thus, membrane treatment technology could be a viable option. Membrane technology started to develop in the 1970s and showed rapid development since the 1990s due to its operational ease, low chemical input, high efficiency, smaller footprint, and low energy. Its use in the water treatment has been rapidly increased due to the development of high-quality membrane and overall decreasing cost (Pendergast and Hoek 2011; Peydayesh et al. 2013; Zhu et al. 2014). Membrane technology can thus play a vital role in removing these contaminants. They have a smaller footprint as compared to conventional treatment and give better removal of contaminants than conventional treatment process (Mulder 1991; Cui and Muralidhara 2010). Despite all the advantages of conventional membrane technology, there are problems associated with it, and there is a possibility to reduce the capital and running cost. The most common problems associated with membrane processes are the life span and the fouling of the membrane. Fouling is considered as the Achilles heel for the membrane technology (Flemming et al. 1997). As feedwater gets filtered into permeate, some of the inorganic, colloidal, organic, and biological contaminants get accumulated on the membrane surface forming, causing fouling and hindering its efficiency. Due to fouling alone, the life span of the membrane gets heavily affected (Zularisam et al. 2010). This limitation can be attributed to the fabrication methods and materials currently in use for membranes, which is still seen as an art rather than a science (Werber et al. 2016b). Thus, there is a need for new-generation membranes that can surpass the limitations of the conventional membrane technologies, with less or no decrement in the efficiency of the treatment.

The advancements in material sciences and membrane fabrication processes can mitigate the limitations of membrane technology by the use of new-generation membrane materials like aquaporin, graphene, metal nanoparticles, etc. (Werber et al. 2016b). Further, the use of a hybrid membrane and catalytic membrane can also help in overcoming the limitations of the conventional membrane systems. Hybrid systems increase purity, productivity, yield, and reducing operating costs and environmental pollution (Ang et al. 2015). By eliminating the thermodynamic equilibrium, the catalytic membranes enhance the reaction conversion, which results in the improvement of the overall process like conversion, yield, residence time, net reaction rate, and selectivity (Westermann and Melin 2009; Wang et al. 2015). Thus, the focus of this report is to provide a basic understanding of the membrane technology, its advantages over conventional treatments, problems relating to its operation, and new-generation membrane materials in order to develop new-generation membranes for water and wastewater treatments with better efficiency.

7.2 Membrane Technology vs. Conventional Water Treatment

The effectiveness of conventional water treatment processes has become limited over the last few decades. The conventional treatment system is not able to meet the current stringent water quality rules and regulations. The reuse of wastewater is also becoming critical, especially in arid and semiarid areas where water is being imported at very high costs. The growing population and increasing water scarcity have imparted the need for the reuse of industrial and municipal wastewater containing many toxic contaminants that cannot be removed by the conventional treatment processes (Heberer 2002; Bohn et al. 2009; Inyinbor et al. 2018). Thus, membrane treatment technology will be a better solution. In addition, improvement in the quality of membrane along with the developing market associated with the membrane processes is paving the way for membrane treatment with respect to their cost. According to some estimates, the cost of new membrane filtration is expected to be more or less than those of conventional treatment processes for small facilities (up to 20,000 m^3/day capacity) (Wiesner et al. 1994). Membranes can be used for the better separation of various types of contaminants without the use of chemicals, and additionally, decontamination can also be performed (Pendergast and Hoek 2011; Peydayesh et al. 2013; Zhu et al. 2014). The emerging contaminants like pharmaceuticals and personal care products have increased the concern for their removal from the water due to their harmful effects at low concentrations; it has also demanded the need for advanced treatments like membrane technology. Reports have suggested that conventional treatment plants are not much effective against these emerging contaminants (Inyinbor et al. 2018). Membrane technology can thus play a vital role in removing these contaminants. Membrane-based treatment plants also have a smaller footprint as compared to conventional treatment (Cui and Muralidhara 2010). The automated controls and alarms in membrane technology can provide trouble-free operation with less operation attention, thus providing ease of operation in comparison with conventional treatments.

7.3 Membrane Filtration

Membrane filtration is a separation process where particles and molecules are being separated from the mixture on the basis of size and chemical properties through a semipermeable membrane. In desalination and wastewater treatment, the contaminated water is allowed to pass through the membrane under some driving force, the treated water called permeate, and a portion of feed that gets retained along with the contaminants and salts called retentate (Mulder 1991). The driving force could be pressure gradients, concentration gradients, electrical potential gradient, or the temperature gradient. More specifically, membrane filtration can be further classified in terms of the size range of permeating species, the mechanisms of rejection, the driving forces employed, the chemical structure and composition of membranes, and the geometry of construction (Mulder 1991). The most important type of membrane filtration is the pressure-driven process, which includes microfiltration (MF), ultrafiltration (UF), nanofiltration (NF), and reverse osmosis (RO) (Bolisetty et al. 2019).

7.3.1 Types of Membranes

Membranes can be divided into two fundamental groups—biological and synthetic membranes.

(a) Biological membranes are the plasma membranes that surround the cells and have a complex structure. The characteristic of these types of membranes is that they have a lipid bilayer structure having the hydrophobic and hydrophilic parts. This bilayer structure does not allow a variety of molecules to enter the cell; however, specific transport of essential molecules like amino acids and sugars is accomplished by proteins located outside (extrinsic protein) and inside (intrinsic protein) the lipid layer (Watson 2015). These proteins behave like a carrier, and thus the molecule transport is carrier-mediated.

(b) Synthetic membranes are the artificially created membrane used for separation processes. They can further be classified as organic or polymeric and inorganic membranes based on the materials used. The most used and important membrane materials are organic materials, i.e., polymers.

Most of the synthetic membranes are made up of organic polymers (e.g., polysulfone, polyethersulfone, polyvinylidene difluoride, polyacrylonitrile). Polymers are the macromolecules with high molecular weight composed of repeated units called monomers. The number of monomers attached to form the long-chain polymer is called the degree of polymerization. Some of the common polymers used for MF and UF are polycarbonate, polypropylene, polyamide, polysulfone, polytetrafluoroethylene, polyacrylonitrile, etc. For MF and UF, the materials used are often the same, but different membrane fabrication conditions result in different pore sizes. Membrane selectivity can be improved by modifying the different properties of the polymers. Since chemically and thermally stable polymers are the desirable membrane materials, polymers with high glass transition temperatures (T_g), melting points, and crystallinity are preferred. T_g is the state of polymer defining property of a polymer below which the polymer is in the glassy state and above which the polymer goes to a rubbery state (Mohanty and Purkait 2011). Glassy polymers allow smaller molecules to pass faster than the larger hydrocarbon molecules, whereas the rubbery polymers have higher permeabilities for hydrocarbons than the much smaller molecules. Some of the commonly used polymers, along with their thermal characteristics, are shown in Table 7.1 (Scott 1995; Ida 2014). Although hydrophobic polymers have excellent thermal and chemical stability, hydrophilic polymers surpass them due to their reduced adsorption property, because of which hydrophilic membranes show less susceptibility toward fouling than hydrophobic counterparts.

Membranes are also formed from inorganic materials (e.g., ceramics, metals). Among the inorganic materials, ceramics are the most popular and frequently used as membrane material. They are often used for MF due to their microporous structure, and properties like thermal stability and resistivity toward chemicals. However, their high cost and fragility have limited their widespread use (Sagle and Freeman 2004).

Table 7.1 Commonly used polymers along with their thermal characteristics (Scott 1995; Ida 2014)

Polymer	T_g (°C)	Melting temperature (°C)
Polyethylene	−60 to −90	137−143.5
Polypropylene	−10	167−170
Teflon	−113	327
Cellulose acetate	69	230
Poly (ether sulfone)	225	>300
Polysulfone	190	>300
Polyacrylonitrile	80−104	319

7.3.2 Types of Pressure-Driven Membrane

A membrane process is defined by the properties of the synthetic membrane, type of driving force, and various other features (Mulder 1991). Based on the driving force, membrane separation processes can be distinguished as pressure-driven (such as MF), concentration-driven (such as osmosis), electric potential gradient-driven (such as electrodialysis), and temperature gradient-driven processes (such as membrane distillation). Pressure-driven membranes are the most widely used membrane process, such as in desalination and in many foods, and chemical industries. In this process, the pressure is used as the driving force for the production of permeate and retentate from the feed after successful separation. Figure 7.1 shows common types of pressure-driven membrane processes in the dead-end mode of filtration.

Microfiltration (MF): MF is the membrane process that is capable of separating particles of size 1000 nm and high molecular weight (>100 kDa). The retention potential of the membrane depends on the pore size. The pore size of the MF membranes is in the range of 100–10,000 nm. Due to the large pore size, low pressure is required for the separation. It is commonly used to separate suspended solids and microorganisms (bacteria, algae, etc.). The macroscopic particles and compounds, smaller than the pore size of the membrane (such as sodium and chloride ions, dissolved organics, viruses), cannot be separated or removed by using MF. The feed is generally passed through the membrane at around 1–3 m/s velocity at low to moderate pressure of 10–300 kPa (Purkait and Singh 2018). Due to its pore size, it is generally used in combination with other membrane processes like UF or RO and plays a crucial role in the pharmaceutical, dairy, and beverage industries (Sadr Ghayeni et al. 1998; Rodriguez-Mozaz et al. 2015; Hayes 2018).

Ultrafiltration (UF): It is the pressure-driven membrane process having a pore size from 1 to 100 nm. The separation mechanism is the sieving type where feed components are separated from the water based on their size. The larger and wider particles are retained on the UF membrane, whereas smaller and narrower particles

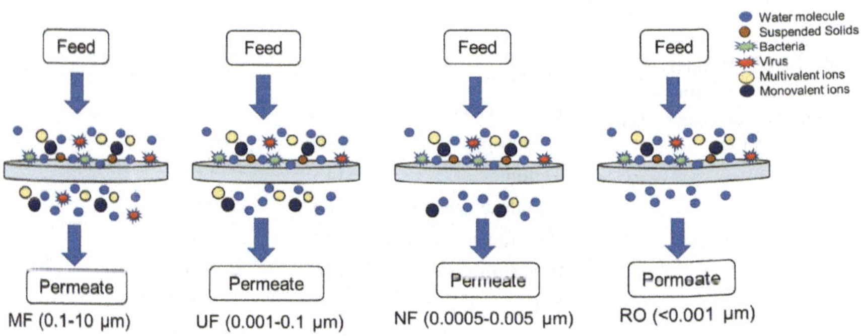

Fig. 7.1 Common pressure-driven membranes (modified from Purkait and Singh 2018)

can pass through it. The operating pressure required in this process for UF is quite lower than other pressure-driven processes like NF and RO. This is due to the porous membrane used in UF with pore sizes larger than the NF and RO. They can be operated with a pressure of 200–1000 kPa and can provide high flux. The conditions for both MF and UF uses are very much similar, with pore size being the only dissimilarity. It can be used to remove micron to submicron-sized components like organic molecules having higher molecular weight, colloids, and suspended solids from the feed (Aoustin et al. 2001; Tomaszewska and Mozia 2002; Falsanisi et al. 2010).

Nanofiltration (NF): NF is among the newest pressure-driven processes. It is considered better than RO in many fields and applications due to its lower energy requirement and higher flux. The properties of NF lie in between UF and RO processes. The size of the pore for NF membrane ranges from 0.5 to 5 nm. The NF membranes are capable of rejecting smaller and uncharged particles and multivalent ions, but they fail to remove monovalent ions. NF is considered as a charged UF process in terms of UF and as a low-pressure RO process in terms of RO. Therefore, NF has higher organic matter rejection than UF and lower pressure requirements than RO. Thus, colloids and large particles will be removed by the sieving mechanism, while ions will be removed by charge effect separation mechanism—Donnan effect (Roy et al. 2017).

Reverse Osmosis (RO): As the name suggests, it is the reverse or opposite of the osmosis process. In osmosis, water (solvent) flows from lower to higher solute concentrations through a semipermeable membrane when placed between the two solutions. But if the water is forced to flow from a higher solute concentration to a lower solute concentration, then it is called RO. In this process, the driving force is the externally applied pressure, which is greater than the osmotic pressure of the system. RO separation process is similar to other pressure-driven membrane processes, but instead of straining, diffusion is the mode of separation. RO membranes can remove even the smallest entity (like monovalent ions, dissolved organic content, etc.) from the water, which other membranes are not capable of removing. Among the various applications of RO, like in wastewater treatment, food, and dairy industry, desalination is the most important aspect of the RO membrane for which it is widely used (Shenvi et al. 2015; Caldera et al. 2016).

7.3.3 Fabrication of Membranes

The membrane is the most important part of every membrane process. Some of the various methods by which polymeric membranes can be fabricated are phase inversion method, stretching, deep coating, track-etching, interfacial polymerization method, etc. (Purkait and Singh 2018). The phase inversion method is the easiest and the most widely used method for preparing various types of polymeric membranes. The commonly used driving factors for this method are temperature, non-solvent, evaporation, and non-solvent vapor (Ismail et al. 2020). In the thermally induced phase method, the prominent role is played by the liquid–liquid demixing. There is a decrease in temperature, and the quality of the solvent also decreases. It remains homogenous at high temperature and separates into polymer-rich and polymer lean phases at low temperatures. Thus, the phase separation induced at low temperature will form a porous structure. In non-solvent-induced phase separation type of membrane preparation, two-phase separation (liquid–liquid demixing and solid–liquid demixing) takes place. The major steps involved in this process are shown in Fig. 7.2a. A thin film of the polymeric solution is casted over a glass plate using a casting knife (Fig. 7.2b) which is then immersed in a non-solvent bath to get the porous membrane (Fig. 7.2c). Liquid–liquid demixing forms the porous structure of the membrane, whereas solid–liquid demixing happens in crystalline membrane parts. With the immersion of the casted thin film in a non-solvent bath, precipitation of the membrane solution takes place. The liquid–liquid demixing results in a dense skin layer and a porous sublayer, whereas instantaneous demixing results in the porous skin layer (Purkait and Singh 2018).

The drying-induced phase separation method uses a volatile solvent for the dissolution of the polymer and a less or nonvolatile and non-solvent. With the evaporation of the solvent, solidification of the polymer-rich phase with pores filled with non-solvent is formed. Thin casted films along with the fast rate of evaporation of solvents result in a dense layer with macrovoids filled with a non-solvent. The vapor-induced phase separation method is the combination of the wet and dry casting process. Here, the casted film is exposed to the vapor of the non-solvent for a certain period, after which it is immersed in a coagulation bath to complete the process. The morphology of the membranes can be controlled by controlling the non-solvent vapor penetration into casting films (Venault et al. 2013; Ismail et al. 2020).

In the stretching method of fabrication, the membrane is prepared by stretching the extruded polymeric film perpendicular to the extrusion direction. The stretching is the cause of the formation of pores (in the range of 0.1–3.0 μm). Generally, this results in highly porous membranes. Crystalline or semicrystalline polymeric

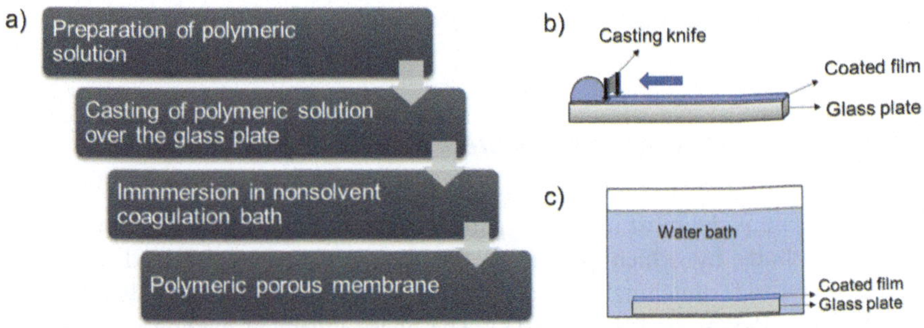

Fig. 7.2 Phase Inversion Process: **a** general scheme of non-solvent-induced phase inversion method of membrane fabrication, **b** casting of polymer solution over a glass plate, and **c** immersion of film in the non-solvent bath to get a porous membrane (modified from Purkait and Singh 2018)

materials are the only suitable materials for this method (Lalia et al. 2013; Purkait and Singh 2018).

The deep coating is a composite membrane preparation method, where a thin selective layer is coated over porous membrane support. Microporous membrane support is coated with a polymer solution prepared in a volatile water-insoluble solvent. The thickness of the coated layer is around 50–100 μm, which reduces to 0.5–2 μm upon evaporation of the solvent. The successful preparation of the desired membrane depends on the preparation method used and the polymer solution (Purkait and Singh 2018).

The track-etching method uses the application of particle radiation (perpendicular to polymeric film) to form tracks by damaging the film at the points of impact. This method is suitable for forming cylindrical pores with uniformity in a membrane. For uniform cylindrical pores, the radiated film is immersed in a bath (acidic or alkaline), which usually results in membranes with pore sizes of 0.02–10 μm. Higher the irradiation, the more the membrane becomes porous (although the thicker film is required); similarly, with an increase in etching time, pore diameter increases (Lalia et al. 2013; Purkait and Singh 2018).

The interfacial polymerization method is used for preparing anisotropic membrane. In this method, microporous support is used over which a thin layer of a polymer is deposited (Cadotte 1975). Porous support material is treated with a solution of amine, and the amine film formed is then dissolved in organic solvents (such as hexane). This results in the formation of a polymer of very thin size over the support surface. This method is typically used for making membranes for RO and NF. This method is considered a very important method of RO and NF membrane fabrication. Cadotte et al. (1980) developed the first thin-film composite membranes using interfacial polymerization method and showed breakthrough in membrane performance for RO applications (Cadotte et al. 1988). The initial protocol used for membrane fabrication using interfacial polymerization is described below:

(a) At first, soaking of the microporous polysulfone support is done in an aqueous polymeric amine solution.
(b) The amine impregnated membrane is then immersed in a solution of diisocyanate in hexane.
(c) Finally, the cross-linking of the membrane is done by performing heat treatment (110 °C).

A wide variety of membranes have been developed by optimizing the properties of the substrate and skin layer (Petersen 1993; Lau et al. 2012). A thin-film composite polyurea membrane formed by interfacial polymerization was found to have better salt rejection and water flux property than an integrally skinned asymmetric cellulose acetate membrane, confirming the authenticity of the method (Purkait et al. 2018).

7.3.4 Problems of Membrane Technology

Fouling: As feedwater gets filtered, some of the particles and molecules get accumulated on the surface or inside pores of the membrane. This accumulation of particles and molecules results in a gelatinous layer over the membrane surface, reducing the efficiency of membrane filtration. This phenomenon is called fouling. It is the Achilles heel for the membrane technology (Flemming et al. 1997). The feed composition is the main factor that determines the extent and rate of fouling. Fouling can be of the following four types (Zularisam et al. 2010; Gao et al. 2011; Guo et al. 2012; Jiang et al. 2017):

(a) *Inorganic fouling*: When the concentration of certain inorganic species exceeds the solubility limit, they start to precipitate and thus cause scaling problems, resulting in the fouling of the membrane.
(b) *Colloidal fouling*: If the fouling is caused by the deposition of organic or inorganic colloidal particles (polysaccharide, proteins, clay, etc.) in the membrane pores, then the fouling is termed as colloidal fouling.
(c) *Organic fouling*: If the feedwater contains a high amount of organic compounds (like extracellular organic debris, dissolved organic carbon, etc.), then they may get accumulated or deposited on the surface of the membrane. Such accumulation results in organic fouling.
(d) *Biofouling*: Microbial species in the membrane system have tendencies to adhere to the surfaces and form a gelatinous layer called biofilm, which drastically reduces the efficiency of the membrane, called biofouling. Biofouling is the worst fouling as microorganisms can attach to the surface and grow with time, and after full growth, they can contaminate other places in the membrane system.

Due to membrane fouling alone, the life span of membranes is low. Generally, it is for a few years. The low life span of the widely used polymeric membrane is due to its low resistance toward harsh chemical, pressure, and temperature factors. The recent development in material science and membrane technology has provided some new-generation membranes which can withstand those adverse condition factors along with a better life span.

Selectivity–permeability trade-off: The flux through the membrane is given by Eq. (1) (Baker 2004):

$$J_w = A(\Delta P - \Delta \pi_m) \tag{1}$$

where
J_w = volumetric water flux
A = water permeability coefficient
ΔP = applied hydraulic transmembrane pressure
$\Delta \pi_m$ = osmotic pressure difference (across the active layer).

In most of the water treatment applications where MF and UF are used, $\Delta\pi_m$ is almost negligible, whereas, for NF and RO, it is a very crucial parameter. For membranes (porous only), flow passing through the active layer is considered and modeled as a laminar flow through cylindrical pores. Water permeability (A) through the membrane is given by the Hagen–Poiseuille equation (Baker 2004), as shown below:

$$A = \left(\varepsilon r_p^2\right)/(8\,\mu\,\delta_m) \tag{2}$$

where
ε = surface porosity
r_p = pore radius
μ = viscosity of the solution
δ_m = thickness of the active layer.

This simplified relationship best describes an idealized membrane with uniformly sized cylindrical pores (i.e., an isoporous membrane). In reality, membranes that are formed by phase inversion have non-cylindrical and tortuous pores that vary substantially in size. The water permeability coefficient (*A*) is related to layer thickness (δ_m), the molar volume of water (V_w), the diffusive water permeability (P_w), gas constant (R_g), and absolute temperature (*T*) by the following equation (Geise et al. 2014):

$$A = P_w V_w/(\delta_m R_g T) \tag{3}$$

Similarly, Fickian diffusion is used for modeling solute movement:

$$J_s = P_s \Delta c_m/\delta_m = B\Delta c_m \tag{4}$$

where
P_s = diffusive solute permeability
B = solute permeability coefficient
Δc_m = solute concentration difference across the active layer.

Coefficients A and B together define the selective layer performance of non-porous membranes. The ideal membranes have both high water permeability and selectivity, a combination which is very difficult to achieve. From the recent studies, it has been suggested that water separation membranes exhibit a trade-off generally termed as permeability–selectivity trade-off, which is similar to the Robeson plot. The Robeson plot is extensively used in the field of gas separation using polymeric membranes for visualizing selectivity and permeability of the membranes (Mehta and Zydney 2005; Robeson 2008; Geise et al. 2011). This trade-off implies that with a small increase in permeability of highly permeating species, a huge increase in permeability of highly retained species will be seen. Thus, instead of working for attaining higher water permeability, the more effective way for membrane materials research would be to work on improving the selectivity of water over other dissolved solutes (Werber et al. 2016b).

7.4 New-Generation Membranes

From the literature, it is evident that the few limitations of membrane technology need to be resolved for better quality water and the long life of the membranes (Werber et al. 2016a, b). At present, there is very little control over solute and water permeabilities, which is one of the key limitations that is hampering the performance of membranes. Development in the membrane fabrication processes and the advancement in the field of material science in terms of new materials can help in crossing the present limitations (Werber et al. 2016b). Among the various materials developed so far, the most anticipated new-generation membrane materials are aquaporins and graphene, while the promising membrane systems being the hybrid membrane and catalytic membrane systems.

7.4.1 Aquaporin

Aquaporins are a class of membrane proteins that help in the transportation of water in and out of the cell membrane. Some categories of the aquaporins are extremely selective, rejecting even ions that easily permeate through current high-quality membranes (Murata et al. 2000; Sui et al. 2001; Hub and De Groot 2008). Intensive research is going on in mimicking these natural biopolymers to obtain a synthetic membrane having high flux and selectivity. Synthetic aquaporin membranes are biomimetic membranes having high water flux, a wide spectrum of solute rejection, stability to external conditions, and easy reconstitution property(Shen et al. 2014; Song et al. 2019). The most extraordinary property of the orthodox membrane (the membrane that allows only water to pass) is rejecting almost all solutes. Although the proper mechanism of water transportation in aquaporins is not fully understood, certain studies have suggested the dipole interaction, electrostatic force, and hydration forces are the main reasons for the selectivity of the aquaporin membrane (Tang et al. 2015).

7.4.2 Graphene

Graphene is considered one of the wonder materials which has its use in different fields, from electronics to water treatment. It has high strength and electron mobility, which makes it a noble material for preparing membranes having antibacterial and antifouling surfaces. Thus, they can be used on the surface of conventional membrane materials for enhancing their efficiency. Graphene is an allotrope of carbon, which is a single layer of carbon atoms arranged in a hexagonal array (Geim and Novoselov 2007; Tadyszak et al. 2018). Its unique electrical, thermal, and mechanical properties make it a suitable and robust material for water treatment. Different forms of graphene that are commonly used are graphene oxides and reduced graphene oxides. Nair et al. showed that laminates formed from graphene oxide provide unhindered permeation of water while being impermeable to other liquids, vapors, and gases (Nair et al. 2012). After that, many applications of graphene oxide membranes have been reported for water purification (Huang et al. 2013; Han et al. 2013; Hu and Mi 2014). The enhanced water transportation was also reported for graphene-based NF (Han et al. 2013) and graphene oxide UF membranes (Huang et al. 2013) where higher water flux of about 4–6 orders of magnitude compared to theoretically predicted flux was measured in the filtration experiments (Huang et al. 2013).

7.4.3 Hybrid Membranes

Integrated or hybrid membrane processes are the integration of membrane processes with conventional unit operations (like coagulation, flocculation, sedimentation) or integration of different membrane processes to increase the performance of the treatment (Van der Bruggen et al. 2003). Many hybrid systems have been established for water treatment due to the major obstacle of membrane filtration, i.e., fouling (Li et al. 2008). With hybrid systems, a pre-treatment before filtration provides a barrier to the contaminants in the feed, thus reducing the possibility of membrane fouling. The main goal of the systems is to increase the purity, productivity, and yield of the product. The hybrid membrane systems also reduce operating costs and environmental pollution (Ang et al. 2015). Due to the advancements in membrane technology, many researchers have suggested the use of membranes to replace conventional treatment plants as they may be inefficient in removing trace organic contaminants, emerging contaminants, etc., and may not be able to meet the standards set by World Health Organization (WHO) guidelines for drinking water quality (WHO 2008). Currently, membrane technology is a good alternative for conventional water treatment technologies as membrane filtration provides several advantages like very little or no chemical requirements, high-quality water, less sludge production, and smaller footprint (Xia et al. 2004). In addition to these, membrane filtration can reject different types of pollutants simultaneously, which cannot be achieved by the conventional treatment processes. For example, the NF membrane can be used to remove hardness, color, and pesticides in the feedwater (Berg et al. 1997; Yeh et al. 2000). Various established hybrid systems are available based on the feedwater quality and the problems of conventional plants that need to be countered. The major benefits of hybrid membrane systems are (Ang et al. 2015):

(a) Production of improved water quality that cannot be produced by the stand-alone processes
(b) Energy savings
(c) Environmentally friendly because of reduced waste disposal
(d) Reduced capital and operating costs as higher efficiency and productivity achieved.

7.4.4 Catalytic Membranes

When separation and chemical reaction are combined into a single membrane unit, it is called a catalytic membrane. A properly designed catalytic membrane reactor improves the yield, reactant conversion, and separation costs. By eliminating the thermodynamic equilibrium, the catalytic membranes enhance the reaction conversion, which results in the improvement of the overall effects like conversion, yield, residence time, net reaction rate, and selectivity (Westermann and Melin 2009; Wang et al. 2015). Several parameters like separation selectivity, productivity, mechanical and chemical perfection needed in operational conditions, membrane lifetime, and cost govern the type of catalytic membrane to be used (Gu et al. 2015). Catalytic membranes can be classified into organic (or polymeric) and inorganic depending on the type of membrane and chemical process requirements. The polymeric membranes can be catalyzed by making the catalyst as a part of the membrane backbone (membrane-embedded catalysts) or by applying catalysts on the surface of the membrane by grafting of deposition processes, where the membrane acts as a catalyst carrier (membrane-assisted catalysts) (Bhattacharya and Misra 2004; MacAnás et al. 2010; Smuleac et al. 2011; Vanherck et al. 2012).

Membrane-embedded catalysts: The polymeric membranes can be catalyzed by the blending processes, which is the physical mixing of two or more polymers to get the membrane of requisite properties. It can be done by blending polymers with liquid catalysts (sulfuric acid) or with solid catalysts (TiO2, Al2O3, resins, etc.) by using various blending techniques (MacAnás et al. 2010). Membranes blended with nanomaterials have shown to provide various benefits like (Smuleac et al. 2011; Vanherck et al. 2012):

(a) Lower corrosion and agglomeration of the nanomaterials
(b) Control over the size of nanopores along with different interactions between nanomaterials and polymers
(c) Catalyzation of nanomaterials by embedding them in the membrane can enhance the permeability of molecules through the membrane and can also provide mechanical stability
(d) The polymeric membranes with nanomaterials in or over them enhance the electrical and optical responses of the membrane.

Embedding the membrane matrix with metal oxide has shown a positive impact on membrane characteristics by enhancing the fouling resistance capability of the membrane. The addition of Al_2O_3 to PVDF membranes has shown improvement in the antifouling property of the membrane (Liu et al. 2013). Also, during advanced oxidation processes, the metal oxides produce OH radicals, which help in catalyzing the degradation of organic pollutants (Xu et al. 2010; Zhao et al. 2010).

Membrane-assisted catalysts: The catalyzation is done by either grafting or precipitation process (Bhattacharya and Misra 2004) for the preparation of membrane-assisted catalysts. Grafting is the process used to modify the surface of the polymers by forming covalent bonds of monomers with or onto the polymeric chain. The factors that affect or control the grafting process are the nature of monomer, additives, solvent, temperature, etc. The nature of the backbone of polymers governs the acceptance or rejection of grafting reaction. The temperature plays a very vital role during grafting step; it controls the graft copolymerization kinetics as well as yield (which increases with the increase in temperature) (Bhattacharya and Misra 2004; Estrada-Villegas and Bucio 2013). The precipitation (deposition) process depends on the polymerization of a mixture of the oligomers, which forms a coating that adheres to the substrates by physical forces only. A novel method for depositing the nanomaterials on the membrane surface by using roll-to-roll fabrication process was studied by Wang et al. (2016) which showed a reduction in fabrication costs due to the use of roll-to-roll process (Merkel et al. 2010; Sawamura et al. 2015). This process can also be applied for the alignment of nanoparticle on the membrane for the support and pattern printing (Cakmak et al. 2015).

Nowadays, photocatalytic degradation of contaminants has also garnered a lot of attention as an alternative method for the removal of pollutants in aqueous or gaseous media. Generally, advanced oxidation processes are used for degrading contaminants, which depend on the oxidizing agents added (such as hydrogen peroxide) in the presence of ultraviolet radiation (UV). This method of advanced oxidation can be used for degrading different toxic compounds, particularly chlorinated compounds (Jia et al. 2015), and thus has been suggested for the decontamination of wastewater as well as drinking water for human uses. Photocatalytic polymeric composite membranes can be prepared by photopolymerization and photo-cross-linking processes. The porosity of the photocatalytic membrane has effects on its performance. Some studies have found improved membrane porosity by incorporating carbon nanoparticles in the membrane and also by using metal–organic frameworks (Dumee et al. 2010; Lee et al. 2014). Thus, the use of new-generation materials and advanced membrane fabrication processes can be explored for enhancing and controlling the membrane porosity.

7.5 Conclusion

The effectiveness of conventional water treatment processes has become limited due to stringent water quality rules and regulations, emerging non-conventional contaminants, space limitations, need for better removal of pathogens and contaminants. Thus, advanced technology, like membrane treatment technology with other hybrid processes, will be required. Membrane technology started to develop in the 1970s and showed rapid development since the 1990s due to its operational ease, low chemical input, high efficiency, energy, and space-saving. Its use in water treatment is rapidly increasing due to the development of high-quality membrane as well as decreasing cost. However, problems like fouling, life span, and selectivity–permeability trade-off are limiting their efficient worldwide use. Thus, there is a need for a new-generation membrane that can surpass the limitations of the conventional membrane technologies, with less or no decrement in the efficiency of the treatment. Development in the field of material sciences and membrane fabrication processes has allowed us to overcome the limitation of membrane technology by the use of new-generation membrane materials like aquaporin, graphene, carbon nanotubes, metal–organic frameworks, etc. New-generation hybrid and catalytic membranes with catalytic degradation of contaminants along with the antimicrobial activity can also enhance the efficiency of membrane treatment. Preparation of electroconductive membrane for better efflux can also be attempted with enhanced conductivity, antifouling, and antimicrobial property. Incorporation of the metallic and bimetallic catalysts on the membrane surface can also be used to accelerate the contaminant degradation processes. Thus, further enhancement of the performance of membrane technology can help the world with the current problems of potable water scarcity, water pollution, and water demand.

References

Ang WL, Mohammad AW, Hilal N, Leo CP (2015) A review on the applicability of integrated/hybrid membrane processes in water treatment and desalination plants. Desalination 363:2–18. https://doi.org/10.1016/j.desal.2014.03.008

Aoustin E, Schäfer AI, Fane AG, Waite TD (2001) Ultrafiltration of natural organic matter. Sep Purif Technol 22–23:63–78. https://doi.org/10.1016/S1383-5866(00)00143-X

Baker RW (2004) Overview of membrane science and technology. Membr Technol Appl 3:1–14

Berg P, Hagmeyer G, Gimbel R (1997) Removal of pesticides and other micropollutants by nanofiltration. Desalination 113:205–208. https://doi.org/10.1016/S0011-9164(97)00130-6

Bhattacharya A, Misra BN (2004) Grafting: A versatile means to modify polymers: Techniques, factors and applications. Prog Polym Sci 29:767–814

Shannon MA, Bohn PW, Elimelech M et al (2009) Science and technology for water purification in the coming decades. In: Nanoscience andTechnology: A Collection of Reviews from Nature Journals. World Scientific Publishing Co., 337–346. https://doi.org/10.1142/9789814287005_0035

Bolisetty S, Peydayesh M, Mezzenga R (2019) Sustainable technologies for water purification from heavy metals: review and analysis. Chem Soc Rev 48:463–487. https://doi.org/10.1039/c8cs00493e

Cadotte J, Forester R, Kim M et al (1988) Nanofiltration membranes broaden the use of membrane separation technology. Desalination 70:77–88. https://doi.org/10.1016/0011-9164(88)85045-8

Cadotte JE, Petersen RJ, Larson RE, Erickson EE (1980) A new thin-film composite seawater reverse osmosis membrane. Desalination 32:25–31. https://doi.org/10.1016/S0011-9164(00)86003-8

Cakmak M, Batra S, Yalcin B (2015) Field assisted self-assembly for preferential through thickness ("z-Direction") alignment of particles and phases by electric, magnetic, and thermal fields using a novel roll-to-roll processing line. Polym Eng Sci 55:34–46. https://doi.org/10.1002/pen.23861

Caldera U, Bogdanov D, Breyer C (2016) Local cost of seawater RO desalination based on solar PV and wind energy: A global estimate. Desalination 385:207–216. https://doi.org/10.1016/j.desal.2016.02.004

Cui ZF, Muralidhara HS (2010) Membrane technology: a practical guide to membrane technology and applications in food and bioprocessing. Elsevier

Dumee L, Velleman L, Sears K et al (2010) Control of Porosity and Pore Size of Metal Reinforced Carbon Nanotube Membranes. Membranes (Basel) 1:25–36. https://doi.org/10.3390/membranes1010025

Eliasson J (2015) The rising pressure of global water shortages. Nature 517:6

Estrada-Villegas GM, Bucio E (2013) Comparative study of grafting a polyampholyte in a fluoropolymer membrane by gamma radiation in one or two-steps. Radiat Phys Chem 92:61–65. https://doi.org/10.1016/j.radphyschem.2013.07.015

Falsanisi D, Liberti L, Notarnicola M (2010) Ultrafiltration (UF) Pilot Plant for Municipal Wastewater Reuse in Agriculture: Impact of the Operation Mode on Process Performance. Water 2:872–885. https://doi.org/10.3390/w2040872

Flemming H, Schaule G, Griebe T et al (1997) Biofouling—The Achilles heel of membrane processes. Desalination 113(2–3), 215–225. https://doi.org/10.1016/S0011-9164(97)00132-X

Gao W, Liang H, Ma J et al (2011) Membrane fouling control in ultrafiltration technology for drinking water production: A review. Desalination 272:1–8

Geim AK, Novoselov KS (2007) The rise of graphene. Nat Mater 6:183–191. https://doi.org/10.1038/nmat1849

Geise GM, Lee HS, Miller DJ et al (2010) Water purification by membranes: The role of polymer science. J Polym Sci, Part B: Polym Phys 48:1685–1718. https://doi.org/10.1002/polb.22037

Geise GM, Park HB, Sagle AC et al (2011) Water permeability and water/salt selectivity trade-off in polymers for desalination. J Memb Sci 369:130–138. https://doi.org/10.1016/j.memsci.2010.11.054

Geise GM, Paul DR, Freeman BD (2014) Fundamental water and salt transport properties of polymeric materials. Prog Polym Sci 39:1–42

Gu Y, Favier I, Pradel C et al (2015) High catalytic efficiency of palladium nanoparticles immobilized in a polymer membrane containing poly(ionic liquid) in Suzuki-Miyaura cross-coupling reaction. J Memb Sci 492:331–339. https://doi.org/10.1016/j.memsci.2015.05.051

Guo W, Ngo HH, Li J (2012) A mini-review on membrane fouling. Bioresour Technol 122:27–34. https://doi.org/10.1016/j.biortech.2012.04.089

Han Y, Xu Z, Gao C (2013) Ultrathin graphene nanofiltration membrane for water purification. Adv Funct Mater 23:3693–3700. https://doi.org/10.1002/adfm.201202601

Hayes M (2018) Novel Proteins for Food, Pharmaceuticals and Agriculture. Wiley, Chichester, UK

Heberer T (2002) Occurrence, fate, and removal of pharmaceutical residues in the aquatic environment: a review of recent research data. Toxicol Lett 131:5–17. https://doi.org/10.1016/S0378-4274(02)00041-3

Hu M, Mi B (2014) Layer-by-layer assembly of graphene oxide membranes via electrostatic interaction. J Memb Sci 469:80–87. https://doi.org/10.1016/j.memsci.2014.06.036

Huang H, Song Z, Wei N et al (2013) Ultrafast viscous water flow through nanostrand-channelled graphene oxide membranes. Nat Commun 4:1–9. https://doi.org/10.1038/ncomms3979

Hub JS, De Groot BL (2008) Mechanism of selectivity in aquaporins and aquaglyceroporins. Proc Natl Acad Sci U S A 105:1198–1203. https://doi.org/10.1073/pnas.0707662104

Ida S (2014) PES (Poly(ether sulfone)), Polysulfone. Encyclopedia of Polymeric Nanomaterials. Springer, Berlin Heidelberg, pp 1–8

Inyinbor AA, Bello OS, Fadiji AE, Inyinbor HE (2018) Threats from antibiotics: A serious environmental concern. J Environ Chem Eng 6:784–793. https://doi.org/10.1016/j.jece.2017.12.056

Ismail N, Venault A, Mikkola JP et al (2020) Investigating the potential of membranes formed by the vapor induced phase separation process. J. Memb. Sci. 597:

Cadotte JE (1975) Reverse osmosis membrane. U. S. PATENT NO. 3, 926, 798, 4 P, 4 TAB, 2 REF

Jia S, Han H, Zhuang H et al (2015) Advanced treatment of biologically pretreated coal gasification wastewater by a novel integration of catalytic ultrasound oxidation and membrane bioreactor. Bioresour Technol 189:426–429. https://doi.org/10.1016/j.biortech.2015.04.062

Jiang S, Li Y, Ladewig BP (2017) A review of reverse osmosis membrane fouling and control strategies. Sci Total Environ 595:567–583

Lalia BS, Kochkodan V, Hashaikeh R, Hilal N (2013) A review on membrane fabrication: Structure, properties and performance relationship. Desalination 326:77–95

Lau WJ, Ismail AF, Misdan N, Kassim MA (2012) A recent progress in thin film composite membrane: A review. Desalination 287:190–199. https://doi.org/10.1016/j.desal.2011.04.004

Lee JY, Tang CY, Huo F (2014) Fabrication of porous matrix membrane (PMM) using metal-organic framework as green template for water treatment. Sci Rep 4:1–5. https://doi.org/10.1038/srep03740

Li NN, Fane AG, Ho WSW, Matsuura T (2008) Advanced membrane technology and applications. In: Li NN, Fane AG, Ho WW, Matsuura T (eds) Advanced membrane technology and applications. Wiley

Liu B, Chen C, Li T et al (2013) High performance ultrafiltration membrane composed of PVDF blended with its derivative copolymer PVDF-g-PEGMA. J Memb Sci 445:66–75. https://doi.org/10.1016/j.memsci.2013.05.043

MacAnás J, Ouyang L, Bruening ML et al (2010) Development of polymeric hollow fiber membranes containing catalytic metal nanoparticles. Catal Today 156:181–186. https://doi.org/10.1016/j.cattod.2010.02.036

Mehta A, Zydney AL (2005) Permeability and selectivity analysis for ultrafiltration membranes. J Memb Sci 249:245–249. https://doi.org/10.1016/j.memsci.2004.09.040

Merkel TC, Lin H, Wei X, Baker R (2010) Power plant post-combustion carbon dioxide capture: An opportunity for membranes. J Memb Sci 359:126–139. https://doi.org/10.1016/j.memsci.2009.10.041

Mohanty K, Purkait MK (2011) Membrane technologies and applications. CRC press

Mulder M, Mulder J (1991) Basic principles of membrane technology, Springer Netherlands

Murata K, Mitsuoka K, Hiral T et al (2000) Structural determinants of water permeation through aquaporin-1. Nature 407:599–605. https://doi.org/10.1038/35036519

Nair RR, Wu HA, Jayaram PN et al (2012) Unimpeded permeation of water through helium-leak-tight graphene-based membranes. Science (335):442–444. https://doi.org/10.1126/science.1211694

Nawab J, Khan S, Ali S et al (2016) Health risk assessment of heavy metals and bacterial contamination in drinking water sources: a case study of Malakand Agency. Pakistan. Environ Monit Assess 188:286. https://doi.org/10.1007/s10661-016-5296-1

Nawab J, Khan S, Khan MA et al (2017) Potentially Toxic Metals and Biological Contamination in Drinking Water Sources in Chromite Mining-Impacted Areas of Pakistan: A Comparative Study. Expo Heal 9:275–287. https://doi.org/10.1007/s12403-016-0240-8

Pendergast MM, Hoek EMV (2011) A review of water treatment membrane nanotechnologies. Energy Environ Sci 4:1946–1971. https://doi.org/10.1039/c0ee00541j

Petersen RJ (1993) Composite reverse osmosis and nanofiltration membranes. J. Memb. Sci. 83:81–150

Peydayesh M, Esfandyari GR, Mohammadi T, Alamdari EK (2013) Pertraction of cadmium and zinc ions using a supported liquid membrane impregnated with different carriers. Chem Pap 67:389–397. https://doi.org/10.2478/s11696-013-0310-3

Purkait MK, Singh R (2018) Membrane Technology in Separation Science. CRC Press

Purkait MK, Sinha MK, Mondal P, Singh R (2018) Introduction to Membranes. In: Interface Science and Technology. Elsevier 25:1–37. https://doi.org/10.1016/B978-0-12-813961-5.00001-2

Robeson LM (2008) The upper bound revisited. J Memb Sci 320:390–400. https://doi.org/10.1016/j.memsci.2008.04.030

Rodriguez-Mozaz S, Ricart M, Köck-Schulmeyer M et al (2015) Pharmaceuticals and pesticides in reclaimed water: Efficiency assessment of a microfiltration-reverse osmosis (MF-RO) pilot plant. J Hazard Mater 282:165–173. https://doi.org/10.1016/j.jhazmat.2014.09.015

Roy Y, Warsinger DM, Lienhard JH (2017) Effect of temperature on ion transport in nanofiltration membranes: Diffusion, convection and electromigration. Desalination 420:241–257. https://doi.org/10.1016/j.desal.2017.07.020

Sadr Ghayeni SB, Beatson PJ, Schneider RP, Fane AG (1998) Water reclamation from municipal wastewater using combined microfiltration-reverse osmosis (ME-RO): preliminary performance data and microbiological aspects of system operation. Desalination 116:65–80. https://doi.org/10.1016/S0011-9164(98)00058-7

Sagle A, Freeman B (2004) Fundamentals of membranes for water treatment. The future of desalination in Texas, 2(363), p. 137

Santhosh C, Velmurugan V, Jacob G et al (2016) Role of nanomaterials in water treatment applications: A review. Chem Eng J 306:1116–1137

Sawamura KI, Furuhata T, Sekine Y et al (2015) Zeolite Membrane for Dehydration of Isopropylalcohol-Water Mixture by Vapor Permeation. ACS Appl Mater Interfaces 7:13728–13730. https://doi.org/10.1021/acsami.5b04085

Scott K (1995) Handbook of industrial membranes. Elsevier

Shen Y, Saboe PO, Sines IT et al (2014) Biomimetic membranes: A review. J Memb Sci 454:359–381. https://doi.org/10.1016/j.memsci.2013.12.019

Shenvi SS, Isloor AM, Ismail AF (2015) A review on RO membrane technology: Developments and challenges. Desalination 368:10–26. https://doi.org/10.1016/j.desal.2014.12.042

Smuleac V, Varma R, Sikdar S, Bhattacharyya D (2011) Green synthesis of Fe and Fe/Pd bimetallic nanoparticles in membranes for reductive degradation of chlorinated organics. J Memb Sci 379:131–137. https://doi.org/10.1016/j.memsci.2011.05.054

Song W, Tu YM, Oh H et al (2019) Hierarchical Optimization of High-Performance Biomimetic and Bioinspired Membranes. Langmuir 35:589–607. https://doi.org/10.1021/acs.langmuir.8b03655

Sui H, Han BG, Lee JK et al (2001) Structural basis of water-specific transport through the AQP1 water channel. Nature 414:872–878. https://doi.org/10.1038/414872a

Tadyszak K, Wychowaniec JK, Litowczenko J (2018) Biomedical applications of graphene-based structures. Nanomaterials 8:944

Tang C, Wang Z, Petrinić I et al (2015) Biomimetic aquaporin membranes coming of age. Desalination 368:89–105. https://doi.org/10.1016/j.desal.2015.04.026

Tomaszewska M, Mozia S (2002) Removal of organic matter from water by PAC/UF system. Water Res 36:4137–4143. https://doi.org/10.1016/S0043-1354(02)00122-7

UN Water (2012) Managing Water Under Uncertainty and Risk, The United Nations World Water Development Report 4. UN Water Reports, World WaterAssessment Programme. UNESCO, Paris, France

UNESCO-WWAP (2003) Water for People, Water for Life: The United Nations World Water Development Report: Executive Summary.UNESCO

Van der Bruggen B, Hawrijk I, Cornelissen E, Vandecasteele C (2003) Direct nanofiltration of surface water using capillary membranes: Comparison with flat sheet membranes. Sep Purif Technol 31:193–201. https://doi.org/10.1016/S1383-5866(02)00184-3

Vanherck K, Verbiest T, Vankelecom I (2012) Comparison of two synthesis routes to obtain gold nanoparticles in polyimide. J Phys Chem C 116:115–125. https://doi.org/10.1021/jp207244y

Venault A, Chang Y, Wang D-M, Bouyer D (2013) A Review on Polymeric Membranes and Hydrogels Prepared by Vapor-Induced Phase Separation Process. Polym Rev 53:568–626. https://doi.org/10.1080/15583724.2013.828750

Wang B, Jackson EA, Hoff JW, Dutta PK (2016) Fabrication of zeolite/polymer composite membranes in a roller assembly. Microporous Mesoporous Mater 223:247–253. https://doi.org/10.1016/j.micromeso.2015.11.003

Wang B, Sun C, Li Y et al (2015) Rapid synthesis of faujasite/polyethersulfone composite membrane and application for CO_2/N_2 separation. Microporous Mesoporous Mater 208:72–82. https://doi.org/10.1016/j.micromeso.2015.01.020

Watson H (2015) Biological membranes. Essays Biochem 59:43–70. https://doi.org/10.1042/BSE0590043

Werber JR, Deshmukh A, Elimelech M (2016a) The Critical Need for Increased Selectivity, Not Increased Water Permeability, for Desalination Membranes. Environ Sci Technol Lett 3:112–120

Werber JR, Osuji CO, Elimelech M (2016b) Materials for next-generation desalination and water purification membranes. Nat Rev Mater 1:1–15. https://doi.org/10.1038/natrevmats.2016.18

Westermann T, Melin T (2009) Flow-through catalytic membrane reactors-Principles and applications. Chem. Eng. Process. Process Intensif. 48:17–28

WHO (2008) World Health Organization. Guidelines for drinking-water quality: second addendum. Vol. 1, Recommendations

Wiesner MR, Hackney J, Sethi S et al (1994) Cost estimates for membrane filtration and conventional treatment. J/ Am Water Work Assoc 86:33–41. https://doi.org/10.1002/j.1551-8833.1994.tb06284.x

Xia S, Li X, Liu R, Li G (2004) Study of reservoir water treatment by ultrafiltration for drinking water production. Desalination 167:23–26. https://doi.org/10.1016/j.desal.2004.06.109

Xu YJ, Zhuang Y, Fu X (2010) New insight for enhanced photocatalytic activity of TiO2 by doping carbon nanotubes: A case study on degradation of benzene and methyl orange. J Phys Chem C 114:2669–2676. https://doi.org/10.1021/jp909855p

Yeh HH, Tseng IC, Kao SJ et al (2000) Comparison of the finished water quality among an integrated membrane process, conventional and other advanced treatment processes. Desalination 131:237–244. https://doi.org/10.1016/S0011-9164(00)90022-5

Zhao X, Cheng J, Chen S et al (2010) Hydrophilic modification of poly(vinylidene fluoride) (PVDF) by in situ polymerization of methyl methacrylate (MMA) monomer. Colloid Polym Sci 288:1327–1332. https://doi.org/10.1007/s00396-010-2261-3

Zhu WP, Sun SP, Gao J et al (2014) Dual-layer polybenzimidazole/polyethersulfone (PBI/PES) nanofiltration (NF) hollow fiber membranes for heavy metals removal from wastewater. J Memb Sci 456:117–127. https://doi.org/10.1016/j.memsci.2014.01.001

Zularisam AW, Ismail AF, Sakinah M (2010) Application and challenges of membrane in surface water treatment. J Appl Sci 10:380–390. https://doi.org/10.3923/jas.2010.380.390

Chapter 8
Forward Osmosis in Desalination and Wastewater Treatment

Kritika Jashrapuria and Swatantra P. Singh

Abstract Fresh, potable water is one of the key elements for the survival of living beings, and with an increasing population, there is water shortage all over the globe. Hence, this situation demands an effective water treatment technology, and one such water treatment method is membrane-based forward osmosis (FO) technology. Wastewater, seawater, and brackish water have great potential to fulfill the water demands, and the FO membrane technology holds the potential to sustainably treat these sources to produce water of good quality. The FO is based on the principle of osmotic pressure difference across the membrane, which provides the gradient for the movement of water molecules from feed solution to draw solution. The advantages offered by the FO-based treatment method are lower fouling propensity and lesser concentration polarization, and, most importantly, it is an energy-efficient process. However, for the enhancement of the treatment process, the FO is used in conjugation with other separation processes, referred to as the FO hybrid system. In this chapter, FO principle, factors affecting the process, membrane fouling, and hybrid FO treatment technology and its application have been discussed.

Keywords Forward osmosis · Draw solution · Membranes · Fouling · Desalination · Wastewater treatment · Concentration polarization

K. Jashrapuria · S. P. Singh
Environmental Science and Engineering Department (ESED), Indian Institute of Technology Bombay, Mumbai, India

S. P. Singh (✉)
Centre for Research in Nanotechnology and Science (CRNTS), Indian Institute of Technology Bombay, Mumbai, India
e-mail: swatantra@iitb.ac.in

S. P. Singh et al. (eds.), *Pollution Control Technologies*, Energy, Environment, and Sustainability, https://doi.org/10.1007/978-981-16-0858-2_8

8.1 Introduction

An expected 138.6 million km^3 of water is found on the earth, covering about 70% surface of our planet. In general, water exists in three significant forms on earth, which are saline water (97%), freshwater (<0.5%), and ice/snow (2%) (Khawaji et al. 2008). Water is the major reason behind the existence of different types of life on this planet. Unfortunately, a very less fraction of water is accessible as freshwater for direct use by living creatures. Moreover, the perpetual and exponential extension of the population has raised concerns about the sustainability of water resources (Lutchmiah et al. 2014). Therefore, one of the foremost challenges is in meeting the growing water demand at low-energy cost. Drinking water is produced mainly from safe water sources, but due to population increase and economic advancement, exploitation of aquifers and waning groundwater levels have diminished freshwater sources. Thus, a potential alternative source of water is wastewater. The evolution of low-energy separation technologies for clean water production is, therefore, vital and has earned a booming concern in the last few decades (Cath et al. 2006).

Nowadays, membrane technologies are one of the most widely used methods to produce clean water, and among all membrane technologies, FO impersonates its potentials in recent times (Korenak et al. 2017). The FO, an alternative membrane process, also has the potential to treat wastewater, producing high-quality water. The FO is a technical term defining the natural phenomenon of osmosis. The principle of the FO process is based on a natural osmotic pressure, which is used to draw the water molecules across a semipermeable membrane from a feed solution (FS) to the draw solution (DS). The driving force is therefore created naturally by the difference in osmotic potential between the DS and the FS. This offers several benefits over traditional pressure-driven membrane processes such as lower energy demands and diminished membrane fouling (Chekli et al. 2016). In fact, in the FO process, using natural osmotic potential gradient diminishes both capital and handling cost and also is profitable for fouling check compared to pressure-driven membrane processes (Haupt and Lerch 2018).

In spite of having numerous accounts, the FO technology still flags some major technological hurdles. Drawbacks include the lack of proper membranes designed for the FO process. The conventional membranes such as RO membranes are asymmetric and were proved not suitable for the FO process as it exacerbates concentration polarization effects, thus causing a notable drop in process competence (Zhao et al. 2011; McCutcheon and Elimelech 2008). The next barrier is the separation of the water from the DS and the DS recovery, peculiarly when high-quality water is required, such as in case of the drinking water. The separation and restoration of the DS require an additional processing unit, which can consume energy and hence remains a notable difficulty for drinking water applications (Luo et al. 2014). The success of the FO for the drinkable purposes will greatly depend on the easy and efficient separation of the DS from the water after the FO process, once the DS is fully diluted (Mehta et al. 2014). Therefore, the FO has been started using in conjugation with other membrane separation technology. Several hybrid FO systems have been described for various applications, including seawater desalination (Choi et al. 2009), wastewater treatment, or both at the same time. Thus, when the FO either replaces conventional pre-treatment technologies or is used as a post-treatment, it reduces the volume of industrial waste and produces good-quality water that can be reused (Yangali-Quintanilla et al. 2011).

8.2 Forward Osmosis Principle

Osmosis defines the spontaneous diffusion of the solvent molecule through a semipermeable membrane from a solution containing a lower solute concentration to a solution containing a higher solute concentration (Cath et al. 2006). The osmotic pressure (π) of an ideal dilute solution is given by van't Hoff's equation, i.e., equation (1) shown below.

$$\pi = nMRT \tag{1}$$

where n stands for the van't Hoff factor (points to the number of individual particles of compounds dissolved in the solution), M denotes molarity of the solution, R is the gas constant ($R = 0.0821$ L atm mol^{-1} K^{-1}), and T is the absolute temperature (in K) of the solution.

In fact, the FO membrane technology is an engineered process that helps in water treatment by utilizing the osmotic pressure gradient between FS and DS.

8.2.1 *Other Membrane-Based Processes*

In reverse osmosis (RO) process, movement of water molecule takes place from a solution having the higher osmotic potential to a solution having lower osmotic potential under applied hydraulic pressure, while in pressure-assisted forward osmosis (AFO), hydraulic pressure is applied at the feed side; therefore, ΔP is negative. This additional pressure at the feed side magnifies the efficiency of the FO process to increase water flux. AFO adds a medium pressure pump to a conventional FO system. The system takes advantage of additional hydraulic pressure that results in water transport via both mechanisms; i.e., flux is driven by hydraulic pressure and by the osmotic pressure.

The theoretical water flux across the membrane (J_w) is calculated using a variation of Darcy's law, i.e., equation (2) (Cath et al. 2006).

$$Jw = A(\sigma\Delta\pi - \Delta P) \quad (2)$$

where J_w stands for water flux and A referred to as water permeability coefficient of the membrane σ denotes the reflection coefficient, which represents the rejection capability of a membrane. A perfect semipermeable membrane has $\sigma = 1$ (Korenak et al. 2017), $\Delta\pi$ represents the differential osmotic pressure, and ΔP is the applied hydraulic pressure. Thus, the direction of the movement of the solvent molecule is governed majorly by osmotic pressure and applied hydraulic pressure.

The operation conditions of each membrane process are (Fig. 8.1):

1. In the RO process, $\Delta P > \Delta\pi$
2. In the FO process, $\Delta P =$ zero
3. In the AFO process, $\Delta P <$ zero.

8.3 Forward Osmosis Membrane

Ideally, the FO membrane must exhibit the trait of higher water flux and solute refusal, minimum internal concentration polarization (ICP), and minimum external concentration polarization (ECP). Along with these properties, higher mechanical and chemical stability is desirable (Korenak et al. 2017). The RO membrane (such as asymmetric cellulose acetate (CA) RO membrane) is incompatible for the FO because of their hydrophobic nature and comparatively thick support layers (SL), which is about 150 μm (McCutcheon and Elimelech 2008). Thick SLs are directly linked with concentration polarization.

Moreover, in the case of wastewater treatment, the problem of membrane fouling must be checked more vigilantly. Thus, desired characteristics of the FO membrane for wastewater treatment are:

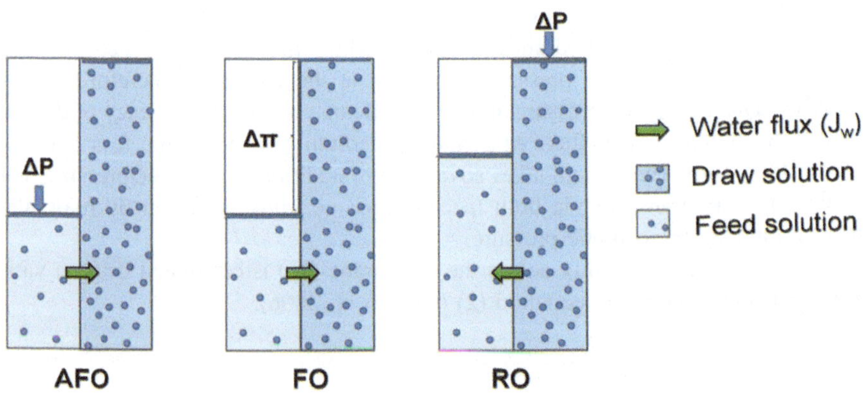

Fig. 8.1 Different osmotic membrane technologies. Adopted and modified from Cath et al. (2006)

1. A compact, ultrathin, membrane with greater solute rejection
2. Hydrophilic thin SL with the ability to reduce ICP
3. Enhanced affinity for water and lesser fouling tendency.

With the advancement in technology, several new commercial FO membranes are available in the market. Details about these membranes are discussed in the next section (Coday et al. 2014).

8.3.1 Membrane Materials

A variety of materials has been used for the preparation of the FO membrane, such as cellulose derivatives, polyamides, polyelectrolytes, and other polymers (Klaysom et al. 2013; Alsvik and Hägg 2013). However, the most advanced development is the biomimetic membrane developed by the incorporation of aquaporin proteins. This membrane has the advantage of higher water flux and higher solute rejection (Nielsen 2009).

Cellulose acetate (CA)- and cellulose triacetate (CTA)-based FO membrane was the foremost membrane type commercially available from Hydration Technologies Incorporation (HTI). These membranes are available in flat sheet and hollow fiber geometries. The extensively used CTA does not adsorb mineral, fatty oil, is extremely resistant to chlorine, and has lesser sensitivity for biological degradation. Additionally, the hydroxyl group of cellulose can also react with other reagents to generate compounds such as cellulose propionate (CP), cellulose acetate butyrate (CAB), or cellulose acetate propionate (CAP). Dual-layer FO hollow fibers derived from CAB and CAP show better performance as compared to CA-based flat sheet or hollow fiber membranes (Korenak et al. 2017). The major disadvantage of these cellulose-based membranes is their hydrolysis in wastewater and their limited stability to pH and temperature (Su et al. 2012). Thus, these cellulose-based membranes were overtaken by thin-film composite (TFC). Most TFC membranes are made with a porous, highly permeable support layer composed of polysulfone, which is coated with a cross-linked aromatic polyamide thin film (Lau et al. 2012). This thin-film coating is also referred as the active layers (ALs). These active layers are responsible for providing the property of solute rejection, while the support layer imparts mechanical stability. Thus, TFC proved better than CTA with respect to stability and permeability (Klaysom et al. 2013).

Other membranes such as polybenzimidazole (PBI)-based hollow membrane have the ability to self-charge in aqueous solution; this imparts the property of higher surface hydrophilicity and lower fouling propensity (Wang et al. 2009). Another approach toward the development of FO membrane is the membrane based on polyelectrolyte. These polyelectrolyte membranes are composed of an alternating film made up of anion and cation electrolyte; this type of deposition leads to layer-by-layer (LbL) configuration fabricated on a suitable support. These types of membranes produce good solute rejection and thermal durability (Saren et al.

2011). Also, the progress in the development of graphene oxide (GO)-based FO membranes is made. Based on the fabrication methods, the GO-based FO membranes can be divided into three types of membrane, which are:

- GO lamellar FO membranes
- GO-blended FO membranes
- GO surface-modified FO membranes.

The development of GO-blended FO membranes leads to improvement in the properties of membranes such as porosity, tortuosity, hydrophilicity, mechanical strength, and thermal stability (Wu et al. 2020). The list of some FO members, along with their producers, is shown in Table 8.1.

The most modern approach in the FO membrane development is aquaporin (AQP)-based membranes. The basic principle behind this biomimetic membrane is the ability of AQP protein to serve as a water channel on the cell membrane. Using AQPs, a membrane having increased water permeability and greater solute rejection is developed. Geometries adopted for these membranes are flat and hollow fiber (Madsen et al. 2015). In a study conducted by Chun et al. (2018), a prototype of an aquaporin-based membrane, i.e., Aquaporin Inside™ membrane (AIM), was compared with commercially available FO membrane in terms of intrinsic filtration properties, membrane surface chemistry, and fouling behavior. The result of the study shows that the AIM shows higher water flux when tested under identical conditions. Also, the membrane integrity of the AIM was maintained after repeated cycles of fouling by a high concentration of gypsum and physical cleaning tests. This indicates the possibility of using the AIM membrane for treating harsh feed solutions.

8.3.2 *Membrane Development*

Methods adopted in the development of the FO membrane have an impact on its efficiency in the filtration, and several other parameters such as fouling and solute rejection (Lutchmiah et al. 2014). Phase inversion technique is most commonly used in the manufacturing of the FO membranes, concentrating on the generation of a compact active layer of the membrane (Klaysom et al. 2013). In general, most of the FO membranes are asymmetrical in nature, with two different layers referred to

Table 8.1 List of FO membrane and their producers. Modified from Korenak et al. (2017)

Company	Membrane type	Configuration	Status
Aquaporin A/S	Biomimetic aquaporin	Hollow fiber and flat sheet	Commercial
Oasys Water	Thin-film composite	Flat sheet	Commercial
Fluid Tech Inc.	CTA	Flat sheet	Commercial
Porifera	Thin-film composite	Flat sheet	Commercial

as an active layer (AL) and a support layer (SL). AL nature is a compact selective layer, whereas SL is used to provide mechanical support (Lutchmiah et al. 2014).

Different methods of membrane development for the optimization of several parameters of FO membranes are:

- Use of functionalized nanoparticles onto the membrane surface in order to reduce fouling and enhance water movement (Tiraferri et al. 2012)
- Development of the SL layer of nanofiber to withstand more stress and provide better mechanical support (Hoover et al. 2011).

8.4 Draw Solution

The concentrated solution in the FO process is referred to as draw solution (DS). The selection of the right kind of DS, with respect to the FO system, is very crucial for the effective working of the entire process. Many types of DS have already been tested, and among all NaCl seems to be most suitable DS due to its property of greater solubility (Cath et al. 2006) and lower cost (Achilli et al. 2010) and also greater osmotic potential. The concentration range of NaCl in DS is 0.3–6 M, but the most common concentration of 0.5 M is used as this concentration mimics the osmotic potential of seawater. The seawater itself cannot be used as DS in most of cases due to the presence of numerous other particles and microorganisms, which promotes biofouling (Lutchmiah et al. 2014). The percentage distribution of different DSs is shown in Fig. 8.2.

8.4.1 Draw Solution for Wastewater Treatment

The desired properties in DS used for wastewater treatment are (Ge et al. 2013)

1. Higher osmotic potential than feed solution (FS)
2. Higher water flux along with lower reverse leakage
3. Non-toxic nature

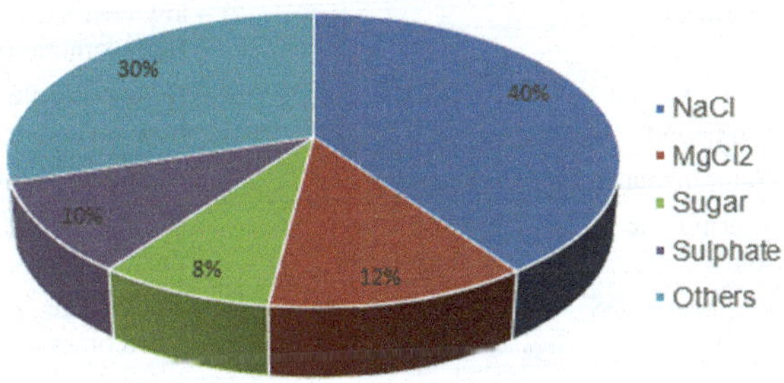

Fig. 8.2 Percentage distribution of a different kind of DS. Adapted and modified from Lutchmiah et al. (2014)

4. Minimum cost and easy to recover
5. It should not affect the sludge quality and microbes when used in the case of bioreactor.

Diffusivity (D) also plays a notable role in leakage as larger molecules diffuse slowly, and thus there are fewer chances of leakage for them but cause higher ICP (Achilli et al. 2010). Wastewater already holds several molecules with varying diffusivity. Thus, these molecules, along with DS, influence the water flux and ICP. For instance, monovalent ion has higher solute leakage as a result of high diffusivity (Cornelissen et al. 2008) but, at the same time, has a lower chance of ICP. Thus, the exact effect depends on the operating condition of the process.

Other factors, such as pH and temperature, must also be taken into account while selecting a DS for the system (Lutchmiah et al. 2014). Any change in one or both may affect the efficiency of the system. One of the problems that occur due to their variation is the scaling of mineral salts when precursor ions like Ca^{2+}, SO_4^{2-}, and CO_3^{2-} are present. Due to the very complex nature of wastewater feed, $MgCl_2$ is considered as ideal DS for the wastewater processing process due to its higher osmotic potential, although the cost is higher as compared to NaCl (Achilli et al. 2010). The list of DS, along with its advantages and disadvantages, is depicted in Table 8.2.

8.4.2 *Draw Solution Recovery System*

Some thermodynamically favorable methods must be adopted to obtain the product and recover the DS. The most frequently used processes are RO, membrane distillation (MD), and thermal recovery. Recovery rate and energy consumption are the two major factors that must be considered while choosing a recovery system for wastewater treatment (Lutchmiah et al. 2014; Luo et al. 2014).

Table 8.2 Different types of DS used in wastewater treatment (Lutchmiah et al. 2014)

DS type	Example	Advantage	Disadvantage
Inorganic substance	Salt	High solubility and low cost High osmotic potential	Scaling/ICP salt leakage Limited storage time
High charged compound	EDTA	High water flux Low reverse leakage	Expensive pH dependency
Natural source	Seawater	Abundant availability	Biofouling
Thermolytic solutes	Ammonium bicarbonate	High solubility Easy recovery	Toxic thermolytic product
Engineered DS	Nanoparticles	No leakage Overcomes scaling	Expensive

8.5 Effect of Temperature and pH

Water flux is directly proportional to the temperature; i.e., an increase in temperature causes higher water flux. An increase of 2.4% of water flux per °C is observed with 0.5 NaCl DS (Cornelissen et al. 2008). At the same time, temperature increase causes a decrease in wastewater viscosity (McCutcheon and Elimelech 2006) along with an increase in solute diffusion rate and lower ICP. However, when FO is used in wastewater application, a rise in temperature causes wastewater to be in more concentrated mode; this causes higher fouling of the membrane. Additionally, the effluent of wastewater contains several microbes, which gets affected by an increase in temperature. Thus, the FO process temperature depends on the characteristic of the wastewater system under consideration. A change in pH affected the draw solution properties, especially DS of inorganic solute and charged compound. In wastewater, heavy metals and suspended solids get precipitated at a higher pH range and get dissolved at a lower pH range.

8.6 Membrane Module Configuration

Module refers to the assembly of the membrane into a unit such that to increase the surface-to-volume ratio along with an increase in cross-flow by reducing particle deposition. Thus, membrane configuration manages how the membrane is packed and arranged inside the modules.

An ideal membrane module configuration must possess the following characteristics (Mccutcheon and Elimelech 2007; Xu et al. 2010):

- Compact, i.e., more surface area-to-volume ratio
- Ultrathin AL active having properties of higher water permeability and solute rejection
- Lowers the effect of concentration polarization
- Lower friction to tangential flow
- Uniform velocity throughout the system
- Should resist fouling
- Simple cleansing and maintenance
- Economically feasible.

Typically, there are four principal types of membrane configurations, and these are (Cath et al. 2006):

- Plate and frame
- Spiral wound
- Tubular
- Hollow fiber.

8.6.1 Plate and Frame

Plate-and-frame modules are made up of flat sheet membranes sealed inside the frames, and such type of arrangement imparts overall mechanical support. These types of modules are less complicated in design, and backwashing can be done efficiently (Bamaga et al. 2011).

In the FO system, these modules are designed in such a manner to maintain cross-flow of feed and draw streams at the same time avoid intermixing of these streams. Practically, it is quite challenging to manage such a cross-flow system. As a result, plate-and-frame modules have the least packing density among all the module design available. Typically, the packing density is in the range of 100–400 m^2/m^3 (Mulder 1997). However, the lower packing ratio acts as an advantage in the case of the feed having high fouling agents such as wastewater streams.

The disadvantages associated with the plate-and-frame module are:

- Low packing density
- Pressure drop
- Requirement of multiple sealings (Balster 2013).

8.6.2 Spiral Wound Module

Spiral wound modules are the membrane module having a higher packing density. In order to achieve such high packing density, the distance between the two layers is reduced to less than 1 mm. Thus, if the wastewater stream is not pre-treated, then the chances of clogging of the spiral wound module are very high, and removal of such clogging can be done by chemicals, air/water mixing, and/or high fluid velocities. In order to remove fouling by backwashing, spiral wound forward osmosis (SWFO) module, having short feed channel length and wide spacers, is suggested (Lutchmiah et al. 2014). Thus due to this limitation, SWFO module application is restricted to wastewater streams having lower concentrations of fouling agents. The HTI spiral wound membrane module is the module configuration that is available in the market (Bamaga et al. 2011).

8.6.3 Tubular Module

Tubular membrane module can be used for continuously operated FO process as compared to other modules due to the following reasons:

- These modules posses self-supported characteristic and can easily withstand high hydraulic pressure.
- Packing density is relatively high.

- Allows flow of fluid on both sides of the membrane.
- Modularization is easier.

In tubular membranes, the inner diameter is about ≥ 2 mm; in such lumen space, turbulent flow can be achieved and therefore fouling and scaling can be reduced (Cath et al. 2006).

8.6.4 *Hollow Fiber Module*

Hollow fiber modules are modified forms of tubular modules having a very high packing density compared to other modules. Thus, these are better for higher volume applications such as desalination. Hollow fiber modules have been developed with an ultrathin RO like skin layer on either the inner or outer surface of the fiber, with a lumen having an inner diameter of around 1 mm with a wall thickness of 0.17–0.215 mm. This module also possesses good mechanical strength (Wang et al. 2010).

The disadvantage associated with this module system is that the internal diameter is very less (around 1 mm). Therefore in such lumen space, laminar flow can only be possible that results in limited mixing at the membrane surface. Under such conditions, scaling or fouling can cause a reduction in mass transport across the membrane (Cath et al. 2006).

8.7 Membrane Fouling

Fouling is the accumulation of matter such as particles, macromolecules onto the surface of the membrane or, in some cases, can be inside pores of the membrane. The matter which gets deposited onto or inside the membrane is referred to as foulant. This accumulation results in a decrease in water flux, which can be either temporary or permanent (Seidel and Elimelech 2002). Broadly, foulants are categorized into four categories which are (Korenak et al. 2017):

- Organic foulants: This includes macromolecules such as alginate, protein, and other naturally occurring organic matter.
- Inorganic foulant: This category includes several sparingly mineral salts which undergo crystallization.
- Biofouling: This involves bacterial cells and other microbes along with attachment, which together leads to the formation of biofilm.
- Colloidal particles: The colloidal particles get accumulated on the membrane surface. In this type of fouling condition, the rate of deposition depends on the roughness of the membrane surface (Vrijenhoek et al. 2001).

Fouling results in a decrease in water flux, deteriorating water quality, and an increase in maintenance cost, and in severe cases, these foulants may degrade the membrane (Rana and Matsuura 2010). Fouling is one of the major issues in membrane-based processes and thus affects the economics of the process. But osmotically driven membrane processes show different fouling patterns as compared to pressure-driven membrane processes as fouling in the pressure-driven process is more as compared to FO process (Korenak et al. 2017). It has been found that the major reason behind the flux decline in FO is reverse salt diffusion from the DS side to the FS side, which leads to accelerated cake-enhanced osmotic pressure (CEOP). Thus by choosing the appropriate DS, which is having less back diffusion property, it is easier to overcome this problem. For instance, on comparing NaCl and dextrose as DS, it was observed that in case of NaCl, much more flux decline is observed due to higher reverse diffusion while in case of dextrose due to larger hydrated radius reverse diffusion is less as shown in Fig. 8.3 (Lee et al. 2010).

Most of the time, fouling in FO is reversible, while in the case of RO, irreversible type of fouling is observed (Xie et al. 2015). Interestingly, membrane in the FO process gets more clogged due to organic contaminants as compared to inorganic contaminants and these organic foulants in the absence of hydraulic pressure make loose fouling layer that can be easily removed by backwashing and in case of RO compact layer of foulant is formed due to additional hydraulic pressure. Thus, controlling fouling and membrane cleaning in FO is more attainable than RO (Mi and Elimelech 2010; Lee et al. 2010).

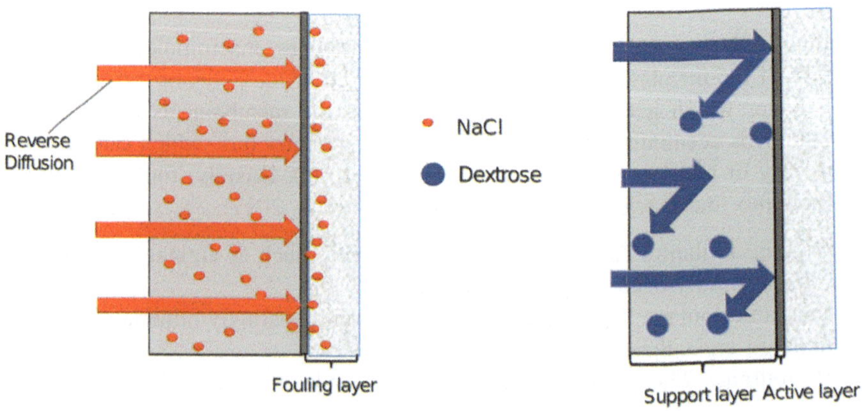

Fig. 8.3 Reverse diffusion comparison of NaCl and dextrose. Adopted and modified from Lee et al. (2010)

8.8 Contribution of FO in Desalination and Wastewater Treatment

FO has a role in a wide range of applications such as power generation, seawater/brackish water desalination, wastewater treatment, and food processing. The two major applications regarding FO are water desalination (Choi et al. 2009) and wastewater treatment (Zhao et al. 2012; Valladares Linares et al. 2014). As the principle of FO is based on the movement of water molecules from FS to DS side, however, to obtain the final product from DS, one additional post-processing step must be performed, and this step causes an increase in the overall energy demand of the FO process. Thus to overcome this problem, hybrid FO system is developed. In a hybrid FO system, FO is integrated with other separation processes, which can be either physical process or chemical process. The hybrid FO system is used widely in saline water and wastewater treatment and has replaced conventional treatment technologies. For instance, the FO-RO Hybrid Desalination Research Center (FOHC, Kookmin University, Korea) has started a project entitled 'ONE Desal' (osmosis-based, no fouling, energy-efficient desalination) as shown in Fig. 8.4. With joint efforts of academic and industrial units, this project aims at setting the first FO-RO hybrid plant having the capacity of 1000 m^3/day and treats wastewater and seawater simultaneously. Thus, the major objectives of this project were (Chekli et al. 2016):

- Development of a hybrid system for laboratory scale as well as for pilot scale
- Development of low-cost treatment technology.

8.8.1 Direct Desalination

In this method, freshwater is obtained from seawater or brackish water using a high osmotic potential DS such as NH_4HCO_3 as shown in Fig. 8.5 (Chekli et al. 2012). The major setback with this system is that a separate process is required to recover DS (Li et al. 2013). For this purpose, the temperature is increased above 60 °C that

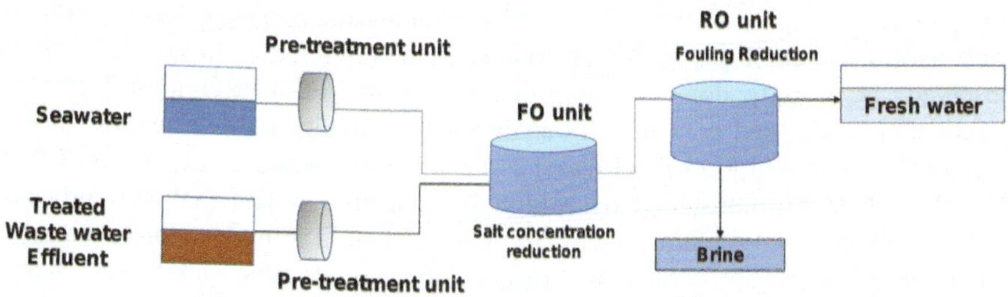

Fig. 8.4 Concept of FO-RO hybrid system. Adopted and modified from Sohn (2017)

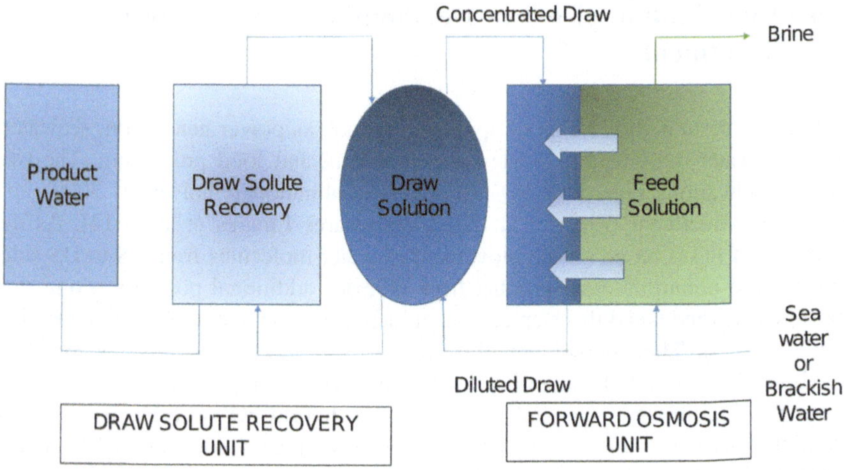

Fig. 8.5 Desalination using direct FO process. Adopted and modified from Korenak et al. (2017)

leads to the production of CO_2 and NH_3 gases. These CO_2 and NH_3 are further reused for the production of DS in the next cycle (McGinnis and Elimelech 2007). Overall, the efficiency of this type of system with respect to cost and energy is low (Valladares Linares et al. 2014).

8.8.2 Indirect System

The indirect systems, a hybrid FO-RO system is used for desalination, and the FO here acts as a pre-treatment step. In the first hybrid FO-RO system, FO system is combined with a low-pressure RO (LPRO) system (Chekli et al. 2016). This hybrid system is developed for the recovery of glucose. For the same purpose, when RO was used alone, the system suffered from very high fouling, whereas when the FO was used as a pre-treatment process, the significant reduction in fouling was observed (Yaeli 1992). Thus, the same principle is used for desalination of saline water in which brackish water is used as DS and wastewater used as FS. Now the diluted seawater can be further processed by low-pressure reverse osmosis (LPRO). This hybrid system leads to lesser fouling, and a reduction in overall cost is observed. In a study done, the performance of the FO-LPRO hybrid for the desalination of the Red seawater was analyzed. The results attested that the energy expenditure of hybrid FO-LPRO was 1.3–1.5 kWh/m^3, which is exactly 50% less than that of energy consumed by a high pressure-driven RO (SWRO) process, which is about 2.5–4 kWh/m^3. Thus, the hybrid system is much more preferable due to two advantages, which are lower fouling propensity and lesser energy consumption (Yangali-Quintanilla et al. 2011).

8.8.3 *Wastewater Treatment*

The FO process in integration with other separation process is widely used for the treatment of wastewater. Generally, wastewater has low osmotic pressure and high fouling tendency and thus the FO with membrane distillation (MD) process examined for processing of municipal wastewater, where water flux is accomplished in a continuous operation recovery rate up to 80% (Xie et al. 2013). Also, a photovoltaic powered FO-electrodialysis (FO-ED) process was experimented to check its efficiency in getting clean water from secondary wastewater effluent. The results indicate the higher removal of total organic carbon (TOC) (Zhang et al. 2013). The FO process is also useful in the treatment of other kinds of wastewater, such as oily wastewater, industrial and nuclear wastewater, landfill leachate, oil–water separation (Duong and Chung 2014), and for these kinds of wastewater treatment osmotic membrane bioreactor (OsMBR) is used. These types of bioreactors are preferred—wastewater clarification methods for reusing of non-potable water. Earlier pressure-driven processes were integrated with these membrane bioreactors, but the FO process replaced these processes due to having the property of lower fouling propensity, no need of additional hydraulic pressure, production of equally good-quality effluent, and easy removal of contaminants (Chekli et al. 2016).

OsMBR is generally of submerged membrane bioreactor type of bioreactor in which wastewater is used as FS, whereas concentrated NaCl solution or sea brine water is used as DS in the system (Yap et al. 2012). The major drawback of this type of treatment process is the decline of water flux with time due to higher rejection of FS-dissolved solutes by the FO membrane and reverse diffusion of DS solutes. These two phenomena lead to decline in osmotic pressure gradient resulting in water flux decline. A study conducted on the salt accumulation model revealed that the decline in water flux could be optimized by the integration of a membrane having a low ratio of salt permeability to water permeability (Xiao et al. 2011).

8.9 Advantages of the FO Process in Desalination and Wastewater Treatment

1. The hybrid FO systems have the potential to consume lesser energy as compared to high pressure-driven process alone, especially in the case where the feed source has extreme characteristics, for instance, the FO-LPRO system for the purpose of desalination.
2. The FO process is successively used as a pre-treatment process with an objective of lowering of fouling propensity and higher fouling reversibility. Thus, the FO surpasses the conventional pre-treatment processes.
3. In the FO-RO system, the simultaneous treatment of seawater and wastewater can be done, eliminating the need for post-treatment of DS. Additionally, due to

the integration of FO, fouling is reduced in the RO membrane. Thus, the overall efficiency of the system increased.
4. No requirement of additional hydraulic pressure as water molecules move along the concentration gradient.
5. Membrane cleaning is easier as, in most cases, the type of fouling observed in the FO is reversible; thus, contaminants can be removed easily.

8.10 Conclusion

The FO-based system used for the treatment of water and wastewater system shows the promising result and has several advantages compared to conventional water treatment processes such as low fouling tendency and absence of additional osmotic hydraulic pressure; thus, it is an energy-efficient process. Drinking water is becoming a marginal resource, which ultimately leads to water stress. Thus to deal with this global concern, FO-based water treatment method has been proposed. Several factors play a role in the efficiency of the FO-based treatment such as membrane type and its orientation and its fouling propensity, type of DS, and their compatibility with water to be treated. Ideally, a membrane with low fouling tendency and high solute rejection is preferred.

Although the FO technology holds great potential and advantages in a variety of application, still it suffers from a few issues which need to be dealt with. One such issue is the recovery of DS for which another energy-dependent post-treatment step must be applied. Thus to avoid this extra energy demand, FO process is integrated with other separation processes, and this type of integrated system is referred to as the hybrid FO system. The hybrid FO system enhances the efficiency of the separation system by a reduction in the fouling and improvement in the product water quality, and, most significantly, decreases the overall energy cost of the separation process. The potential applications of the FO are desalination of seawater, recovery of the useable form of water from wastewater, and treatment of high salinity brines. Therefore, nowadays, studies are more focused on the development of new membranes such as aquaporin and improvement of existing FO membrane by surface modification. Along with that, the focus is on the development of more efficient DSs and to optimize their compatibilities with various wastewaters. Thus in order to move FO from bench-scale to full-scale application, more focus on identifying specific areas should be done where the FO makes both practical and economic senses.

References

Achilli A, Cath TY, Childress AE (2010) Selection of inorganic-based draw solutions for forward osmosis applications. J Memb Sci 364:233–241. https://doi.org/10.1016/j.memsci.2010.08.010

Alsvik IL, Hägg MB (2013) Pressure retarded osmosis and forward osmosis membranes: Materials and methods. Polymers (Basel 5:303–327. https://doi.org/10.3390/polym5010303

Balster J (2013) Plate and Frame Membrane Module. Encyclopedia of Membranes, 1–3. https://doi.org/10.1007/978-3-642-40872-4_1584-1

Bamaga OA, Yokochi A, Zabara B, Babaqi AS (2011) Hybrid FO/RO desalination system: Preliminary assessment of osmotic energy recovery and designs of new FO membrane module configurations. Desalination 268(1–3):163–169. https://doi.org/10.1016/j.desal.2010.10.013

Cath TY, Childress AE, Elimelech M (2006) Forward osmosis: Principles, applications, and recent developments. J Memb Sci 281:70–87. https://doi.org/10.1016/j.memsci.2006.05.048

Chekli L, Phuntsho S, Kim JE, Kim J, Choi JY, Choi JS, Kim S, Kim JH, Hong S, Sohn J, Shon HK (2016) A comprehensive review of hybrid forward osmosis systems: Performance, applications and future prospects. J Memb Sci 497:430–449. https://doi.org/10.1016/j.memsci.2015.09.041

Chekli L, Phuntsho S, Shon HK, Vigneswaran S, Kandasamy J, Chanan A (2012) A review of draw solutes in forward osmosis process and their use in modern applications. Desalin. Water Treat. 43:167–184. https://doi.org/10.1080/19443994.2012.672168

Choi YJ, Choi JS, Oh HJ, Lee S, Yang DR, Kim JH (2009) Toward a combined system of forward osmosis and reverse osmosis for seawater desalination. Desalination 247(1–3):239–246. https://doi.org/10.1016/j.desal.2008.12.028

Chun Y, Qing L, Sun G, Bilad M, Fane A, Chong T (2018) Prototype aquaporin-based forward osmosis membrane: Filtration properties and fouling resistance. Desalination 445:75–84. https://doi.org/10.1016/j.desal.2018.07.030

Coday BD, Xu P, Beaudry EG, Herron J, Lampi K, Hancock NT, Cath TY (2014) The sweet spot of forward osmosis: Treatment of produced water, drilling wastewater, and other complex and difficult liquid streams. Desalination 333:23–35. https://doi.org/10.1016/j.desal.2013.11.014

Cornelissen ER, Harmsen D, de Korte KF, Ruiken CJ, Qin JJ, Oo H, Wessels LP (2008) Membrane fouling and process performance of forward osmosis membranes on activated sludge. J Memb Sci 319:158–168. https://doi.org/10.1016/j.memsci.2008.03.048

Duong PHH, Chung TS (2014) Application of thin film composite membranes with forward osmosis technology for the separation of emulsified oil-water. J Memb Sci 452:117–126. https://doi.org/10.1016/j.memsci.2013.10.030

Ge Q, Ling M, Chung TS (2013) Draw solutions for forward osmosis processes: Developments, challenges, and prospects for the future. J Memb Sci 442:225–237. https://doi.org/10.1016/j.memsci.2013.03.046

Haupt A, Lerch A (2018) Forward osmosis application in manufacturing industries: A short review. Membranes (Basel) 8:47. https://doi.org/10.3390/membranes8030047

Hoover LA, Phillip WA, Tiraferri A, Yip NY, Elimelech M (2011) Forward with osmosis: Emerging applications for greater sustainability. Environ Sci Technol 45:9824–9830. https://doi.org/10.1021/es202576h

Khawaji AD, Kutubkhanah IK, Wie JM (2008) Advances in seawater desalination technologies. Desalination 221(1–3):47–69. https://doi.org/10.1016/j.desal.2007.01.067

Klaysom C, Cath TY, Depuydt T, Vankelecom IFJ (2013) Forward and pressure retarded osmosis: Potential solutions for global challenges in energy and water supply. Chem Soc Rev 42:6959–6989. https://doi.org/10.1039/c3cs60051c

Korenak J, Basu S, Balakrishnan M, Hélix-Nielsen C, Petrinic I (2017) Forward osmosis in wastewater treatment processes. Acta Chim Slov 64:83–94. https://doi.org/10.17344/acsi.2016.2852

Lau WJ, Ismail AF, Misdan N, Kassim MA (2012) A recent progress in thin film composite membrane: A review. Desalination 287:190–199. https://doi.org/10.1016/j.desal.2011.04.004

Lee S, Boo C, Elimelech M, Hong S (2010) Comparison of fouling behavior in forward osmosis (FO) and reverse osmosis (RO). J Memb Sci 365:34–39. https://doi.org/10.1016/j.memsci.2010.08.036

Li D, Zhang X, Simon GP, Wang H (2013) Forward osmosis desalination using polymer hydrogels as a draw agent: Influence of draw agent, feed solution and membrane on process performance. Water Res 47:209–215. https://doi.org/10.1016/j.watres.2012.09.049

Luo H, Wang Q, Zhang TC, Tao T, Zhou A, Chen L, Bie X (2014) A review on the recovery methods of draw solutes in forward osmosis. J Water Process Eng 4(C):212–223. https://doi.org/10.1016/j.jwpe.2014.10.006

Lutchmiah K, Verliefde ARD, Roest K, Rietveld LC, Cornelissen ER (2014) Forward osmosis for application in wastewater treatment: A review. Water Res 58:179–197. https://doi.org/10.1016/j.watres.2014.03.045

Madsen HT, Bajraktari N, Hélix-Nielsen C, Van der Bruggen B, Søgaard EG (2015) Use of biomimetic forward osmosis membrane for trace organics removal. J Membr Sci 476:469–474. https://doi.org/10.1016/j.memsci.2014.11.055

McCutcheon JR, Elimelech M (2006) Influence of concentrative and dilutive internal concentration polarization on flux behavior in forward osmosis. J Membr Sci 284(1–2): 237–247. https://doi.org/10.1016/j.memsci.2006.07.049

Mccutcheon J, Elimelech M (2007) Modeling water flux in forward osmosis: Implications for improved membrane design. AIChE J 53:1736–1744

McCutcheon JR, Elimelech M (2008) Influence of membrane support layer hydrophobicity on water flux in osmotically driven membrane processes. J Memb Sci 318:458–466. https://doi.org/10.1016/j.memsci.2008.03.021

McGinnis RL, Elimelech M (2007) Energy requirements of ammonia-carbon dioxide forward osmosis desalination. Desalination 207:370–382. https://doi.org/10.1016/j.desal.2006.08.012

Mehta D, Gupta L, Dhingra R (2014) Forward Osmosis in India: Status and Comparison with Other Desalination Technologies. Int Sch Res Not 2014:1–9. https://doi.org/10.1155/2014/175464

Mi B, Elimelech M (2010) Organic fouling of forward osmosis membranes: Fouling reversibility and cleaning without chemical reagents. J Membr Sci 348(1–2):337–345. https://doi.org/10.1016/j.memsci.2009.11.021

Mulder M (1997) Basic principles of membrane technology, 2nd edn. Kluwer Academic Publishers, Dordrecht

Nielsen CH (2009) Biomimetic membranes for sensor and separation applications. Anal Bioanal Chem 395(3):697–718. https://doi.org/10.1007/s00216-009-2960-0

Yaeli J, (1992) Method and apparatus for processing liquid solutions of suspensions particularly useful in the desalination of saline water. U.S. Patent 5,098,575. https://doi.org/10.1016/0375-6505(85)90011-2

Rana D, Matsuura T (2010) Surface modifications for antifouling membranes. Chem Rev 110:2448–2471. https://doi.org/10.1021/cr800208y

Saren Q, Qiu CQ, Tang CY (2011) Synthesis and characterization of novel forward osmosis membranes based on layer-by-layer assembly. Environ Sci Technol 45:5201–5208. https://doi.org/10.1021/es200115w

Seidel A, Elimelech M (2002) Coupling between chemical and physical interactions in natural organic matter (NOM) fouling of nanofiltration membranes: Implications for fouling control. J Memb Sci 203:245–255. https://doi.org/10.1016/S0376-7388(02)00013-3

Sohn J (2017) FO-RO Hybrid Desalination Project: Progress and Challenges. http://en.bwf.kr/data/files/B1_Jinsik_Sohn.pdf

Su J, Ong R, Wang P, Chung T, Helmer B, de Wit J (2012) Advanced FO membranes from newly synthesized CAP polymer for wastewater reclamation through an integrated FO-MD hybrid system. AIChE J 59:1245–1254

Tiraferri A, Kang Y, Giannelis EP, Elimelech M (2012) Highly hydrophilic thin-film composite forward osmosis membranes functionalized with surface-tailored nanoparticles. Appl Mater Interfaces 4(9):5044–5053. https://doi.org/10.1021/am301532g

Valladares Linares R, Li Z, Sarp S, Bucs SS, Amy G, Vrouwenvelder JS (2014) Forward osmosis niches in seawater desalination and wastewater reuse. Water Res 66(4):122–139. https://doi.org/10.1016/j.watres.2014.08.021

Vrijenhoek EM, Hong S, Elimelech M (2001) Influence of membrane surface properties on initial rate of colloidal fouling of reverse osmosis and nanofiltration membranes. J Membr Sci 188 (1):115–128. https://doi.org/10.1016/S0376-7388(01)00376-3

Wang KY, Yang Q, Chung TS, Rajagopalan R (2009) Enhanced forward osmosis from chemically modified polybenzimidazole (PBI) nanofiltration hollow fiber membranes with a thin wall. Chem Eng Sci 64:1577–1584. https://doi.org/10.1016/j.ces.2008.12.032

Wang R, Shi L, Tang CY, Chou S, Qiu C, Fane AG (2010) Characterization of novel forward osmosis hollow fiber membranes. J Membr Sci 355(1–2):158–167. https://doi.org/10.1016/j.memsci.2010.03.017

Wu W, Shi Y, Liu G, Fan X, Yu Y (2020) Recent development of graphene oxide based forward osmosis membrane for water treatment: A critical review. Desalination 491: https://doi.org/10.1016/j.desal.2020.114452

Xiao D, Tang CY, Zhang J, Lay WCL, Wang R, Fane AG (2011) Modeling salt accumulation in osmotic membrane bioreactors: Implications for FO membrane selection and system operation. J Memb Sci 366:314–324. https://doi.org/10.1016/j.memsci.2010.10.023

Xie M, Lee J, Nghiem LD, Elimelech M (2015) Role of pressure in organic fouling in forward osmosis and reverse osmosis. J Membr Sci 493:748–754. https://doi.org/10.1016/j.memsci.2015.07.033

Xie M, Nghiem LD, Price WE, Elimelech M (2013) A forward osmosis-membrane distillation hybrid process for direct sewer mining: System performance and limitations. Environ Sci Technol 47:13486–13493. https://doi.org/10.1021/es404056e

Xu Y, Peng X, Tang CY, Fu QS, Nie S (2010) Effect of draw solution concentration and operating conditions on forward osmosis and pressure retarded osmosis performance in a spiral wound module. J Membr Sci 348(1–2):298–309. https://doi.org/10.1016/j.memsci.2009.11.01

Yangali-Quintanilla V, Li Z, Valladares R, Li Q, Amy G (2011) Indirect desalination of Red Sea water with forward osmosis and low pressure reverse osmosis for water reuse. Desalination 280:160–166. https://doi.org/10.1016/j.desal.2011.06.06

Yap WJ, Zhang J, Lay WCL, Cao B, Fane AG, Liu Y (2012) State of the art of osmotic membrane bioreactors for water reclamation. Bioresour Technol 122:217–222. https://doi.org/10.1016/j.biortech.2012.03.060

Zhang Y, Pinoy L, Meesschaert B, Van Der Bruggen B (2013) A natural driven membrane process for brackish and wastewater treatment: Photovoltaic powered ED and FO hybrid system. Environ Sci Technol 47:10548–10555. https://doi.org/10.1021/es402534m

Zhao S, Zou L, Mulcahy D (2011) Effects of membrane orientation on process performance in forward osmosis applications. J Memb Sci 382:308–315. https://doi.org/10.1016/j.memsci.2011.08.020

Zhao S, Zou L, Tang CY, Mulcahy D (2012) Recent developments in forward osmosis: Opportunities and challenges. J Memb Sci 396:1–21. https://doi.org/10.1016/j.memsci.2011.12.023

Chapter 9
Food Processing Wastewater Treatment: Current Practices and Future Challenges

Darakhshan Nayyar, Tabish Nawaz, Saba Noore, and Anubhav Pratap Singh

Abstract The quality and quantity of wastewater released from food processing industries vary widely with the products and production processes. These wastewater streams are rich in BOD, suspended solids, and oily substances. Traditionally, biological treatment methods have been utilized in food processing wastewater treatment in meeting the discharge regulations. However, the issues of size of the wastewater treatment systems, their process efficiency, and space availability need to be addressed in the context of stricter standards and sustainability. Moreover, the idea of resource recovery coupled with wastewater treatment is also transforming the traditional wastewater treatment systems, and the focus is increasingly shifting toward recovering nutrients, energy, and water in addition to meeting the discharge standards. This has resulted in the development of novel technologies in the context of food processing wastewater treatment. In this book chapter, the quality and quantity of wastewater generated by major industries are discussed. Subsequently, current wastewater treatment technology that is utilized in key food processing facilities is highlighted. The key challenges to each industry, treatment system in the context of its wastewater are identified and future opportunities are explored. The waste streams and treatment technology have also been identified with respect to their resource recovery potential.

D. Nayyar · T. Nawaz (✉)
Environmental Science and Engineering Department, Indian Institute of Technology Bombay, Mumbai 400076, Maharashtra, India
e-mail: tnawaz@iitb.ac.in

S. Noore
Department of Food Engineering, CSIR-Central Food Technological Research Institute, Mysore 570 020, India

A. P. Singh
Faculty of Land and Food Systems (LFS), University of British Columbia, Vancouver Campus 213-2205 East Mall, Vancouver, BC V6T 1Z4, Canada

S. Singh et al. (eds.), *Pollution Control Technologies*, Energy, Environment, and Sustainability, https://doi.org/10.1007/978-981-16-0858-2_9

9.1 Introduction

Food and water are intricately linked to each other. Water, in particular, impacts global food security, as it is an essential component of the food production, right from the beginning of crop production, through the raw food material production and packaging to finally in our kitchens where the food is prepared in its final form. The food industry, ostensibly a water-intensive sector, consumes 1–3L, 9.13L, 0.97L, and 2.05L of water per kilogram or liter of product from dairy, meat, bakery, and oil sector, respectively (Munderere et al. 2016). www.wwdmag.com 2020. These applications generate a huge quantity of wastewater with quality differing widely from one food industry to another depending upon the products formed and processes employed.

However, certain commonalities can also be seen on a broader examination, such as high biological oxygen demand (BOD), suspended solids (SS), and smelly odor that are typically noticed in the wastewater from the food industry. Based on these commonalities, industries can be clubbed together while analyzing their wastewater characteristics, environmental impact, and treatment methodology. Some common food industries that have typically high BOD and SS levels are brewery, starch industries, whereas beverage, vegetable oil, dairy, confectionary show high levels of BOD and chemical oxygen demand levels (COD), with oil industry also showing high levels of oil in their wastewater and acidic pH values (env.go.jp 2020). Additionally, there are industries like confectionary, sugar, fruit juice, and breweries that operate at their peak on seasonal basis. These kinds of industries pose another challenge as wastewater quality does not remain consistent throughout the year. Therefore, the treatment strategy must be optimized on seasonal basis to save the treatment operation cost from overdoing during the off-seasons. Meat and poultry industries though they both process meat but produce wastewater which are remarkably different. However, a major source of pollutant load in both the industries comes from the blood released during slaughtering stage. Therefore, it can be said that while broadly wastewater from different industries have high BOD and COD, which make them suitable for biological treatment, there also exist peculiarity in each food industry that is unique to the process employed and products formed that each require deeper investigation into the wastewater characteristics and the technological requirement that addresses the treatment challenge to each kind of wastewater. Apart from process wastewater, the general-purpose water used in food processing facilities used for cleaning and sanitizing the process equipment also contains detergent and disinfectant in them, which can compromise the treatment efficacy in the biological treatment system for the wastewater (Malollari et al. 2019). Therefore, a food processing facility must employ an ancillary treatment scheme apart from the biological treatment methods which are typically applied.

In terms of the availability of technology for processing food industry wastewater, technologies can be characterized primarily into physicochemical methods, biological methods, and novel emerging methods. The plan, design, and treatment mechanisms often need to be determined according to specific contaminant concentration profile of the food processing wastewater and local regulations. Currently, coagulation, adsorption, and oxidation-based processes dominate the food industry, as they are easy to carry out. However, new biological methods are increasingly getting attention due to their widespread popularity. Further, emerging methods based on novel concepts, such as membrane filtration and biofilm technology, are being developed.

In this chapter, typical attributes of wastewater from common food industries are analyzed and discussed. Food industries considered in this chapter include dairy, meat processing facilities like slaughter house, abattoir and poultry farms, alcoholic beverage, edible oil, confectionary, and sugar industries. Next, different treatment technology has been identified and discussed.

9.2 Literature Review Methodology

This article is based on the relevant and recent literature gathered from various scientific publications' databases in the area of wastewater from food-related industries. The databases used for this study included Google Scholar, Science Direct, Scopus, and Web of Science. The entire literature review methodology can be summarized in following steps as follows: (a) peer reviewed references of books, book chapters, papers, and review articles were searched from year 2005–2020 using a combination of keywords. The idea is to include only the recent state-of-the-art knowledge in one and half decade. The main keywords used were: food industry wastewater, dairy wastewater, poultry wastewater, meat processing facility wastewater, cattle rearing wastewater, confectionery wastewater, sugar industry wastewater, olive oil wastewater, corn ethanol wastewater, cooking oil wastewater, characterization, quality, (b) from the total number of search results provided by a given database, relevant searches were selected according to the subject of review and then finally most apt ones were downloaded. The summary of search results from databases is shown in Table 9.1, (c) key findings and relevant results from those papers were studied and evaluated, and (d) subsequently, the papers were discussed in this review article. It was found that Google Scholar provided maximum number of total search results Fig. 9.1.

Table 9.1 Summary of search results from databases

S. no	Keywords	Scientific database	Total search results	Relevant number of search results	Number of documents downloaded
1.	Characteristics/quality of dairy industry wastewater	Google Scholar	17,300	1230	22
		Science Direct	3334	21	11
		Scopus	3505	2	2
		Web of Science	4	2	2
2.	Characteristics/quality of poultry industry wastewater	Google Scholar	16,800	23	10
		Science Direct	2389	30	5
		Scopus	2096	2	2
		Web of Science	3	0	0
3.	Characteristics/quality of meat processing industry wastewater	Google Scholar	16,900	72	25
		Science Direct	2672	36	25
		Scopus	13	3	3
		Web of Science	5	4	1
4.	Characteristics/quality of confectionary wastewater, sugar industry wastewater	Google Scholar	3580	18	6
		Science Direct	691	9	5
		Scopus	0	0	0
		Web of Science	0	0	0
5.	Characteristics/quality of fermentation industry, brewery industry wastewater, vegetable oil and edible oil industry wastewater	Google Scholar	3930	16	10
		Science Direct	1	1	1
		Scopus	0	0	0
		Web of Science	0	0	0

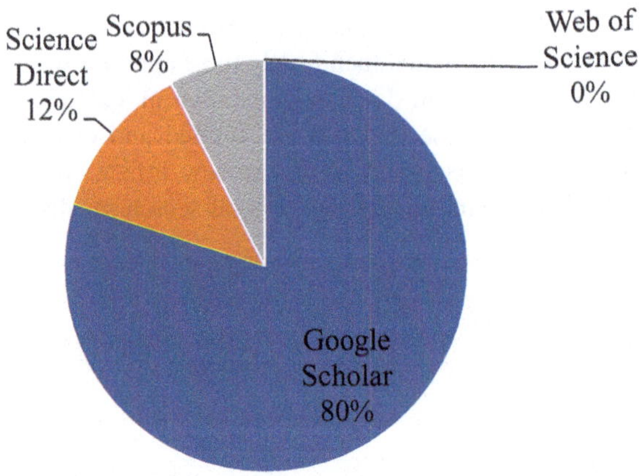

Fig. 9.1 Percentage contribution of databases in the total search results

9.3 Characteristics of Food Industry Wastewater

Food industry cluster consists of different number of industries such as dairy, poultry, cattle meat processing, distillery, brewery, winery, confectionary, sugar, olive oil, and cooking oil. Food industry consumes a higher amount of water per tonne compared to other industries (Abdullh et al. 2016). This industry alone consumes approximately 30% of the total water utilized by manufacturing industries (Meneses et al. 2017). Wastewaters from these industries come from a diverse range of operations consisting of production, cleaning, sanitizing, cooling, transport, etc. Typically, food industry effluent is non-toxic to the environment and aquatic life (Aderibigbe et al. 2017), richer in organic content, with COD and BOD being 10–100 times of domestic wastewater (Heponiemi and Lassi 2012). The quantity, concentration, and composition of wastewater effluent from the food industry depend primarily on the product type being handled, the processing system, operating methods, treatment plant design, extent of water treatment, and the quantity of water collected (Christian 2010). Wastewater characteristics from a particular industry vary significantly and determine the kind of treatment it demands. Food industries are grouped into six major sectors, and their wastewater characteristics are discussed as follows.

9.3.1 *Dairy Industry Wastewater*

A dairy is a milk processing center where the raw milk is processed either for immediate use or is transformed into cheese, whey, butter, ice cream, etc. (Pathwardhan 2008). Dairy industry deals with different products and therefore has various production lines. Depending upon the amount of water used at different steps like production, processing, packaging, cleaning, transport, storage, the volume of wastewater generated may vary significantly (Christian 2010). Table 9.2 shows wastewater characteristics from dairy industries. According to a study, dairy industry produces 1.7 million liters of wastewater per liter of milk (Pathak et al. 2015). Global milk production in 2018 was 843 million tonnes, and the major contributing countries were India, Turkey, the European Union, Pakistan, and USA (Dairy Market Review 2019). This ratio draws attention toward the scope of water recovery and water reuse which can be done through better housekeeping, production, and treatment processes. Even though the environmental impact due to its wastewater is not very severe, dairies must be checked for its organic content (Christian 2010) as it is rich in organic matter. Dairy wastewater contains a high concentration of BOD (0.35–48 g/L), COD (0.43–95 g/L), suspended solids (0.25–580.26 g/L), and nutrients (10^{-5}–1.12 g/L and 0.008–0.5 g/L of total nitrogen and phosphorus, respectively). This wastewater poses a high polluting risk for water bodies if discharged untreated (Ashekuzzaman et al. 2019). A higher level of BOD and COD is due to existence of a greater quantity of proteins, carbohydrates, lipids, and fats in the waste stream from milk and its derivatives. When this protein-rich wastewater is dumped into water bodies, bacteria convert these proteins into ammonium, nitrate, and nitrite. Nitrite in higher concentration is toxic to both human and animals and must be treated before its disposal. The BOD/COD ratio of wastewater gives an idea about the suitable treatment option among aerobic or anaerobic processes. pH is supposed to be acidic mainly due to the formation of lactic acid from lactate but a significant variation in pH 5–9.8 is observed as cleaning water could contain different sterilizing and alkaline detergents. The presence of dissolved salts in wastewater can lead to higher chloride content. Whey is one of the most contaminating constituents of all due to its high level of casein, lactate, nitrogen, nitrates, and odor producing nature (Wang and Serventi 2019). Cattles, poultries, slaughterhouses, and meat industries inject animals with antibodies, pharmaceutical drugs to prevent them from diseases, as a growth stimulus and for better feed efficiency. Regardless of how it is injected, the milk from these cattle has a potential risk of containing antibodies in concentration higher than the permissible limits. These drugs, antibodies cannot be degraded completely in conventional treatment units and further act as secondary pollutants in water bodies where they are dumped and adversely affect aquatic flora and fauna. With increasing urbanization, industrialization, and diminishing water resources, industries are under pressure to plan safe disposal of waste, recycle and reuse water, recover essential nutrients, etc. The use of dairy waste for anaerobic bihydrogen production can be exploited as a source of energy (Tikariha and Sahu 2014). As dairy sludge contains less heavy metal and is richer in biodegradable matter, it can be used both as a soil nutrient and fertilizer for acid soil (Tikariha and Sahu 2014).

9.3.2 Slaughterhouses and Cattle Industry Wastewater

Slaughterhouse is a center where animals are slaughtered, meat is dressed, finished products such as fresh carcass and a variety of meat are produced. This meat producing sector produces different kinds of wastes from various stages such as slaughtering of animals, cleaning, and meat processing units. The cattle meat industry is one of the largest water-consuming industries with worldwide animal production using 2422 cubic gigameters (Gm^3) of water per year and the slaughterhouses alone demanding one-third of this volume (Alfonso-Muniozguren et al. 2018). Studies suggest that meat processing industry consumes approximately 24% of the total fresh water used by food and beverage sector and roughly 29% of that of agricultural consumption globally (Malollari et al. 2019). Table 9.3 shows wastewater characteristics from slaughterhouse and cattle industries. Slaughterhouse waste is an obnoxious waste due to complex wastewater characteristics and foul odor. The slaughterhouse effluent includes a variety of wastes such as manure, blood, hair, dirt, flesh, grease, and offal (Bustillo-Lecompte and Mehrvar 2015). Blood, stomach, and intestinal mucus are the major source of contamination. This wastewater also has pathogenic, non-pathogenic microorganism, disinfectants, and

Table 9.2 Wastewater characteristics from dairy industries

Sr. no.	Name of industry	Wastewater characteristics												
		pH	Conductance	COD (mg/L)	BOD (mg/L)	TDS (mg/L)	TSS (mg/L)	Alkalinity (mg/L)	Chlorides (mg/L)	TN* (mg/L)	TP* (mg/L)	FOG* (mg/L)	Turbidity (NTU)	References
1	Dairy industry	8.34		10251.2	4840.6		5802.6		616	663	153.6			Christian (2010)
		5.5–7.5		1500–3000	350–600	800–1200	250–600						15–30	Sarkar et al. (2006)
		10	2600 mho		1500					0.01	35			Pathak et al. (2015)
		6.1–7.7		4958	9033	180.2–445.4			24.85–92.91	NH_4-2.10–6.50 No_2-0.0375–0.7000, No_3—10.24–15.52	18–26.42		35.9–97.1	Tikariha and Sahu (2014)
		6–11		1150–9200				350–1340			8–68			Aziz et al. (2019)
		6–7.4		430–95,000	1600–48,000					17–1120	10–500			Wang and Serventi (2019)
		9.8		1448	650	1222								Wang and Serventi (2019)
2	Dairy rinse water	6.7 ± 0.24	174 ± 26 µS/cm	2230 ± 141						99.7 ± 13.2	6.5 ± 1.2	575 ± 140	401 ± 28	Briao et al. (2019)

*TN = Total nitrogen, TP = Total phosphorus, FOG = Fat, oil, and grease

Table 9.3 Wastewater characteristics from slaughterhouse and cattle industries

S. no.	Name of industry	Wastewater characteristics											
		pH	Conductance	COD (mg/L)	BOD (mg/L)	TSS (mg/L)	Alkalinity (mg/L)	Chlorides (mg/L)	TN* (mg/L)	TP* (mg/L)	FOG* (mg/L)	Turbidity (NTU)	References
1	Abattoir wastewater			4200–8500	1600–3000	1300–3400	350–800		114–148		100–200		Pathak et al. (2015)
2	Southern meats (Sheep)			3100–11,500	1600–5500	1150–5700	340–700		180–440				Pathak et al. (2015)
3	Meat industry	6.6		3230	1760	9300					960	1200	Rennio et al. (2008)
4	Slaughterhouse industry	5–7.8		1100–15,000	600–3900	220–6400	350–1340		50–840	15–200	40–1385		Aziz et al. (2019)
		6.4–6.6		1900–2200	2000–2400	1300–1700							Wu and Doan (2005)
		4.9–8.10		1250–15,900	610–4635	300–2800			50–841	25–200			Bustillo-Lecompte et al. (2016)
		10.23–10.25	10.17 ± 0.01	25,558 ± 5007	2098 ± 40	1677 ± 244							Almuktar et al. (2018)
		6.3–6.6		2000–6200	1300–2300	850–6300				15–40	40–c600	90–130	Caixetaa et al. (2002)
		8.02		1683.6	863.4	640.2		382.6	2743.6	328.4			Christian (2010)
		7			1500–2500	800							Abdullh et al. (2016)

detergents from cleaning operations (Bustillo-Lecompte and Mehrvar 2015). It has high BOD (0.6–3.9 g/L), COD (1.1–15.9 g/L), suspended solids (0.22–9.3 g/L), color, turbidity, nutrients, heavy metals, etc. With intensification of food-producing animal industry, use of antibiotics has become an integral part of it. According to an estimate, 0.5 million kg of antibiotics is allotted to cattle industry, 1.4 million kg to swine, 1.0 million kg to poultry, and 0.4 million kg to other animals (National Research Council 1980). 30–90% of pharmaceutical provided in animal feed is defecated in different forms and are also found in wastewater of such industries. Such wastewater may contaminate soil, get leached into water bodies, and have a possibility to enter food chain (Wei et al. 2011). Heavy metals like cadmium, lead, arsenic, and zinc can also leach into wastewater through animal manure, fecal content and this can further enter into food chain and may bioaccumulate. Meat industry waste can be utilized for generation of biogas (de Sena et al. 2018).

9.3.3 Poultry Industry Wastewater

Wastewater produced from poultry industry comes from several processes like live poultry droppings, slaughtering, feather and innards removal. High strength, large quantity wastewater of 8–5 L is produced for every bird slaughtered (Rajab et al. 2017). By-product wastewater also contains a large number of pathogenic microorganisms such as Coliforms, Salmonella sp., Staphylococcus sp., and Clostridium sp. (Eryuruk et al. 2018).

Poultry wastewater is generally laden with both organic and inorganic content. It is enriched in nutrients (nitrogen and phosphorus), suspended solids, COD, and BOD. Table 9.4 shows wastewater characteristics from poultry industries. High organic matter is due to a large number of contaminants that includes blood, fat from skin, eviscerated organs, oil used in defeathering boiling water, feathers, dirt and manure. Faeces and bird's urine also contribute to high nitrogen content. Grease content varies from 0.5 to 1 kg per thousand birds (Pathwardhan 2008). A study highlighted the use of ultrafiltration (UF) for water recycling and value-added protein recovery (Avula et al. 2009). As UF recovers most solid by-products and does not leave solid sludge, it can be used as a potential solution to reduce the extent of environmental pollution.

9.3.4 Alcohol Fermentation Industry Wastewater

Distilleries, wineries, and breweries are different types of alcoholic industries that make products through fermentation of various raw materials such as grains, fruits, and vegetables. Table 9.5 shows wastewater characteristics from wineries, distilleries, and breweries. Distilleries and breweries use raw materials like grains, malted barley, and molasses. Wineries use fruit juice particularly grapes as raw

Table 9.4 Wastewater characteristics from poultry industries

S. no.	Name of industry	Wastewater characteristics										
		pH	Conductance	COD (mg/L)	BOD (mg/L)	TDS (mg/L)	TSS (mg/L)	Alkalinity (mg/L)	TN (mg/L)	FOG (mg/L)	Turbidity (NTU)	References
1	Poultry industry	6.5–9		600–1050	500–800		450–800					Abdullh et al. (2016)
		6.8 ± 0.2	795 ± 109 µS/cm	2711 ± 487	930 ± 96	917 ± 135	835 ± 162	160 ± 21	153 ± 32	281 ± 63	>1000	Rajab et al. (2017)
		6.2–7.5		800–1050	275–500		160–280			100–190		Sardari et al. (2018)

Table 9.5 Wastewater characteristics from breweries, wineries, and distilleries

S. no.	Name of industry	Wastewater characteristics								
		pH	COD (mg/L)	BOD (mg/L)	TDS (mg/L)	TSS (mg/L)	Chlorides (mg/L)	TN (mg/L)	TP (mg/L)	References
1	Brewery industry	4.5–12	2000–6000	1200–3600		200–1000		25–80	10–50	Hidalgo and Martin-Marroquín (2019)
			750–80,000	500–64,000		100–3000		1–8	17–216	Vymazal (2014)
		4–8	1700	850		90				Abdullh et al. (2016)
2	Winery industry	3.5–4.1	13,230–23,170	9260–18,360		650–1340		332–415	96.3–176	Pathak et al. (2015)
			500–45,000	500–40,000		1000–17,000		0.001–2	5–77	(Vymazal 2014)
		2.5–11	500–15,000	300–12,000		10–800		1–40	1–40	Hidalgo and Martin-Marroquín (2019)
3	Distillery industry		750–80,000	6000–65,000		1000–17,000		7000	740–2500	Vymazal (2014)
		3–4.5	110,000–190,000	50,000–60,000	90,000–150,000	13,000–15,000	8000–8500	5000–7000	2500–2700	Mohana et al. (2009)
4	Beer production Unit	4.5–12	2000–6000	1200–3500		200–1000		25–80	10–50	Valta et al. (2014)

material (Pathwardhan 2008). As a result, the wastewater is composed of unutilized raw materials, chemicals used in the process, biological agents, and so on. Distillery wastewater is also called as spent wash. According to an estimate, 8–15 L of spent wash is produced per liter of alcohol (Mohana et al. 2009). Typical spent wash characteristics are dark brownish color, acidic pH (3–4.5) due to fermentation, foul odor due to sulfate, high BOD, and COD concentration. Presence of melanoidin formed through millers process gives its wastewater dark brownish color (Bezuneh 2016). Spent wash is also laden with inorganic substances like nitrogen, phosphorus, potassium, sulfate, and calcium. Apart from that, it also contains recalcitrant compounds like melanoidins, caramel, anthocyanins, tannins, etc., which are toxic to microbes that dwell in processing systems (Mohana et al. 2009). High nitrogen and phosphorus content can result in eutrophication of water bodies if discharged untreated. Some value-added products can be extracted from spent wash through incineration of evaporated spent wash (Pathwardhan 2008). Approximately 3.85 tonnes of potash or 575 tonnes of potassium chloride and 1.27 tonnes of potassium sulfate can be recovered from spent wash (Pathwardhan 2008).

Brewery industry wastewater contains residual amount of unprocessed materials including solids, sugary compounds, yeasts, and so on. It has higher COD. A recent study investigated exploitation of brewery waste to produce biogas laden with H_2 and CH_4, where CH_4-rich biogas can further be used in power generation (Arantes et al. 2017). Brewery waste is generally less strong than spent wash from distilleries.

Winery wastewater is mainly contributed through cleaning and washing operations at different processes. Winery effluent contains biodegradable organic content, high electrical conductivity, and varying total suspended solids. Chemical oxygen demand (COD) varies from 0.500 to 45 g/L, biological oxygen demand (BOD) is about 0.4–0.9 of the COD value, recalcitrant constituents comprise about 0.1–5% of total COD, pH varies from 3.5 to 7, and total suspended solids (TSS) from 0.012 to 7.3 g/L (Serrano et al. 2011). If this obnoxious wastewater is released in water bodies untreated, it can adversely affect flora and fauna. High acidity of effluent can affect plant adversely by depriving essential nutrients to them and also by reducing growth of vital soil bacteria (Ioannou et al. 2015).

9.3.5 *Edible Oil Industry Wastewater*

Raw materials like soya, palm, olive, sunflower, and cottonseeds are utilized to produce edible oil. Palm oil has the highest consumption throughout world occupying 37% of the total vegetable oil consumption (Ahmad et al. 2020). Large volume of wastewater is produced which is rich in organic waste of seeds, chaffs, and inorganic matter. Often, wastewater effluent from edible oil industries has high COD, BOD, fats, phosphate, and sulfate. If this effluent is discharged untreated, many environmental problems can arise such as ground and surface pollution,

Table 9.6 Wastewater characteristics from edible oil industries

S. no.	Name of industry	Wastewater characteristics									
		pH	Conductance	COD (mg/L)	BOD (mg/L)	TDS (mg/L)	TSS (mg/L)	TN (mg/L)	TP (mg/L)	FOG (mg/L)	References
1	Olive mill industry	5.30 ± 0.21	20.10 ± 1.55 mscm	178,400 ± 8950	25,600 ± 5280		17,000 ± 3040	975.30 ± 0.12	260 ± 0.15		Hachicha et al. (2009)
2	Palm oil mill industry	4.15–4.45		45,500–65,000	21,500–28,500	15,500–29,000	15,660–23,560	500–800	94–131	1077–7582	Wu et al. (2010)
		3.9		114,800		32,000	44,000				Ahmad et al. (2020)
3	Vegetable oil industry	5.04		640	42.3	72	68.1				Ahmad et al. (2020)
		6.9		933	634	298.8	389.2				Ahmad et al. (2020)
		6.9		676	509	2346					Zajda and Aleksander-Kwaterczak (2019)
4	Oil refinery industry	6.9–10		125–1095			9–93				Aziz et al. (2019)

phytotoxicity, alteration of soil quality, and odor nuisance (Ahmad et al. 2020). Table 9.6 shows typical wastewater characteristics from edible oil industries.

9.3.6 *Confectionary, Fruit Juice & Sugar Industry Wastewater*

Confectionary industry is considered as an important food industry all over the world. It produces sweets including chocolate, candy, chewing gum, bubble gum, that are made primarily of sugar, cocoa, fats, flavors and emulsifiers. The cleaning operation in the confectionary industry is the main process that contributes to wastewater effluent. For this reason, its effluent witnesses daily and seasonal variability in quantity as well as composition. According to a study, confectionary plants release 300–500 m^3 of wastewater per month (Zajda and Aleksander-Kwaterczak 2019). The organic content in the waste stream is due to sugar, fats, dyes, etc. Thus, wastewater has high value of BOD and COD and is predominantly biodegradable. The confectionery sewage often has a discharge from washing and disinfectants which is responsible for varying pH value and increased content of nitrogen and phosphorus compounds.

The sugar industry is a kind of biorefinery where sugarcane or beet is used as raw material and various commodities like raw sugar, molasses, ethanol, bagasse, and electricity are produced. It is an energy as well as a water-intensive process. Sugar industry being seasonal in operation is dependent on the availability of sugarcane or beet in a year. Due to its seasonal nature, it poses a peculiar problem of pollution. A useful exploitation of bagasse is for the generation of electricity. Fruit processing industry is also a seasonal industry. Its wastewater is primarily rich in organic matter, but due to low pH value, poor nutrients, and fluctuating volume, problem in treatment arises. The characteristics of wastewater from confectionary, fruit juice, and sugar industries are summarized in Table 9.7.

9.4 Food Industry Wastewater Treatment Technology

Most wastewater treatment methods may be broadly characterized into (a) physical and chemical treatment processes, (b) biological treatment processes, and (c) novel treatment processes. While one might argue that physical and chemical processes could be subcategorized further, it must be mentioned that physical and chemical processes are often imparted together in order to enhance process efficiency. Physical processes generally involve screens, filter media, membranes, solid beds, etc., for separating solids from flowing wastewater using either gravity or floatation. Chemical processes generally rely on effecting chemical changes in pollutants, like floc formulation, or particle mass enhancement, that might make their removal

Table 9.7 Wastewater characteristics from confectionary, fruit juice, and sugar industries

S. no.	Name of industry	Wastewater characteristics											
		pH	Conductance	COD (mg/L)	BOD (mg/L)	TDS (mg/L)	TSS (mg/L)	Alkalinity (mg/L)	Chlorides (mg/L)	TN (mg/L)	TP (mg/L)	FOG (mg/L)	References
1	Sugar industry	6.61	1.659 m mho	1529.01	910			975	112.465				Lakdawala and Patel (2015)
		7		8170	2600						17.4		Puszczało and Marszałek (2019)
2	Confectionary industry	6.5		3150	1340	473	1037				8.64*10–3	1730	El-Kassas et al. (2015)
		5.05 ± 0.35	22 ± 1.70 ms/cm	100,580 ± 6250	24,300 ± 4650		8100 ± 2700			66.14 ± 0.35	350.50 ± 0.20		Hachicha et al. (2009)
3	Candy manufacturing			2500	8000								Zajda and Aleksander-Kwaterczak (2019)
4	Chocolate manufacturing	7.4	750 µS/cm	3608									Zajda and Aleksander-Kwaterczak (2019)
5	Sour cherry juice wastewater			1000–8000						3.3–55	0.104–10		Ozbas et al. (2006)
6	Apple juice wastewater			124.8						5694	49.14		Ozbas et al. (2006)

easier. Biological treatment processes use biomass (microorganisms, plants, etc.) that either degrade organic effluents or convert them into other less problematic products. Rapid research and development in processing technologies has also led to emergence of various novel wastewater treatment approaches which may not be categorized into either of the above categories. Most food industries do not rely on a single method or principle and incorporate a plethora of technologies for treating their wastewater before disposal (Qasim and Mane 2013). Current practice often involves application of process engineering fundamentals to develop a site-specific process configuration for each food plant.

9.4.1 Physical and Chemical Processes

9.4.1.1 Coagulation–Flocculation-Based Processes

Coagulation–flocculation technology is a simple, economic, and eco-friendly process based on coagulants such as alum, ferric chloride, lignosulfonate (LSA), bentonite (BEN), carboxymethylcellulose (CMC), and chitosan to precipitate suspended matter and dissolved organic compounds in wastewater (Xu et al. 2001), followed by their removal due to orthokinetic flocculation produced due to velocity gradients. Coagulation technology for treatment has been reported as early as 2000 BC in the Egyptian civilization, wherein almond smears were used to clarify river water. Around the same time, Sanskrit literature suggests using crushed nuts from Nirmali tree (Struchnos potatorum) for treating water (Deshmukh et al. 2013). Coagulation is also of importance in other wastewater treatment approaches, like chemical phosphorus removal method. Qasim and Mane (Qasim and Mane 2013) observed that turbidity, hardness, COD, electrical conductivity decrease on treating wastewater from dairy, sweets, snacks, and ice cream industries, respectively, with alum. Rivas et al. report that cheese whey wastewater (CWW) treated with $FeSO_4$, $FeCl_3$, and alum resulted in reduction in COD, BOD, total Kjeldahl nitrogen and phosphorus. Capponi et al. (2006) developed a modified column floatation method, wherein a colloidal precipitate (carrier, ferric hydroxide) diffused counter-current into a column with rising bubbles inside industry wastewater, followed by separation of the flocs (coagula). The effect and impact of the coagulation–flocculation technique is often impacted by various properties including coagulant types, properties, dosage, and pH (Wang and Serventi 2019). 57% reduction of COD was observed on using chitosan (at 0.01–0.05 g/L concentration) in dairy industry wastewater by Sarkar et al. (Ioannou et al. 2015). Gunes et al. (2018) compared coagulation–flocculation-based processes with Fenton-like processes (discussed later) for industrial drum cleaning wastewater and found coagulation–flocculation-based processes (88% COD removal) to outperform coagulation–oxidation-based processes (68% COD removal). Kurup et al. found that sodium lignosulfonate could not only cheaply reduce BOD, but also aid in proteins and lipid recovery from dairy wastewater due to their ability to form complexes with cationic colloidal particles at low pH (Kurup et al. 2019).

9.4.1.2 Adsorption-Based Processes

Adsorption-based wastewater treatment techniques rely on collection and adherence of dissolved substances and effluents on to an adsorbent (solid) surface. For adsorption to occur, attractive forces on the surface of the adsorbing material must be greater than the intramolecular forces of the substance being adsorbed (Qasim and Mane 2013). Activated carbon is widely used adsorbent material for industry wastewater. Amuda and Ibrahim (Amuda and Ibrahim 2006) reported using various activated carbon from coconut shell (acid-activated carbon, barium chloride-activated coconut, and commercial-activated carbon) to treat beverage industry wastewater and found acid-activated carbon to achieve maximum (92%) COD removal. Qasim and Mane (Qasim and Mane 2013) also used activated carbon and reported in COD and TDS upon treatment of dairy, sweet, and ice cream industries wastewater with activated carbon reduction. The choice of the adsorbent material has been a matter of research in recent years. Often nowadays, this choice is also governed by other sustainability principles. Thereby, one of the notable aspects has been the emergence of cheap agricultural by-products as attractive adsorption materials in recent literature (Adenuga et al. 2019). Adenuga et al. (2019) used *Calophyllum inophyllumi* spent seedcake to remove up to 70% of metals (Pb, Cd, and Zn). Murali et al. (Murali et al. 2013) used water hyacinth to reduce the COD of dairy industry by up to 90%. Kulkarni et al. (Kulkarni et al. 2011) used bagasse fly ash, a by-product from sugar industry, for treating distillery wastewater resulting in more than 85% COD removal. Umukoro et al. used peel charcoal for removal of copper and zinc ions for industrial wastewaters. Moringa pods have been used by Matouq et al. (2015) to remove more than 90% chromium and copper ions from model wastewater. More recently, nanomaterials such as carbon nanotubes are being developed and employed for treating industrial wastewater (Xu et al. 2018; Siddiqui et al. 2019). In addition to this, adsorption is one of the most common membrane-based method used industrially. Moreover, it is effective, efficient, and reliable. Activated carbon, alumina, and silica are the most common sources that serve as reliable, effective, and efficient adsorbents for wastewater treatment (Crini 2005). It also acts as an environment protectants due to its natural characteristic (Ciesielski et al. 2003). Besides zeolites are often used for wastewater purification and filtration. Zeolites contain a porous structure filled with alkaline cations, which renders them with high cationic exchange activity and good sieving properties. As they are widely present in nature, using zeolites is also considered very cost-effective. Due to its high cation exchange ability and sieve property of molecule, it is utilized in separation and purification of wastewater (Wang and Peng 2010).

Table 9.8 Comparison among various treatment methods of wastewater treatment

S. no.	Treatments	Influent impurities removal (%)				
		COD	BOD	Color	Salinity	References
1.	*Advance oxidation process*					
a.	Electrochemical coagulation	90	57	87	–	Xu et al. (2001), Ghanbari and Moradi (2015), and Bazrafshan et al. (2015)
b.	Fenton-based processes	87	65	84	47	Mandal et al. (2010) and Šostar-Turk et al. (2005)
c.	Ozonation	59.9		82	–	Roa-Morales et al. (2007)
d.	Photocatalysis	63	43	98	–	
2.	*Biological treatments (bioremediation/biodegradation)*					
a.	Anaerobic	34–47	15	26	25	Mandal et al. (2010) and Castillo et al. 1999)
b.	Aerobic	52–56	–	–	–	Castillo et al. (1999)
3.	*Emerging food industry wastewater treatments*					
a.	Biofilm technology	86	–	94	–	Kesaano and Sims (2014) and Yun et al. (2006)
b.	Membrane filtration	80	70	–	–	Šostar-Turk et al. (2005)

9.4.1.3 Advanced Oxidation Processes

The organic waste of contaminated water from food and other industries usually contains substances including polyphenols, tannins, and lignins, that are difficult to degrade or break down easily, therefore advanced oxidation process has obtained popularization in the field of wastewater treatment (Mosteo et al. 2008). It primarily deals with the formation of hydroxyl radicals (OH) using chemical, photocatalytic, and photochemical reactors, wherein hazardous waste is degraded into simpler and non-toxic particles. In addition to this, biorecalcitrant can also be degraded by using strong oxidizing agent. During the process, free radicals (OH) react with organic molecules probably by hydrogen abstraction, direct electron transfer or by mineralization process, resulting in the production of water, inorganic acids, and CO_2 (Legrini et al. 1993). Principally, various methods including ozonation, sonication, electrochemical oxidation, ultraviolet (UV) photolysis, and Fenton-based are followed during this treatment. Refractory organic pollutants rely extremely on advance oxidation processes (Jamil et al. 2011; Oller et al. 2011). Table 9.8 shows the comparison among various treatment methods of wastewater treatment.

Electrochemical Coagulation

Electrochemical coagulation implies development of OH at the active portion of anode, which helps in removal of organic contaminations from the wastewater (Rizzo 2011). In this process, various coagulants are used including lignosulfonate (LSA), bentonite (BEN), carboxymethylcellulose, and ferric chloride (FeCl), to get back the improved wastewater that is discharged from food processing industries as shown in Fig. 9.2. Xu et al. (2001) reported that in egg processing plant, there was an improvement in COD, TS, and turbidity of the water by 90%, 95%, and 97%, respectively. Similarly when Fe and Al electrode were used for electrocoagulation to treat wastewater discharged from slaughterhouse, COD removal was observed to be 94.4% and 81.1%, respectively. Roa-morales et al. (Roa-Morales et al. 2007) observed that the application of aluminum electrocoagulation with hydrogen peroxide minimized the levels of COD, TS, and BOD by 90%, 95%, and 96%, respectively, in wastewater from confectionary industries at pH 4. Likewise, Sangeetha et al. (Sangeetha et al. 2014) reported that COD level in wastewater collected from sago industries was limited to 67.86% at 100 mg/L of alum (coagulant) dosage. However, huge amount of electricity is consumed during this treatment and is extremely expensive for industries. Maithlo et al. (2019) clearly indicated a substitute solution as a metal-air fuel cell electrocoagulation (MAFCEC) which can be useful for removing arsenic and other pollutants from wastewater with minimal energy consumption. Heavy metal ions (Al-Qodah and Al-Shannag 2017), oil (An et al. 2017) can also be removed from wastewater by practicing this treatment. Chellam and sari (Chellam and Sari 2016) reported that aluminum electrocoagulation is used to control fouling smell and virus in waste water.

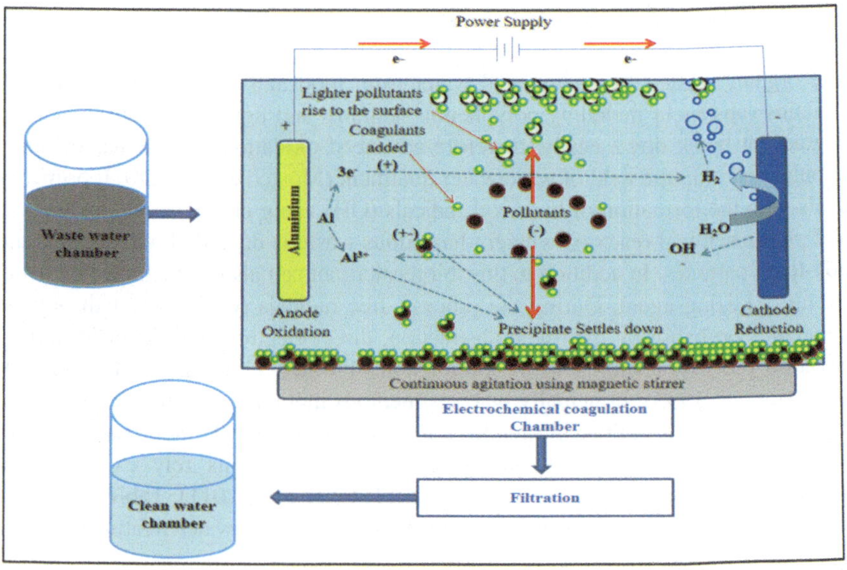

Fig. 9.2 Working principle of electrochemical coagulation

Fenton-Based Processes

In this process, oxidation takes place between ions of ferrous and hydrogen peroxide commonly known as Fentons's reagent (Gogate and Pandiit 2004). The working principle of Fenton-based wastewater treatments is demonstrated in Fig. 9.3 that includes processes like acid regulation, catalyst mixing, oxidation reaction, neutralization, and solid–liquid separation. This process is known to oxidize organic compounds at ambient temperature (Bigda 1995) and break them into environmental-friendly components such as H_2O_2 into H_2O and O_2, parhaps its stringent pH range, high H_2O_2 consumption as well as increasing ferric sludge adversely affect the oxidation process (Yuan et al. 2013; Wang et al. 2016). To overcome these challenges, catalysts such as heterogeneous and homogeneous were taken into consideration to substitute Fe^{2+}, including ferric oxide (Matta et al. 2007), iron minerals (Shinya and Bergwall 2007), and nanozero-valent iron (Babuponnusami and Muthukumar 2012). Research survey indicates that Fento-based processes are normally used for food processing industries like baker's yeast industry, where COD level was reduced to 88% after 20 min at pH 4 (Pala and Erden 2005). Besides, Lee and Shoda (2008) reported that Fenton treatment was useful in reduction of color and COD level (95.4% and 88%, respectively) in wastewater collected from industries such as livestock. Similarly, 80% COD level was reduced with 85% of total phenol elimination using Fenton's reagents against wastewater collected from olive oil mill (Martins and Quinta-Ferreira 2010).

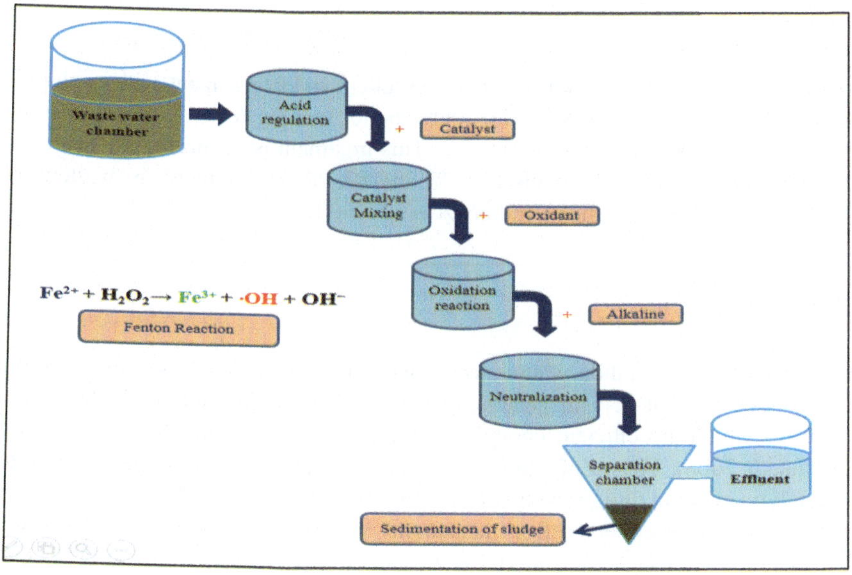

Fig. 9.3 Schematic representation of Fenton-based wastewater treatment

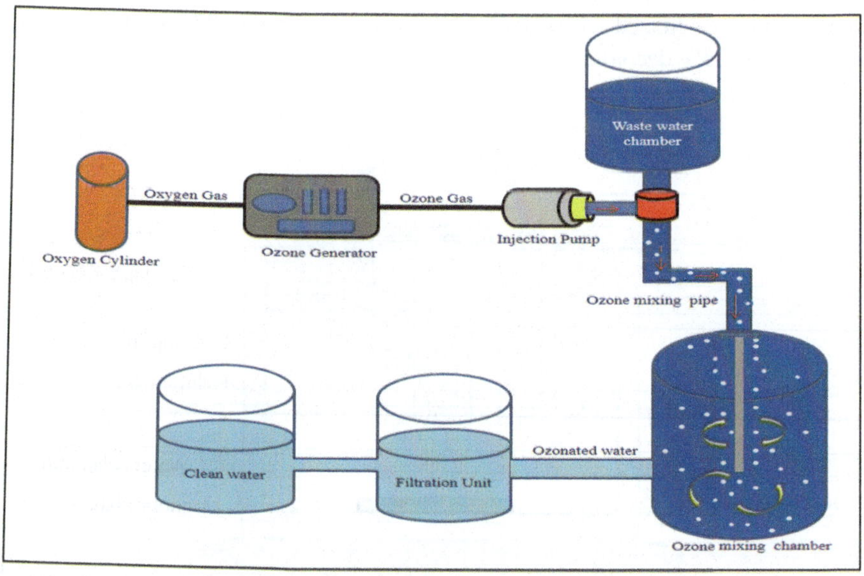

Fig. 9.4 Schematic illustration of ozone-based wastewater treatment

Ozonation

Ozone (O_3) is universally known as an excellent oxidizing agent with negligible level of toxic residue. O_3 reacts with double bonds (conjugated) which are generally related with colors as shown in Fig. 9.4. This treatment is applicable for reduction of COD and polyphenols levels (59.8% and 82.4%, respectively) in wastewater collected from olive oil mill (Andreozzi et al. 2008).

Photocatalysis

Photocatalysis is a method where electromagnetic radiations are used to excite the semiconductors that help in generating valence band holes and conduction band electrons which are efficient enough to eradicate contamination from wastewater (Stasinakis 2008). TiO_2, ZnO, SnO, WO_3, ZrO_2, CeO, and YO_3 are generally used as catalytic material during the treatment (Fig. 9.5). Kang et al. (2011) reported that greasy wastewater collected from restaurants were taken for photocatalysis using UV radiation with TiO_2 (150 mg/L) as a catalytic material for 10 min at pH 7. Oil, COD, and BOD level of the treated water was reduced to 70%, 63%, and 43%, respectively. Additionally, when wastewater discharged from winery was treated with TiO_2/H_2O_2 nearly 84% of COD level was reduced (Agustina et al. 2008). Further, Qin et al. (2011) reported that the wastewater collected from molasses fermentation was treated with calcined YO_3 which resulted in 92.98% reduction of COD and 98.23% decolorization of the water.

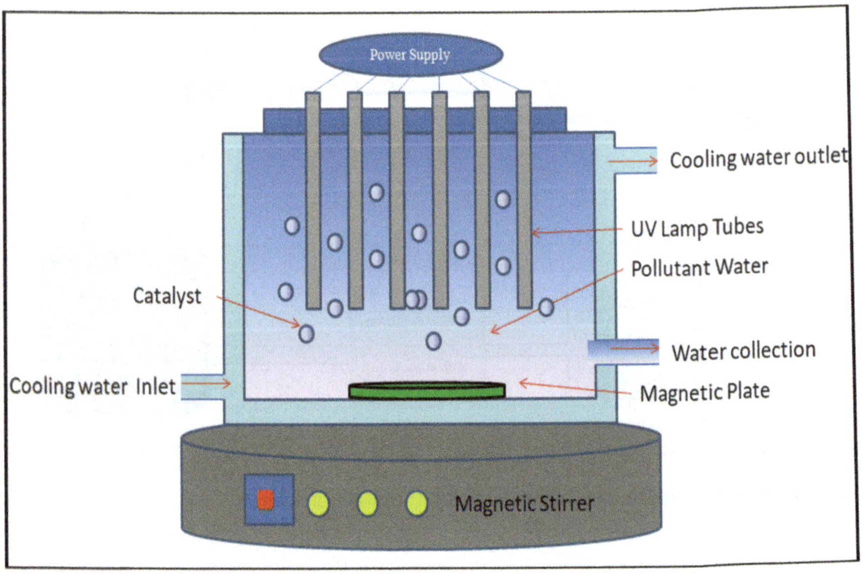

Fig. 9.5 Schematic diagram of photocatalysis

Hybrid Processes

Hybrid processes are generally utilized in limiting cost, increasing efficiency, and minimizing the reaction time. Péerez et al. (2007) collected wastewater from coffee and treated it with the combination of two different methods including ozonation and photocatalysis ($UV/H_2O_2/O_3$). The results indicated reduction of COD level by 87% (35 min/ pH 2.0). Similarly in 2009, De Sena et al. (2009) reported that wastewater collected from meat industry when treated with the combination of photo-Fento (DAF/UV/H2O2) treatment, resulted in dropping of BOD_5, COD, TS, and VS levels by 82.9%, 91.1%, 72.5%, and 77.0%, respectively, whereas wastewater collected from olive mill subjected to photo-ozonation treatment (UV/O_3), resulted in 91% reduction of COD (Lafi et al. 2009).

9.4.2 *Biological Treatments (Bioremediation/ Biodegradation)*

Presently, plethora of chemical and physicochemical practices are followed to eradicate pollutants from the industrial wastewater. Nevertheless total removal of the pollutants was yet not achieved and due to the use of excessive chemicals, new toxic substances (by-products) were produced which further contaminated soil, water, and other natural sources (Rana et al. 2014). Moreover, these treatments require chemicals which are expensive for the industries to manage, whereas biological treatment methods include plants and microorganisms which are cost-effective and environmental-friendly. In addition to this, biological treatments do not produce any toxic by-products that may be hazardous to environment. Microorganisms have a great application in breaking down of complex substances that are difficult to degrade (Fig. 9.6). *Eichhornia crassipes and Panicum maximum* were used as a spillage to degrade pollutants discharged from food processing industries. Results reflected a remarkable decline in TDS (3.1–0.351 g/L), COD (22.5–0.150 g/L), BOD_5 (20.3–0.123 g/L), SS (0.1–0.008 g/L), phosphate (0.101 from 0.0024 g/L), and nitrates (0.012–0.008 g/L) levels (Mosteo et al. 2008).

9.4.2.1 Anaerobic Treatments

Anaerobic treatment can be elaborated as a practice wherein microorganisms decay organic matter into CO_2 and CH_4. During this treatment, pollutants are reduced into smaller particles and biogas is produced that can further be used for the production of energy. Energy consumed during this treatment is very low but high amount of pollutants is degraded with minimum sludge (Nandy et al. 2002). Ganesh et al.

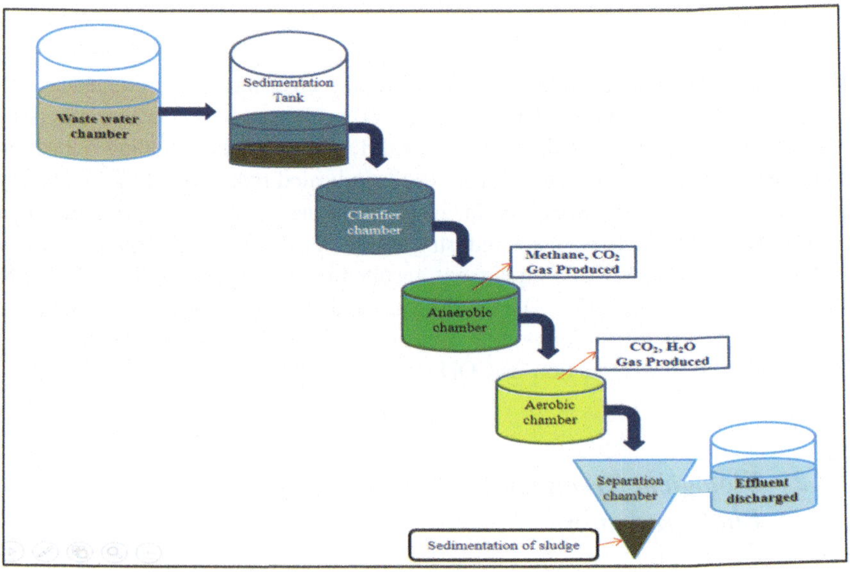

Fig. 9.6 Schematic illustration of biological treatments (anaerobic–aerobic treatments for influents)

(2010) reported that wastewater collected from winery industry was anaerobical treated and 80% of COD level was reduced.

9.4.2.2 Aerobic Treatments

Aerobic treatments usually depend on oxygen to break down organic substances into simpler form. 89% of COD was reduced during aerobic treatment in cheese wastewater (Monier and Abdel-Latif 2013). Due to the formation of excessive sludge, its utilization is limited. Table 9.9 shows comparison between anaerobic and aerobic wastewater treatments.

Table 9.9 Comparison between anaerobic and aerobic waste water treatments (Mandal et al. 2010)

Parameters	Aerobic	Anaerobic
Efficiency of organic removal	High	High
Treated water quality	Excellent	Moderate to poor
Sludge production	High	Low
Energy consumed	High	Low to moderate
Odor	Less	More

9.4.3 *Emerging Food Industry Wastewater Treatments*

9.4.3.1 Biofilm Technology

It is a cluster of microorganisms where each cell is attached to another. Due to its sticky nature, it gets attached to the surface very easily. These cells which act as support get embedded on a matrix produced from extracellular polymeric substances (EPS). This biofilm helps in destroying pathogens and plethora of other nutrients including nitrogen-containing compounds, phosphorus (P) and carbonaceous substances present in wastewater. After the treatment of biofilms, water is fit to be used for agricultural purposes as illustrated in Fig. 9.7. Additionally, this method is environmental-friendly and consumes less space with high active biomass concentration (Wilderer et al. 2015).

9.4.3.2 Membrane Filtration

Membrane separation method is widely used to clean wastewater collected from various food industries. Generally it acts on the principal of spraying out the pollutants and other heavy metals from water. As shown in Fig. 9.8, there are five different types of membrane filtration method practised such as ultrafiltration (UF), microfiltration (MF), nanofiltration (NF), reverse osmosis (RO), and electrodialysis (ED) (Sarode et al. 2018).

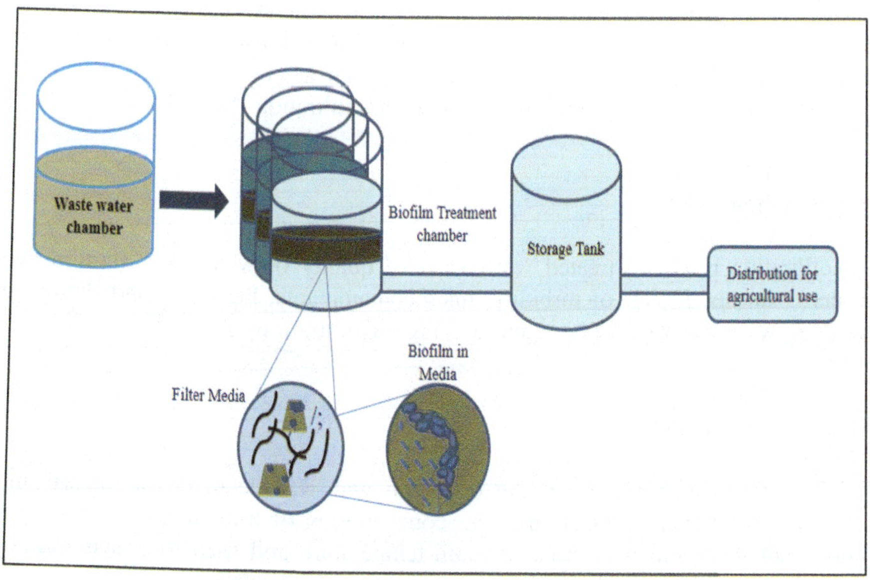

Fig. 9.7 Schematic illustration of biofilm treatments for influents

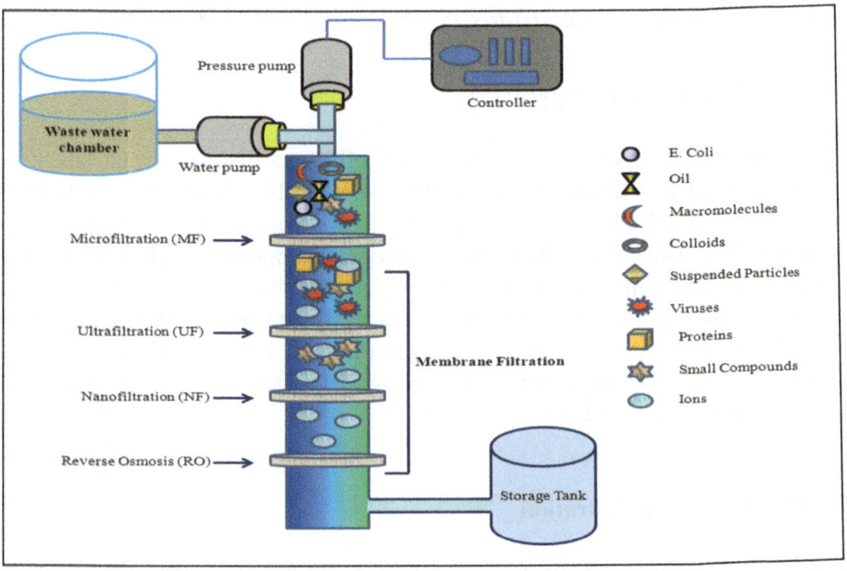

Fig. 9.8 Schematic illustration of membrane filtration for influents

Ultrafiltration (UF)

In order to remove colloidal and dissolved material, transmembrane pressure (TMP) is mostly used. UF is available in two types including polymer-enhanced and micellar-enhanced configurations of ultrasonication. All the pollutants related to metal ions, organic and colored dyes are removed using micellar-enhanced ultrafiltration (MEUF) while minor particles which are not possible for conventional UF are removed by polymer-enhanced ultrafiltration (PEUF).

Nanofiltration (NF)

Nanofiltration produces treated wastewater of quality between ultrafiltration and reverse osmosis. Levels of turbidity, microcontaminants, hardness, and dissolved solids in wastewater could be reduced using nanofiltration.

Reverse Osmosis (RO)

RO is a chemical engineering unit operation with a wide range of known and potential commercial applications, e.g., concentration of fruit juices. It is a separation technique, which operates at room temperature and used to concentrate or purify liquids without a phase change leading to energy efficiency.

9.5 Conclusions and Future Trends

In this book chapter, different types of food industries, their typical wastewater characteristics with scope of resource recovery and reuse are discussed briefly. The discharge from these industries is rich in nutrients, carbonaceous, and nitrogenous matter which can further cause problems like eutrophication, algal bloom, and interference with aquatic life, which when dumped untreated. Therefore, various types of treatments are practised to purify wastewater that are mostly liberated from plethora of industries. This industry being water intensive can be further explored for better options of water reuse, water recovery before sending its wastewater for treatment. As water is extremely important for the survival of all living beings on this planet, reuse of treated wastewater is the most suitable route to counter water-related environmental challenges. A wide range of treatment technologies both conventional and emerging ones was also discussed. With the growing population, environmental pollution, and depleting natural resources, need of the hour is to look for more sustainable treatment solutions with smaller carbon footprint and lesser waste disposal. Inhouse wastewater treatment of industries can also be used for reduced time and less overall cost. The use of emerging technologies including membrane filtration and biofilm technology can lead to the better future in the field of wastewater treatment as it is capable of removing impurities. The development of life-long and energy-saving membranes can decrease the operating cost and therefore reduce the overall investment in treatment plants. Besides the presence of these exquisite technologies and future developments to treat wastewater, the industries require to reduce the generation of wastewater by adapting new technologies and improving awareness in conservation of water resources.

References

Abdullh MN, Abdelhalim WS, Abdelhalim HS (2016) Industrial wastewater treatment of food industry using best techniques. Int J Eng Sci Inven 5(8):15–28

Adenuga AA, Amos OD, Oyekunle JOO, Umukoro EH (2019) Adsorption performance and mechanism of a low-cost biosorbent from spent seedcake of calophyllum inophyllum in simultaneous cleanup of potentially toxic metals from industrial wastewater. J Environ Chem Eng. https://doi.org/10.1016/j.jece.2019.103317

Aderibigbe DO, Giwa AA, Bello IA (2017) Characterization and treatment of wastewater from food processing industry: A review. Imam J Appl Sci 2:27–36

Agustina TE, Ang HM, Pareek VK (2008) Treatment of winery wastewater using a photocatalytic/photolytic reactor. Chem Eng J 135:151–156

Ahmad T, Belwal T, Li L, Ramola S, Aadil RM, Abdullah Xu Y, Zisheng L (2020) Utilization of wastewater from edible oil industry, turning waste into valuable products: A review. Trends in Food Science & Technology. https://doi.org/10.1016/j.tifs.2020.02.017

Alfonso-Muniozguren P, Lee J, Bussemaker M, Chadeesingh R, Jones C, Oakley D, Saroj D (2018) A combined activated sludge-filtration-ozonation process for abattoir wastewater treatment. J Water Process Eng 25:157–163. https://doi.org/10.1016/j.jwpe.2018.07.009

Almuktar SAAAN, Abed SN, Scholz M (2018) Wetlands for wastewater treatment and subsequent recycling of treated effluent: a review. Environ Sci Pollut Res Int 25(24):23595–23623. https://doi.org/10.1016/j.pce.2019.07.005

Al-Qodah Z, Al-Shannag M (2017) Heavy metal ions removal from wastewater using electrocoagulation processes: A comprehensive review. Separation Science and Technolog 52(17):2649–2676

Amuda OS, Ibrahim AO (2006) Industrial wastewater treatment using natural material as adsorbent. Afr J Biotech 5:1483–1487

An C, Huang G, Yao Y, Zhao S (2017) Emerging usage of electrocoagulation technology for oil removal from wastewater: A review. Sci Total Environ 579:537–556

Andreozzi R, Canterino M, Di Somma I, Lo Giudice R, Marotta R, Pinto G, Pollio A (2008) Effect of combined physico-chemical processes on the phytotoxicity of olive mill wastewaters. Water Res 42:1684–1692

Arantes MK, Alves HJ, Sequinel R, Silva EA (2017) Treatment of brewery wastewater and its use for biological production of methane and hydrogen. Int J Hydrogen Energy. https://doi.org/10.1016/j.ijhydene.2017.08.206

Ashekuzzaman SM, Forrestal PJ, Richards KG, Fenton O (2019) Dairy industry derived wastewater treatment sludge: Generation, type and characterization of nutrients and metals for agricultural reuse. J Clean Prod 203:1266–1275. https://doi.org/10.1016/j.jclepro.2019.05.025

Avula RY, Nelson HM, Singh R (2009) Recycling of poultry process wastewater by ultrafiltration. Innov Food Sci & Emerg Technol 10:1–8. https://doi.org/10.1016/j.ifset.2008.08.005

Aziz A, Basheer F, Sengar A, Irfanullah Khan, SF, Farooqi IQ (2019) Biological wastewater treatment (anaerobic-aerobic) technologies for safe discharge of treated slaughterhouse and meat processing wastewater. Sci Total Environ 686:681–708. https://doi.org/10.1016/j.scitotenv.2019.05.295

Babuponnusami A, Muthukumar K (2012) Removal of phenol by heterogeneous photo electro Fenton-like process using nano-zero valent iron. Sep Purif Technol 98:130–135

Bazrafshan E, Alipour MR, Mahvi AH (2015) Textile wastewater treatment by application of combined chemical coagulation, electrocoagulation, and adsorption processes. Desalination and Water Treatment 57(20):9203–9215

Bezuneh TT (2016) The role of Microorganisms in Distillery Wastewater Treatment: Review. J Bioremediat Biodegrad 7:375. https://doi.org/10.4172/2155-6199.1000375

Bigda RJ (1995) Consider Fenton's chemistry for wastewater treatment. Chem Eng Prog 91:62–66

Briao VB, Salla AC, Miorando T, Hemkemeier M (2019) Water recovery from dairy rinse water by reverse osmosis: Giving value to water and milk solids. Resour, Conserv Recycl 140:313–323. https://doi.org/10.1016/j.resconrec.2018

Bustillo-Lecompte CF, Mehrvar M (2015) Slaughterhouse wastewater characteristics, treatment, and management in the meat processing industry: A review on trends and advances. Environ Manage 8:153–174. https://doi.org/10.1016/j.jenvman.2015.07.008

Bustillo-Lecompte C, Mehrvar M, Quiñones-Bolaños E (2016) Slaughterhouse wastewater characterization and treatment: An economic and public health necessity of the meat processing industry in ontario, Canada. J Geosci Environ Prot 4:175–186. https://doi.org/10.4236/gep.2016.44021

Caixetaa CET, Cammarotab CM, Xavier AM (2002) Slaughterhouse wastewater treatment: evaluation of a new three-phase separation system in a UASB reactor. Biores Technol 81:61–69. https://doi.org/10.1016/s0960-8524(01)00070-0

Capponi F, Sartori M, Souza ML, Rubio J (2006) Modified column flotation of adsorbing iron hydroxide colloidal precipitates. Int J Miner Process 79:167–173. https://doi.org/10.1016/j.minpro.2006.02.002

Castillo A, Llabres P, Mata-Alvarez J (1999) A kinetic study of a combined anaerobic–aerobic system for treatment of domestic sewage. Water Res 33:1742–1747

Chellam S, Sari MA (2016) Aluminum electrocoagulation as pretreatment during microfiltration of surface water containing NOM: A review of fouling, NOM, DBP, and virus control. J Hazardous Materials 304:490–501

Christian O (2010) Characteristics of the untreated wastewater produced by food industry. Fascicula: Protecția Mediului, XV:709-714

Ciesielski W, Lii C-Y, Yen M-T, Tomasik P (2003) Interactions of starch with salts of metals from the transition groups. Carbohyd Polym 51(1):47–56

Crini G (2005) Recent developments in polysaccharide-based materials used as adsorbents in wastewater treatment. Prog Polym Sci 30(1):38–70

Dairy Market Review. Food and Agricultural Organisation-FAO, Rome (2019)

de Sena RF, Claudino A, Moretti K, Bonfanti IC, Moreira RF, Jose HJ (2008) Biofuel application of biomass obtained from a meat industry wastewater plant through the flotation process—A case study. Resources, Conservation and Recycling 52:557–569. https://doi.org/10.1016/j.resconrec.2007.07.002

De Sena FR, Tambosi JL, Geneva AK, Moreira R, Schrider HF, Jose HJ (2009) Treatment of meat industry wastewater using dissolved air floatation and advanced oxidation processes monitored by GC-MS and LC-MS. Chem Eng J 152:151–157

de Sena RF, Claudino A, Moretti K, Bonfanti ÍCP, Moreira RFPM, José HJ (2018) Biofuel application of biomass obtained from a meat industry wastewater plant through the flotation process—a case study. Resour Conserv Recycl 52:557–569. https://doi.org/10.1016/j.resconrec.2007.07.002

Deshmukh DB, Pimpalkar S, Rakhunde R, Joshi D (2013) Evaluation Performance of Natural Strychnos Potatorum over the Synthetic Coagulant Alum, for the Treatment of Turbid Water. Int J Innov Res Sci Eng Technol 2:6183–6189

El-Kassas HY, Heneash AMM, Hussein NR (2015) Cultivation of Arthrospira (Spirulina) platensis using confectionary wastes for aquaculture feeding. J Genet Eng Biotechnol 13:145–155. http://dx.doi.org/10.1016/j.jgeb.2015.08.003

Eryuruk K, Tezcan Un U, Bakır Ogutveren U (2018) Electrochemical treatment of wastewaters from poultry slaughtering and processing by using iron electrodes. J Clean Prod 172:1089–1095. https://doi.org/10.1016/j.jclepro.2017

Ganesh R, Rajinikanth R, Thanikal JV, Ramanujam RA, Torrijos M (2010) Anaerobic treatment of winery wastewater in fixed bed reactors. Bioprocess Biosyst Eng 33:619–628

Ghanbari F, Moradi M (2015) A comparative study of electrocoagulation, electrochemical Fenton, electro-Fenton and peroxi-coagulation for decolorization of real textile wastewater: Electrical energy consumption and biodegradability improvement. J Environ Chem Eng 3(1):499–506

Gogate PR, Pandiit AB (2004) A review of imperative technologies for wastewater treatment II: hybrid methods. J Adv Environ Res 8:501–551

Güneş E, Demir E, Güneş Y, Hanedar A (2018) Characterization and treatment alternatives of industrial container and drum cleaning wastewater: Comparison of Fenton-like process and combined coagulation/oxidation processes. Separation and Purification Technology, 209. https://doi.org/10.1016/j.seppur.2018.07.060

Hachicha S, Sellami F, Cegarra J, Hachicha R, Drira N, Medhioub K, Ammar E (2009) Biological activity during co-composting of sludge issued from the OMW evaporation ponds with poultry manure—Physico-chemical characterization of the processed organic matter. J Hazard Mater 162(1):402–409. https://doi.org/10.1016/j.jhazmat.2008.05.053

Heponiemi A, Lassi U (2012) Advanced oxidation processes in food industry wastewater treatment—A review

Hidalgo D, Martin-Marroquín JM (2019) Adding sustainability to the beverage industry through nature-based wastewater treatment. Environ Sci. https://doi.org/10.1016/b978-0-12-815259-1.00001-x

https://www.env.go.jp/earth/coop/coop/document/male2_e/007.pdf. Accessed on 20 May 2020

https://www.wwdmag.com/industrial-water-wastes-digest/water-wastewatertreatment-food-industry. Accessed on 20 May 2020

Ioannou LA, Puma GL, Fatta-Kassinos D (2015) Treatment of winery wastewater by physicochemical, biological and advanced processes: A review. J Hazard Mater 286:343–368. https://doi.org/10.1016/j.jhazmat.2014.12.043

Jamil TS, Ghaly MY, El-Seesy IE, Souaya ER, Nasr RA (2011) A comparative study among different photochemical oxidation processes to enhance the biodegradability of paper mill wastewater. J Hazard Mater 185:353–358

Kang JX, Lu L, Zhan W, Li B, Li DS, Ren YZ, Liu DQ (2011) Photocatalytic pretreatment of oily wastewater from the restaurant by a vacuum ultraviolet/TiO_2 system. J Hazard Mater 186:849–854

Kesaano M, Sims RC (2014) Algal biofilm based technology for wastewater treatment. Algal Res 5:231–240

Kulkarni SJ, Patil SV, Bhalerao YP (2011) Flyash adsorption studies for organic matter removal accompanying increase in dissolved oxygen. Int J Chem Eng Appl 2:434–438

Kurup GG, Adhikari B, Zisu B (2019) Recovery of proteins and lipids from dairy wastewater using food grade sodium lignosulphonate. Water Resources and Industry 22:100–114

Lafi WK, Shannak B, Al-Shannag M, Al-Anber Z, Al-Hasan M (2009) Treatment of olive mill wastewater by combined advanced oxidation and biodegradation. Sep Purif Technol 70:141–146

Lakdawala MM, Patel YS (2015) Studies on adsorption capacity of zeolite for removal of chemical and biochemical oxygen demands. Chemistry Journal 1(4):139–143

Lee H, Shoda M (2008) Removal of COD and color from livestock wastewater by the Fenton method. J Hazard Mater 153:1314–1319

Legrini O, Oliveros E, Braun AM (1993) Photochemical processes for water—treatment. Chem Rev 93:671–698

Maitlo HA, Kim JH, Kim K-H, Park JY, Khan A (2019) Metal-air fuel cell electrocoagulation techniques for the treatment of arsenic in water. J Cleaner Production 207:67–84

Malollari I, Pinguli L, Buzo R, Lajqi V, Makolli S, Cani XH (2019) Actual situation of wastewater from food industry and a case study of their treatment. J Environ Prot Ecol 20(1):432–438

Mandal T, Maity S, Dasgupta D, Datta S (2010) Advanced oxidation process and biotreatment: Their roles in combined industrial wastewater treatment. Desalination 250(1):87–94

Martins RC, Quinta-Ferreira RM (2010) Final remediation of post—Biological treated milk whey wastewater by ozone. Int J Chem React Eng 8:A142

Matouq M, Jildeh N, Qtaishat M, Hindiyeh M, Al Syouf MQ (2015) The adsorption kinetics and modeling for heavy metals removal from wastewater by moringa pods. J Environ Chem Eng 3:775–784. https://doi.org/10.1016/j.jece.2015.03.027

Matta R, Hanna K, Chiron S (2007) Fenton-like oxidation of 2,4,6-trinitrotoluene using different iron minerals. Sci Total Environment 385:242–251

Meneses YE, Stratton J, Flores RA (2017) Water reconditioning and reuse in the food processing industry: Current situation and challenges. Trends Food Sci Technol 61:72–79

Mohana S, Acharya BK, Madamwar D (2009) Distillery spent wash: Treatment technologies and potential applications. J Hazard Mater 163:12–25. https://doi.org/10.1016/j.jhazmat.2008.06.079

Monier M, Abdel-Latif D (2013) Modification and characterization of PET fibers for fast removal of Hg (II), Cu (II) and Co (II) metal ions from aqueous solutions. J hazardous materials 250:122–130

Mosteo R, Sarasa J, Ormad MP, Ovelleiro J (2008) Sequential solar photo-Fenton-biological system for the treatment of winery wastewaters. J Agric Food Chem 56(16):7333–7338

Munderere E, Munyaneza O, Wali UG (2016) Industrial water demand for agro-processing and beverage industries in Rwanda. Smart Water 2. https://doi.org/10.1186/s40713-017-0006-x

Murali K, Karuppiah PL, Nithish M, Kumar SS, Raja VS (2013) COD reduction using low cost biosorbent as part of cleaner production. Int J Sci Res Public 3:1–3

Nandy T, Shastry S, Kaul SN (2002) Wastewater management in a cane molasses distillery involving bioresource recovery. J Environ Manage 65:25–38

National Research Council (US) (1980) Committee to Study the Human Health Effects of Subtherapeutic Antibiotic Use in Animal Feeds. National Academies Press (US), Washington, DC

Oller I, Malato S, Sánchez-Pérez JA (2011) Combination of advanced oxidation processes and biological treatments for wastewater decontamination—A review. Sci Total Environ 409:4141–4166

Ozbas EE, Tufeckci N, Yilmaz G, Ovez S (2006) Aerobic and anaerobic treatment of fruit juice industry effluents. J Sci Ind Res 65:830–837

Pala A, Erden G (2005) Decolorization of a baker's yeast industry effluent by fenton oxidation. J Hazard Mater 127:141–148

Pathak N, Hagare P, Guo W, Ngo H (2015) Australian food processing industry and environmental aspects-A review. I: International Conference on Biological, Civil and Environmental Engineering. https://doi.org/10.15242/iicbe.c0215016

Pathwardhan AD (2008) Industrial wastewater treatment. PHI learning Pvt. Ltd., New Delhi

Péerez TZ, Geissler G, Hernandez F (2007) Chemical oxygen demand reduction in coffee wastewater through chemical flocculation and advanced oxidation processes. J Environ Sci 19:300–305

Puszczało E, Marszałek A (2019) Treatment of wastewater from the confectionery industry using pressure membrane processes. Paper Presented at Innovations-Sustainability-Modernity-Openness Conference (ISMO'19) 16, 27. https://doi.org/10.3390/proceedings2019016027

Qasim W, Mane AV (2013) Characterization and treatment of selected food industrial effluents by coagulation and adsorption technologies. Water Resour Ind 4:1–12

Qin Z, Liang Y, Liu Z, Jiang W (2011) Preparation of in YO3 catalyst and its application in photodegradation of molasses fermentation wastewater. J Environ Sci 23:1219–1224

Rajab AR, Salim MR, Sohaili J, Anuar AN, Salmiati, Lakkaboyana SK (2017) Performance of integrated anaerobic/ aerobic sequencing batch reactor treating poultry slaughterhouse wastewater. Chem Eng J 313:967–974. https://doi.org/10.1016/j.cej.2016.10.144

Rana RS, Singh P, Kanderi V, Singh R, Dobhal R, Gupta S (2014) A review on characterization and bioremediation of pharmaceutical industries wastewater: An Indian perspective. Appl Water Sci 7:1–12

Rizzo L (2011) Bioassays as a tool for evaluating advanced oxidation process in water and wastewater treatment. Water Res 45:4311–4340

Roa-Morales G, Campos-Medina E, Aguilera-Cotero J, Bilyeu B, Barrera-Diaz C (2007) Aluminium electrocoagulation with peroxide applied to wastewater from pasta and cookie processing. Sep Purif Technol 54:124–129

Sangeetha V, Sivakumar V, Sudha A, Priyenka Devi KS (2014) Optimization of process parameters for COD removal by coagulation treatment using Box-Behnken design. Int J Eng Technol 6:1053–1058

Sardari K, Askegaard J, Chiao Y, Darvishmanesh S, Kamaz M, Wickramasinghe SR (2018) Electrocoagulation followed by ultrafiltration for treating poultry processing wastewater. J Environ Chem Eng 6:4937–4944. https://doi.org/10.1016/j.jece.2018.07.022

Sarkar B, Chakrabarti PP, Vijaykumar A, Kale V (2006) Wastewater treatment in dairy industries —possibility of reuse. Desalination 195:141–152

Sarode S, Upadhyay P, Khosa MA, Mak T, Shakir A, Song S, Ullah A (2018) Overview of wastewater treatment methods with special focus on biopolymer chitin-chitosan. Int J Biol Macromol

Serrano LG, Varga DD, Ruiz I, Soto M (2011) Winery wastewater treatment in a hybrid constructed wetland. Ecol Eng 37:744–753. https://doi.org/10.1016/j.ecoleng.2010.06.038

Shinya A, Bergwall L (2007) Pyrite oxidation: Review and prevention practices. J Vertebr Paleontol 27:145A

Siddiqui SI, Naushad M, Chaudhry SA (2019) Promising prospects of nanomaterials for arsenic water remediation: A comprehensive review. Process Safety and Environmental Protection 126. https://doi.org/10.1016/j.psep.2019.03.037

Šostar-Turk S, Petrinić I, Simonič M (2005) Laundry wastewater treatment using coagulation and membrane filtration. Resour Conserv Recycl 44(2):185–196

Stasinakis AS (2008) Use of selected advanced oxidation processes (AOPs) for wastewater treatment—A mini review. Global NEST J 10:376–385

Tikariha A, Sahu O (2014) Study of characteristics and treatments of dairy industry waste water. J Appl & Environ Microbiol 2(1):16–22. https://doi.org/10.12691/jaem-2-1-4
Valta K, Kosanovic T, Malamis D, Moustakas K, Loizidou M (2014) Overview of water usage and wastewater management in the food and beverage industry. Desalination and Water Treatment 53:1–13. https://doi.org/10.1080/19443994.2014.934100
Vymazal J (2014) Constructed wetlands for treatment of industrial wastewaters: A review. Ecol Eng 73:724–751. https://doi.org/10.1016/j.ecoleng.2014.09.034
Wang S, Peng Y (2010) Natural zeolites as effective adsorbents in water and wastewater treatment. Chem Eng J 156(1):11–24
Wang Y, Serventi L (2019) Sustainability of dairy and soy processing: A review on wastewater recycling. J Clean Prod 237:117821. https://doi.org/10.1016/j.jclepro.2019.117821
Wang NN, Zheng T, Zhang GS, Wang P (2016) A review on Fenton-like processes for organic wastewater treatment. J Environ Chem Eng 4:762–787
Wei R, Ge F, Huang S, Chen M, Wang R (2011) Occurrence of veterinary antibiotics in animal wastewater and surface water around farms in Jiangsu Province, China. Chemosphere. 82 (10):1408–1414. https://doi.org/10.1016/j.chemosphere.2010.11.067 PMID: 21159362
Wilderer P, Irvine R, Goronszy M (2015) Sequencing Batch Reactor Technology. Water Intelligence Online 6
Wu J, Doan H (2005) Disinfection of recycled red-meat-processing wastewater by ozone. J Chem Technol Biotechnol 80:828–833. https://doi.org/10.1002/jctb.1324
Wu TY, Mohammad AW, Jamaliah MJ, Anuar N (2010) Pollution control technologies for the treatment of palm oil mill effluent (POME) through end-of-pipe processes. J Environ Manag 91:1467–90. https://doi.org/10.1016/j.jenvman.20
Xu LJ, Sheldon BW, Carawan RE, Larick DK, Chao AC (2001) Recovery and characterization of by-products from egg processing plant wastewater using coagulants. Poult Sci 80:57–65
Xu J, Cao Z, Zhang Y, Yuan Z, Lou Z, Xu X, Wang X (2018) A review of functionalized carbon nanotubes and graphene for heavy metal adsorption from water: Preparation, application, and mechanism. Chemosphere 195:351–364. https://doi.org/10.1016/j.chemosphere.12.061
Yuan SH, Gou N, Alshawabkeh AN, Gu AZ (2013) Efficient degradation of contaminants of emerging concerns by a new electro-Fenton process with Ti/MMO cathode. Chemosphere 93:2796–2804
Yun M-A, Yeon K-M, Park J-S, Lee C-H, Chun J, Lim DJ (2006) Characterization of biofilm structure and its effect on membrane permeability in MBR for dye wastewater treatment. Water Res 40(1):45–52
Zajda M, Aleksander-Kwaterczak U (2019) Wastewater treatment methods for effluents from the confectionery industry—an overview. J Ecol Eng 20(9):293–304. https://doi.org/10.12911/22998993/112557

Chapter 10
Energy Analysis of Wastewater Infrastructure

Dina Zaman, Manoj Kumar Tiwari, and Gopika Ashokan

Abstract Urban wastewater systems (WWS) comprising of municipal sewage collection, treatment, and disposal play a vital role in alleviating urban water pollution and maintaining overall public health and sanitation. WWS forms one of the most energy-intensive components of the urban water cycle and requires huge financial investment for its construction, operations, and management. The financial burden of WWS operations can be subsided to a significant extent by implementing strategies for energy reduction, optimization as well as energy production and recovery from wastewater. However, developing an energy-neutral or energy-positive WWS is a challenging task, requiring the application of several interdependent energy pathways. In this chapter, comprehensive energy analysis of WWS is carried out, considering the available routes for reducing, optimizing, producing, and recovering energy from wastewater. Alongside this, available approaches for assessing the total energy consumption and energy efficiency of the WWS have also been evaluated. The use of indicators and indices can be particularly effective to both quantify and compare energy efficiency within different components of a WWS or among different WWS. The diagnostic potential of energy efficiency indices (EEIs) proposed specifically for WWS has been comparatively evaluated in this study. Such an index-based energy evaluation can conveniently characterize the energy efficiency of WWS and facilitating energy benchmarking of systems.

Keywords Municipal wastewater · Energy recovery · Energy efficiency indices · Energy benchmarking · Energy self-sufficiency

---

D. Zaman · M. K. Tiwari (✉) · G. Ashokan
School of Water Resources, Indian Institute of Technology Kharagpur, Kharagpur 721302, India
e-mail: mktiwari@swr.iitkgp.ac.in

S. Singh et al. (eds.), *Pollution Control Technologies*, Energy, Environment, and Sustainability, https://doi.org/10.1007/978-981-16-0858-2_10

10.1 Introduction

Water and energy are the most valuable resources in today's urban world, in the wake of global climate change caused by unprecedented environmental pollution and explosive population growth (Howells et al. 2013; Khalkhali and Mo 2020). With more than 50% of the world's population residing in urban regions, the water-energy nexus has particularly impacted the urban water cycle in cities and the surrounding regions (Bougnom and Piddock 2017). The urban water cycle chiefly consisting of water and wastewater infrastructure is highly energy intensive, either in direct or indirect form (Khalkhali and Mo 2020).

Urban wastewater systems (WWS) are massive infrastructures, often spread over cities that receive wastewater from domestic, commercial, and non-manufacturing sources. It primarily consists of a large network of conduits or pipes for wastewater collection, a wastewater treatment plant (WWTP), and sludge management systems for disposing of the treatment residues. WWS can be categorized as centralized or decentralized, depending on the scope and scale of the systems (Naik and Stenstrom 2016). Centralized WWS is economically suitable in large urban areas, whereas decentralized systems are more economical in low-income, small urban communities away from centralized WWS and with a demand for localized use of reclaimed water.

The primary function of a WWS is to reduce the organic and nutrient load, microbial load, other objectionable solids, and harmful micropollutants, along with odor from the wastewater influent. Urban wastewater management is intricately related to the overall sanitation and public health in a city (Winkler et al. 2017). Untreated sewage leads to detrimental impacts on the geo-aquatic environment in the long term. The ease of access to clean drinking water sources also indirectly depends on adequate wastewater management in the region, as untreated sewage (directly or indirectly) finds its way to the freshwater reserves. The issue of river pollution prevalent in several countries is also chiefly attributed to the release of municipal and industrial wastewater in its streams, without adequate treatment (Liu et al. 2012).

Conventional WWS usually has low-treatment efficiency and high energy consumption. A huge volume of energy is required to convey wastewater through the WWS, from the generation point to the treatment facility, and then through each component of the WWTP (Simon-Várhelyi et al. 2020). Further, the treated effluent and the waste materials separated as sludge need to undergo conditioning and then transported to the site of reuse, recycling, or disposal. Most wastewater collection systems operate under gravity flow (intermittent pumping if required), which makes them fairly energy efficient. Moreover, high infiltration into the sewers may lead to dilution and an increase in wastewater volume, which indirectly leads to high energy consumption in the subsequent processes. Overall, energy consumption in the wastewater collection system and to reach the primary treatment facility is in the range of 0.05–0.50 kWh/m^3 (Gu et al. 2017).

The total energy consumption in the treatment stage is dependent on several intrinsic and extrinsic factors. The volume of wastewater produced, type and concentration of constituent pollutants, type and extent of treatment provided (i.e., primary, secondary, or tertiary), as well as desired effluent water quality standards are the primary factors determining the energy requirement in the treatment process (Liu et al. 2018). Other factors that indirectly influence the energy requirement are the type of treatment plant (i.e., centralized or decentralized), the scale of the system, type of sewage collection system (i.e., separate or combined), need for pumping within the system, and the climatic conditions in the region (Niu et al. 2019). Energy is further consumed to condition or stabilize the sludge generated from WWTP through anaerobic digestion, composting or incineration (Frijns et al. 2013). Similarly, wastewater effluent is further treated and transported in accordance with its ultimate use.

Until recently, wastewater was largely perceived as a burden that needs to be sufficiently treated and responsibly disposed to reduce its detrimental impact on health and the environment. Traditionally, WWS has been built and operated with the myopic vision of pollution control and not as a self-sustaining and profit-garnering enterprise. This viewpoint is being gradually discarded as several routes of resource recovery from wastewater have come into application in recent times (Mo and Zhang 2013). The primary resources recoverable from wastewater are nutrients, water, and energy (Frijns et al. 2013). A paradigm shift in WWS operation from an energy sink to a potential energy source can be achieved only when energy optimization, resource recovery, and cost-benefit ratio is integrated with the primary goal of pollution control in WWS.

Energy optimization in a WWS entails reducing or conserving energy consumption, increasing energy efficiency, substituting energy-intensive components, processing with low-energy counterparts, switching to green energy sources in place of greenhouse gas (GHG) emitting fuels, producing energy from alternative and non-conventional sources, as well as recovering energy from wastewater by-products (Gu et al. 2017). Overall, the energy utilized in a WWS should be justifiable for the subsequent environmental or economic benefits garnered, through process optimization (Smith et al. 2019). Quantification of energy consumption and evaluation of energy efficiency is also a crucial task for a WWS, which can be effectively carried out using energy efficiency indicators (EEIs) (Sabia et al. 2020). EEIs facilitate for comparative evaluation of performance, classification, and benchmarking of WWS, in terms of energy efficiency.

In this chapter, a comprehensive energy efficiency analysis of WWS is carried out, considering case studies from different regions of the world. Approaches for reducing, optimizing, producing, and recovering energy from wastewater are critically explored from several WWS belonging to different geographical regions. Furthermore, quantitative evaluation of energy consumption using EEIs is also scrutinized for WWS from different countries.

10.2 Wastewater-Energy Nexus in a WWS

There has been a paradigm shift in the municipal wastewater treatment philosophy during the past 50 years. The early focus of WWS was primarily to reduce biochemical oxygen demand (BOD) and pathogenic content that may otherwise lead to the outbreak of diseases. Later, the detrimental environmental impact of eutrophication caused by the nitrogen and phosphorus in wastewater effluent discharge was recognized, and treatment units for their removal were included in the WWS. Presently, several emerging contaminants and pollutants have been included in the wastewater stream that requires specific advanced treatment units for their degradation and removal (Zaman et al. 2020). With time, the standards for allowable discharge effluents of WWS are also evolved. Stringent wastewater discharge standards led to manifold increase in energy consumption but also resulted in enormous environmental benefits (Smith et al. 2019). This has led to the exploration of energy reduction, optimization, and recovery from wastewater to compensate for the increased costs. Energy consumption, as well as the potential for energy recovery, is a function of influent characteristics of the wastewater (Yang et al. 2020).

10.2.1 WWS as an Energy Sink

The major sources of energy utilized in the WWS are electricity, natural gas, and other fossil fuels. Around 40% of the total energy consumed is because of electromechanical equipment in the WWTP. The average energy consumption for wastewater treatment in different regions of the world is USA—0.52 kWh/m^3, China—0.31 kWh/m^3, Japan—0.304 kWh/m^3, Korea—0.243 kWh/m^3, and Europe—0.42 to 0.53 kWh/m^3 (Tang et al. 2019). The high energy consumption in WWS also entails a high GHG emission from these systems in the form of carbon dioxide (CO_2), methane (CH_4), and nitrous oxide (N_2O) (Maktabifard et al. 2020). The carbon footprint of WWS is a measure of its impact on climate change, and nearly 70% of the carbon footprint of the urban water cycle is produced by WWS, out of which almost 56% is solely due to energy use (Frijns et al. 2013).

Several factors are responsible for determining the total energy consumption in a WWS. The size of a WWS represented in terms of population equivalent (PE), organic load, or hydraulic load determines the energy demand of the system to a large extent. Although the total energy consumption is lower for small WWTPs, the overall energy efficiency (energy consumption per unit of treated wastewater) and overall operating costs at all treatment stages studied worldwide have exhibited a significant decrease in large systems (Gu et al. 2017). Small WWTPs display higher variation in influent load, unstable treatment conditions and may lack the skilled manpower to operate them due to economic constraints, which may lead to lesser energy efficiency. Moreover, the type of WWTP—centralized and decentralized also impacts the energy demand of the system. Lifecycle assessment studies revealed that decentralized systems are almost five times more energy intensive than centralized systems (Shehabi et al. 2012). Hence, decentralized systems should be opted for only on a case-specific basis, where the local water reuse is feasible. In the near future, changes in the climatic variables like temperature, rain, and snowfall in a particular region due to climate change are also expected to implicate the overall energy usage in WWS (Khalkhali and Mo 2020). Projected climate-related stresses in the next 100 years is expected to increase the seasonal variation in flow rate and lead to frequent operational failures.

The energy consumption in the primary treatment stage remains more or less constant for a particular WWS. On the other hand, the energy demand in the secondary treatment process largely varies depending on the type of treatment and capacity of the WWTP. Activated sludge process (ASP) with extended aeration, sequential nitrification and denitrification, sequencing batch reactor (SBR), and oxidation ditch usually exhibits higher energy demand (Gude 2015). The aeration of the biological tank in a conventional ASP consumes nearly 60% of the total energy requirement of the treatment plant, and more energy is consumed for sludge pumping, processing, and disposal (Guven et al. 2019). On the other hand, trickling filters, lagoons, constructed wetlands, and adsorption-biology systems of wastewater treatment are relatively energy efficient. Energy consumption further increases in advanced treatment processes applied for nutrient removal and sludge conditioning. The energy requirement of advanced biological treatment with nutrient removal and filtration is almost 50% higher than that of conventional WWTP (Panepinto et al. 2016). The typical average energy consumption values in various treatment units are depicted in Fig. 10.1.

Apart from the direct energy requirement, a huge amount of embodied (or indirect) energy is associated with the massive infrastructures built for the WWS. Embodied energy encompasses the energy associated with construction and maintenance of WWS, from material extraction to deconstruction at the end of service life, and is often measured in terms of CO_2 equivalent (Georgiou et al. 2019). A comprehensive representation of energy dynamics in WWS can be

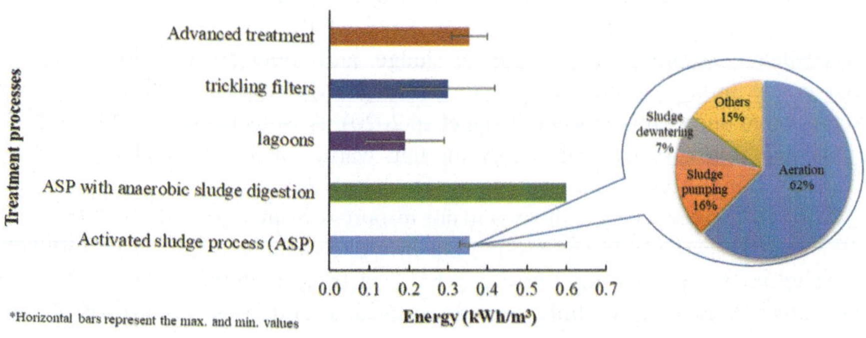

Fig. 10.1 Energy consumption in different WWTPs (Data sourced from Gude 2015)

achieved only by considering both the direct and embodied energy requirements of the system. However, this chapter does not focus on embodied energy consumption in detail, as it requires a different approach for its quantification and analysis.

10.2.2 WWS as an Energy Source

The energy sources in wastewater may be broadly divided as direct and indirect. The energy embedded in the wastewater by virtue of its characteristic properties and which is recoverable by suitable methods during the treatment process is considered as direct energy sources. On the other hand, the recyclable nutrients, materials, and reusable water obtained from the WWTP which indirectly substitute the energy resources spent in the producing them are denoted as indirect energy sources of wastewater (Liu et al. 2018). Resources can be extracted at different stages of treatment and can be either used in its raw form or transformed to more economically viable end products.

Energy remains embedded in the wastewater in different forms and in varying proportion and degree of recoverability, depending upon the characteristics of the wastewater. The internal energy of wastewater is chiefly due to the chemical energy of organic constituents present and can be measured as the chemical oxygen demand (COD) of the influent (Yang et al. 2020). Wastewater also possesses internal energy in the form of thermal energy, denoted by its characteristic temperature (Hao et al. 2019). Another form of recoverable energy is the hydraulic energy of the wastewater—potential energy due to elevation difference or kinetic energy due to a high flow rate (Gude 2015).

During treatment, 60% of the constituent organic substances in wastewater accumulates as sludge. The excess sludge may be treated through landfilling, composting, incineration, or biotransformed to produce biogas during anaerobic sludge digestion (Frijns et al. 2013). Incineration can be used to convert the internal chemical energy in organic sludge to heat and electric energy, along with substantial sludge volume reduction and abatement of soil pollution. Alternatively, the chemical energy in the wastewater content can be effectively transformed into electric energy and heat energy. Around 70% of the energy ingrained in raw wastewater is separated in the form of sludge, and about 50% of this energy is recoverable through anaerobic digestion of sludge. Moreover, phosphorus can also be recovered during this process (Yang et al. 2020). Wastewater can also be used as a potential source of thermal energy through warm water conservation and heat recovery, using shower heat exchangers (Hao et al. 2019).

Apart from direct energy, the two major resources from wastewater are nutrients and water. Essential nutrients are present in wastewater in the form of nitrogen, phosphorus, and potassium, which can be used in agricultural purposes by land application of partially or fully treated wastewater sludge as biofertilizer or soil enhancer (Zaman et al. 2020). Fertilizers can be procured from different stages of a WWS in the form of raw wastewater, semitreated or treated effluent, or treatment by-products like biosolids. Moreover, treated sludge can also be used as a landfill cover in municipal landfills. Other non-conventional pathways for nutrient recycling are struvite precipitation from source-separated urine and sludge digester liquor. Harvesting micro and macroalgae and wetland plants in wastewater effluent is another energy-efficient application of wastewater. Municipal wastewater can be directly used for the biomass production of microalgae by utilizing the carbon, nitrogen, phosphorus, and other micronutrients (Nagarajan et al. 2020). The obtained biomass is then applied as feedstock for biofuel generation. Additionally, treated wastewater may provide as an alternative source of water for various non-potable uses in industries and urban landscaping or irrigation of cash crops in agriculture. In some studies, the COD in wastewater was converted to high-value organic products like biopolymers instead of applying the traditional route of producing biomethane (Hao et al. 2019).

Nutrient recycling as fertilizer and water reuse indirectly offsets energy consumed in the production of chemical fertilizers, as well as treatment, and transportation of irrigation or industrial water and reduces the nutrient removal load on WWTP. Moreover, very negligible energy is required to redirect the wastewater nutrients for such alternative uses. Substituting water usage by treated effluent can significantly reduce the load on freshwater sources, thus snowballing several environmental benefits. However, nutrient recycling and effluent reuse are often more challenging than energy production and recovery in terms of safety, acceptability of all stakeholders involved, technological inadequacies, risk of intersystem contamination, land and finance resource constraints.

10.3 Energy Optimization in WWS

Although 25–40% of the expenses in any WWS is due to energy, optimization of operational parameters in WWS for energy considerations is seldom carried out, at the design stage (Rojas and Zhelev 2012). Energy optimization in the operations of existing systems may reap significant cost reduction with little or no investment. The energy optimization strategies in a WWS can be achieved through three potential routes—(a) Energy reduction in the existing wastewater management processes, (b) Improvement of existing processes and infrastructure for lower energy consumption, and (c) Substitution of energy-intensive processes with energy-efficient practices. These may be achieved through infrastructure-based or process-based modifications, and benefits incurred may be direct or indirect. Infrastructure-based modifications are difficult to implement in an existing WWS, require significant investments, and generally have a longer payback period. On the other hand, modifications in the operations and maintenance of WWS can be implemented with lesser capital investment and a shorter payback period.

The beneficial outcome of energy optimization is better represented in a WWS in terms of energy efficiency rather than energy consumption. Typically, WWS operating at a capacity lower than its design capacity leads to lower energy efficiency (Niu et al. 2019). Even when operating at optimal capacity, smaller WWS exhibits lower energy efficiency than centrally operated large WWS. Further, the higher energy efficiency of WWS is obtained with wastewater influent having higher organic content (Niu et al. 2019). Thus, some of the straightforward infrastructure-based energy optimization strategies for WWS are expanding the sewage collection system to supply the design capacity of wastewater at the WWTP, opting for centralized wastewater treatment facilities instead of decentralized systems wherever possible, and establishing separate sewage collection systems to increase the inlet pollutant load at the WWTP. Further, energy efficiency can also be increased to a considerable extent by the renovation of older WWS and retrofitting the wastewater collection systems for ensuring lower infiltration.

Around 5000 MWh of energy savings along with improvement in effluent quality was obtained in the largest WWTP in Italy, by optimizing the operational parameters of wastewater influent treatment (Borzooei et al. 2019). In another study, it was established that if the COD of wastewater is higher than 500 mg/L, the energy consumption can be reduced by 20% (Niu et al. 2019). Alternatively, municipal wastewater was utilized as irrigation water for seven energy crops, and the biomass produced was anaerobically digested for biomethane production, instead of conventional sewage treatment (Shilpi et al. 2019). The organics and nutrients present in wastewater were synergistically utilized in this process, thereby substituting the energy required for wastewater treatment. A distinctive framework for reducing energy consumption in a WWTP was suggested by Simon-Várhelyi et al. (2020), wherein the periodic influent load was stored during daytime and scheduled for purification during the night. The proposed strategy was implemented in a Romanian WWTP, and an overall energy reduction by 47% and effluent quality improvement by 25% were achieved.

Despite ASP being very energy intensive, it continues to be the most popularly utilized secondary treatment process in a WWS. More than 50% of the total energy is consumed at the aeration stage and 12% in pumping operations, in a typical ASP (Henriques and Catarino 2017). A substantial scope of energy reduction is possible in the aeration stage by optimized air distribution, through variable aeration as per the oxygen uptake rate (Rojas and Zhelev 2012). Installation of control devices on blowers and aerators, using bubble aeration techniques, can also increase the energy efficiency of the aeration stage (Panepinto et al. 2016). Conversely, alternative wastewater treatment technologies to high energy-consuming ASP have to be explored and implemented to reduce the overall energy demands (Gikas 2017).

Sludge treatment is the second most energy-intensive process in the WWTP, after secondary treatment. The operational parameters—aeration flow rate, reaction time, and sludge flow rate in two full-scale autothermal thermophilic aerobic digestion process were optimized using the *eSS* algorithm (Rojas and Zhelev 2012). The utilization of optimized values of these parameters garnered energy improvement of 22% in the first single-stage system and 18% in the two-stage system. In another study, the thermo-alkali pre-treatment resulted in an increase of specific biogas production and specific methane production by 46.2 and 86.1%, respectively, (Borzooei et al. 2019). The third most energy-intensive component of WWS is the wastewater pumping stations. Operational energy consumption of wastewater pumping stations was reduced by 16.7% by implementing a data-driven optimization using supervised and deep reinforcement for predictive control (Filipe et al. 2019).

Operational modifications in the existing WWS may also reduce direct energy consumption or indirectly facilitate lesser energy demand in the subsequent stages. WWTP with pre-settling and sludge digestion facilities consumes 40% less net energy than those without these stages. Pre-treatment processes applied to make the sludge more accessible for anaerobic digestion and enhance methane production are classified as mechanical, biological, chemical, or thermal (Borzooei et al. 2019). Mechanical and chemical pre-treatment processes such as coagulation, flocculation, and microsieving increase the sludge recovery from wastewater. Similarly, thermal pre-treatment of the sludge before anaerobic digestion increased the methane production, proportional to the duration and temperature applied. However, pre-treatment processes signify that additional cost in the form of equipment or chemicals is employed, and it should be ensured that it incurs proportional benefits.

10.4 Energy Production and Recovery in WWS

The internal energy embedded in wastewater can be recovered to a considerable extent by implementing effective energy capturing and energy recovery techniques during wastewater transmission, treatment, or disposal. Energy can be generated onsite by transforming the energy stored in influent organic matter, nutrients, kinetic energy of wastewater flow, residual heat in wastewater, to useable heat or electricity (Gude 2015). The integration of wastewater treatment and energy production in a WWS can reduce the net energy consumption to a great extent. Alongside, the energy produced can also compensate for the carbon footprint of the WWTP (Maktabifard et al. 2020). However, the pre-requisites for onsite energy production from a WWTP are site specific, and each process needs to be specifically analyzed before implementation as it also entails high capital investment. The scope of energy production and recovery from WWS is depicted in Table 10.1.

Energy recovery from wastewater has the highest potential to operate energy efficient and economically sustainable WWS. The electric and heat energy generated on site in a WWTP can be reutilized in the energy-consuming units of the WWTP, in heating and lighting the auxiliary infrastructures of the WWTP, and the excess can be distributed to the nearby region (Gude 2015). The thermal energy recovered from the wastewater can be used for district heating/cooling, drying dewatered sludge, or for powering greenhouses (Hao et al. 2019). The algal biomass produced by utilizing wastewater influent can be used as feedstock for producing biofuels, with high recoverability through hydrothermal liquefaction, pyrolysis, and gasification (Nagarajan et al. 2020).

Several steps can be undertaken to increase the recoverability of energy from wastewater, along with its effective treatment. Higher production of biogas is

Table 10.1 Scope of energy production and recovery from WWS

Process	Input energy component	Output energy component	Advantages	Disadvantages/ limitations
Solar and wind power generation	On-site solar energy and wind energy at the WWTP site	Electric energy	• Use of massive areas dedicated to WWTPs, away from population • Clean energy production for on-site use in the WWTP	• Large capital investment • Restrictions for location of WWTPs • Appropriate climatic conditions required
Chemical energy conversion	Biogas produced from anaerobic processes	Heat and electric energy	• High electricity generation potential • Produces 25–100% of on-site energy • Efficiency can be increased by pre-treatment and co-digestion	• High installation costs • Cannot be implemented effectively in small WWS • Cost-effective only for flow rate >5 MGD/ 18,500 cum/day
Chemical energy conversion	Biogas from anaerobic processes	Substitute natural gas (Bio-SNG) and biopolymers	• More environmentally efficient by-products of biogas • More energy recoverability and less GHG emissions	• Technological barriers • High investment
Incineration	Incineration of biosolids	Heat and electric energy	• High electricity generation potential if water content of biosolids <30% • Eliminates odor generation • A better option for highly populated regions with less option for solids disposal	• Harmful pollutants released during • High capital investment • Inconsistency in quality of process • Additional energy required for dewatering of biosolids
Hydropower generation systems	Kinetic energy or potential energy in wastewater	Hydroelectric energy	• Increase in dissolved oxygen in effluent and natural aeration • Clean source of energy	• Requires significant kinetic energy or potential energy

(continued)

Table 10.1 (continued)

Process	Input energy component	Output energy component	Advantages	Disadvantages/ limitations
Heat recovery systems	Thermal energy from wastewater	Heat energy for cooling or heating	• High energy efficiency • Low operations and maintenance • High recoverability	• Performs optimally only in moderate temperature regions
Microbial fuel cells (MFC)	Microbial metabolic energy in wastewater	Electric energy	• Reduces excess sludge to 20% of conventional systems	• Energy loss during electricity generation • Low organic utilization rate • High capital costs
Microbial electric cells (MEC)	Electric energy	Biochemicals like hydrogen and methane gas	• High gas generation potential for raised wastewater temperature and organic VFA content • Outperforms MFC in global warming potential, resource consumption and carcinogen production	• Unsatisfactory hydrogen recovery • Low COD removal
Microbial desalination cells	Metabolic energy in wastewater	Electric energy	• Simultaneous wastewater treatment and desalination • Additional nutrient removal	• Less energy production than MFC • Membrane fouling
Microalgae technology (harvesting microalgae in wastewater)	Biochemical energy in algal biomass	Bioenergy products like biodiesel	• Negative greenhouse gas emission as it utilizes CO_2 • Reduces water footprint of biodiesel production by 90% • No additional nutrients, except phosphate is required	• High cost of algal cultivation • Complex process of species selection, harvesting, dewatering, and lipid extraction

SNG Synthetic natural gas, *GHG* Greenhouse gas, *VFA* Volatile fatty acids

obtained in an anaerobic MBR than in the conventional ASP with anaerobic digestion. Furthermore, concentrated sewage and high temperature facilitate high energy recoverability from its sludge. Energy recovery is also high if high specific sludge loading rates and low sludge retention time can be maintained in the reactors. Alternatively, sludge from WWTP can be mixed with kitchen/food waste (also known as co-digestion) to increase biogas production (Guven et al. 2019). Co-digestion of wastewater sludge can also be carried out with other lignocellulose biomass. Such operational modifications garner a varying level of reduction in on-site energy demand and increase the energy production potential of the process, depending on site-specific conditions.

10.5 Energy Self-sufficiency in WWS

'Energy self-sufficiency' largely denotes an aspirational state of any system, wherein its various components synergistically subsist and function, without little or no external energy contribution (Müller et al. 2011). Synonyms like 'energy autarky,' 'energy autonomy,' 'energy independence,' and 'energy self-reliance' are also used in lieu of energy self-sufficiency (Engelken et al. 2016). WWS is a tremendously energy-intensive subsystem of the urban water cycle, and designing energy self-sufficient WWS is a rather challenging task. Overall energy self-sufficiency in WWS can be demarcated as absolute or physical self-sufficiency (no requirement of external energy) and relative self-sufficiency (use of locally generated energy). Physical energy self-sufficiency is more or less impractical and costly concerning a WWS, and hence, most WWS desires to attain only relative energy self-sufficiency. Thus, energy self-sufficiency in a WWS can be denoted either as energy-neutral (i.e., net energy consumed = overall energy produced) or energy-positive (i.e., overall energy produced > net energy consumed).

Energy self-sufficiency in a WWS is the culmination of benefits garnered from energy reduction, optimization, energy reuse, and decentralization of energy sources, configured synergistically for overall energy benefits, as depicted in Fig. 10.2. However, establishing energy self-sufficient systems may incur high cost and a longer payback period, which should also be taken into consideration, while choosing the alternatives to be implemented. Increasing the energy efficiency of WWS is the first and foremost step toward energy self-sufficiency. Thereafter, more than one energy-positive process needs to be employed to achieve this objective, and very few studies have explored the scope of energy self-sufficiency in real WWS. In one of the earliest studies, a hypothetical system for complete anaerobic treatment comprising of co-generation of heat and electricity from methane as well as membrane filtration and methane stripping for producing reusable water was proposed (McCarty et al. 2011). However, the model was not substantiated with data on a real system. In case of real WWS, the scope of attaining energy self-sufficiency will also vary among different systems, depending on several economic, socio-political, and regional factors. In a study carried out for Germany, Engelken et al. (2016) established that the key driving factors for municipal organizations to strive for energy self-sufficiency are tax revenues, environmental awareness, and independence from private utilities.

One of the most obvious routes of attaining energy self-sufficiency is the production of renewable energy or incorporating green energy resources at the WWTP

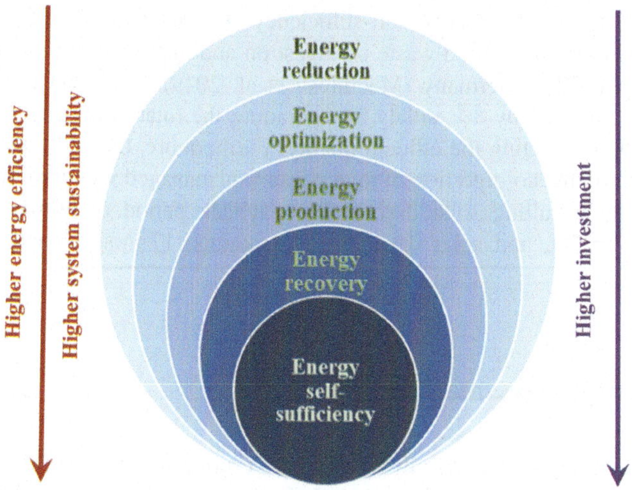

Fig. 10.2 Energy self-sufficiency as a culmination of all energy-positive efforts in a WWS

to decentralize the energy system (Müller et al. 2011). The total energy consumption in a WWTP in Korea was offset by 6.5% (payback time of 7.4 year) by incorporating 100 kW photovoltaics, 10 kW small hydropower plants, and effluent heat energy recovery system as green energy resources (Chae and Kang 2013). Subsequently, it also led to the reduction of GHG emissions by 261 tonnes CO_{2e}/year. Another possible avenue for attaining energy self-sufficiency could be the substitution of the energy-intensive subcomponents in the WWTP. Gikas (2017) proposed an alternative treatment process, which eliminates the need for aeration in WWTP. Nearly 85% of total suspended solids (TSS) and 65% of BOD_5 in wastewater was removed by using advanced microsieving and filtration and followed by trickling filter with encapsulated denitrification. Further, biosolids concentrate obtained from sieving and filtration was dewatered and then partially thermally dried to be ultimately used for thermal and electrical energy. This process utilized 85% less energy than a conventional ASP and produced average energy of 0.085–0.115 kWh/m^3, thus achieving an energy-positive system.

Conventional biological processes of wastewater treatment are highly energy intensive. Liu et al. (2018) attempted to configure energy self-sufficient biological treatment processes by combining more than two complementing methods with optimization of operational parameters. The proposed systems were integrated high-rate activated sludge and mainstream deammonification process, integrated anaerobic anammox process, and anaerobic membrane bioreactor (MBR) coupled with nutrient recovery. Synergistic effects of electrochemical, biological, and bioelectrochemical wastewater treatment technologies were successfully used to achieve energy self-sufficiency in a WWTP (Tang et al. 2019). The high energy consumption in the electrochemical process was provided by energy produced from wastewater using biological and bioelectrochemical processes. In another study, a

24% increase in overall energy self-sufficiency was achieved by implementing combined techniques like food waste co-digestion and improved aeration and solar dryers in a WWTP in Germany (Macintosh et al. 2019). The plant increased its energy self-sufficiency by 8%, solely by upgrading the rotary blowers used during aeration, without affecting the effluent quality. Furthermore, the additional organic load produced by the co-digestion of food waste was managed with minimal energy requirement by installing solar dryers. The payback period of blower upgrade, co-digestion facility, and solar dryer was 10 months, 17 months, and 30 years, respectively.

10.6 Energy Assessment in WWS

Assessment of energy efficiency in a WWS is a crucial step towards the continual performance evaluation of the systems. It has already been established that in dynamic systems like WWS, low energy consumption does not necessarily imply high energy efficiency. Moreover, since each treatment unit requires a different volume of energy for its operation for the removal of varying degrees of target pollutants, energy efficiency is a more representative parameter for comparison. The increasing use of more energy-intensive advanced wastewater treatment processes employed to remove recalcitrant pollutants, adhering to stringent effluent standards or produce reusable effluents, has rendered energy efficiency assessment more crucial for WWS in the present times. However, no two WWS are functionally and operationally alike, due to which comparative assessment among them is not an easy task, in the absence of standard energy benchmarking. To our best knowledge, there is no universally accepted approach for energy evaluation in WWS, as of now. Sporadic literature, addressing different aspects of energy analysis in WWS, is available in research, with no comprehensive study to formulate a unified framework.

Depending on the data availability, period of evaluation, and purpose of the study, different energy assessment methods may be used in a WWS (Longo et al. 2019). One of the most popular methods of energy assessment in a WWS is energy auditing. Energy audits of WWS are useful to obtain the total energy consumption or overall energy efficiency of the individual components or the overall wastewater management process, in a systematic manner (Henriques and Catarino 2017). It provides the baseline energy consumption in the WWS under consideration, for future comparison of energy consumption. It also facilitates in assessing the scope of energy improvement within systems. The long-term analysis of WWS can be carried out using energy analysis and lifecycle analysis (LCA). The energy of a WWS signifies the total direct and indirect energy required for the transformation of wastewater to treated water of the desired standard (Winfrey and Tilley 2016), whereas LCA is a more comprehensive and long-term investigation of the transformation process in WWS.

The LCA of WWS can be carried out from the aspects of environmental impacts, energy impacts, carbon emissions aspects, or cost aspects. Although essential for understanding the 'cradle-to-grave' energy budget of the system, studies on LCA from energy impacts have not been extensively explored for WWS (Lopes et al. 2018). Further, an in-depth analysis of the greenhouse gas emission in the WWS can be specifically carried out using a carbon footprint analysis (CFA) (Maktabifard et al. 2020). CFA is useful for assessing the impact of a WWS on climate change and is often considered as a subset of LCA. All forms of greenhouse gas emissions from a WWS such as CO_2, CH_4, and N_2O are converted to CO_2 equivalents in terms of their global warming potential and expressed as CO_{2e} in a CFA.

Energy benchmarking is another crucial aspect of energy evaluation in WWS. Benchmarking studies are one of the most fruitful means of assessing the evolution of energy consumption in WWS, over a long period, and also facilitate intercomparison among different systems (Lorenzo-Toja et al. 2016). Benchmarking studies are more pragmatic in nature as they highlight the energy hotspots in the system and also reveal the degree of energy efficiency achievable through energy conservation. The most popular approaches employed in energy benchmarking of WWS are normalization, ordinary least squares, data envelopment analysis, and LCA (Longo et al. 2019). However, barring a few developed countries, comprehensive energy benchmarking studies have not been carried out by most countries for WWS (Sabia et al. 2020). Continual energy benchmarking studies in Austrian wastewater utilities have facilitated them to reduce energy consumption in the WWTPs by 30%. A preliminary energy benchmarking exercise for two full-scale WWTP located in East Sussex, England, was also carried out, by segregating the total energy consumption into electrical, manual, mechanical, and chemical energy equivalent (Belloir et al. 2015).

Another vital aspect of energy assessment in a WWS is the representation of the energy parameters obtained from real systems. As energy efficiency is a better measure than energy consumption for WWS, the individual values of absolute energy consumption are transformed into energy efficiency indicators (EEIs). The EEIs provide a realistic and easily understandable platform for reporting and representing raw energy data of WWS and play an effective role in the assessment of energy management systems. Specific energy consumption (SEC), denoted by the total energy consumed per volume of wastewater treated (kWh/m^3), and energy intensity (EI), denoted by the energy consumed per unit of profit incurred, are the most widely used indicators in the WWS. However, due to the variable nature of wastewater parameters, SEC may be a misleading indicator when it is used to compare among different systems. Similarly, with no direct profit implications in most wastewater infrastructure, EI is also not directly applicable in these systems. Thus, SEC and EI were modified to formulate new EEIs for specific usability in wastewater infrastructure (Vaccari et al. 2018; Niu et al. 2019). However, not all EEIs proposed truly represent the energy efficiency of the system (Longo et al. 2019). Most of EEIs signify the overall energy consumption at best or merely a part of it. Moreover, it is explicit that energy efficiency cannot be characterized by a single EEI in case of a WWS, owing to the large number of variable parameters involved in the process.

The most prominent SEC-derived indicator is SEC/PE_{served} that has been utilized in several studies for energy analysis (Longo et al. 2019). Alternatively, the energy efficiency of specific treatment stages can be represented in terms of its constituent pollutants removed, such as $SEC/TSS_{removed}$, $SEC/COD_{removed}$, $SEC/BOD_{removed}$, and $SEC/N/P_{removed}$ (Lackner et al. 2014). At times, the major pollutants removed by the WWTP are combined using weighted sum and $SEC/pollutants_{removed}$ is computed. Although simple and straightforward, these indicators are still inadequate in terms of comprehensive energy efficiency analysis and benchmarking. Nevertheless, these indicators have been used for energy audit and benchmarking studies (Longo et al. 2019; Foladori et al. 2015). Benchmarking using an index-based technique is easily computable, provided a reference benchmark is established. Similarly, an energy audit of WWS with similar operational parameters can also be easily compared based on EEIs.

Seventeen wastewater treatment plants of Portugal were investigated for ascertaining their EEIs (Silva and Rosa 2015). Four specific energy consumption indices relating to the equivalent population, the volume of treated water, amount COD removed, and the amount of total nitrogen removed were utilized to quantify the energy efficiency in the largest WWTP in Italy (Panepinto et al. 2016). Energy intensity of wastewater treatment plants of China for a wide range of influencing factors such as treatment scale, technology, treatment degree, load factor, sludge amount, age, topography, and wastewater collection area was determined by Niu et al. (2019). Moreover, energy intensity values from different WWTPs from around the world were also assimilated from the literature, and a comparative analogy was carried out. Benchmarking of energy consumption of 200 wastewater treatment plants in Italy was conducted by Vaccari et al. (2018), using three EEIs. Different hydraulic stages of the WWTP are accessed using different energy indicators. Out of these, SEC/PE/year proved to be a better indicator of energy efficiency in WWTPs. Some of the SEC-derived EEIs used in case studies from different WWS around the world are summarized in Table 10.2.

Apart from the SEC-derived EEIs in the mentioned studies, several new indicators were introduced. Chae and Kang (2013) formulated an indicator called energy independence, denoted as the percent ratio of green energy production to the total energy consumed. The energy self-sufficiency of a WWS can be expressed in terms of 'energy self-sufficiency index' denoted as the ratio of net energy consumption to the gross energy consumption, and 'degree of energy self-sufficiency' represented as the ratio of locally generated energy to the local energy demand (Engelken et al. 2016). An energy efficiency indicator called 'sustainable value indicator' which correlated removal efficiency to the cost incurred and energy consumed by the system (α—Satisfaction of needs/resources used) was proposed (Henriques and Catarino 2017). This indicator effectively associates WWTP performance with the involved resources. A novel index 'specific energy consumption index' (E_O) which denoted the total energy consumption per unit removal of total oxygen demand (TOD) was proposed by Luo et al. (2019) for evaluating the comparative energy consumption of 2022 WWTPs. This index is a better representation of the true energy consumption of WWTPs with variable influent quality.

Table 10.2 Energy efficiency indicators (EEIs) formulated for energy analysis in WWS

Region	Number of WWTPs	Indicators	Unit	References
China	2022 WWTPs	Specific energy consumption index	kWh/kgTOD	Luo et al. (2019)
Several countries	601 WWTPs	EC per cubic meter of treated influent EC per PE served EC per kg COD removed	kWh/m^3 kWh/PE/year kWh/kgCOD	Longo et al. (2019)
China	85% of all WWTPs	Energy intensity	kWh/kgCOD	Niu et al. (2019)
Italy	241 WWTPs	EC per PE served	kWh/PE/year	Vaccari et al. (2018)
		EC per cubic meter of treated influent	kWh/m^3	
		EC per kg COD removed	kWh/kgCOD	
Italy	One WWTP (largest in Italy)	EC per PE served	SEC/PE	Panepinto et al. (2016)
		EC per volume of treated influent	SEC/m^3	
		EC per kg of COD removed	SEC/kgCOD	
		EC per kg of nitrogen	SEC/kg N	
Italy	5 WWTPs	EC per volume of treated influent	kWh/m^3	Foladori et al. (2015)
		EC per kg of BOD/COD	kWh/kgCOD	
		EC per year per PE served	kWh/PE.year	
		EC per year per designed PE	kWh/PE.year	
		Capacity utilization (CU)	PE_{served}/PE_{design}	
Portugal	17 WWTPs	EC per cubic meter of treated influent EC per kg of BOD_5 removed EC per kg COD removed Energy production from biogas Net use of external energy source Electrical energy costs	kWh/m^3 kWh/BOD_5 kWh/COD kWh/m^3 (or %) kWh/m^3 €/m^3	Silva and Rosa (2015)

EC Energy consumption, *PE* Population equivalent

Another composite indicator proposed by Longo et al. (2019) was 'water treatment energy index' to represent the overall energy efficiency of WWTP. Individual energy consumption indicators were fused into a 'global index' for energy performance evaluation, classification, and benchmarking of WWS (Sabia et al. 2020).

10.7 Conclusions

The global decline in the sustainable supply of water and energy resources and the concerns regarding rising carbon emissions have impelled all urban municipalities to reimagine the operations and functioning of WWS. Modern wastewater management systems need to be established not only to control water pollution but also to transform the wastewater into water and energy resources. However, establishing an efficient WWS requires a huge investment, in terms of finance, energy, infrastructure, and skilled manpower. The failure to recognize wastewater management as a profitable enterprise garnering environmental and economic benefits has also led to neglecting these services, especially in the low-income countries.

Energy balance in a WWS comprises of three crucial aspects—internal energy present in the wastewater, external energy consumed in wastewater treatment and prospects of energy production or recovery from the wastewater. The primary energy constituents of wastewater are chemical energy, thermal energy, and hydraulic energy, which can be recovered as electricity, heat, and biofuels. Apart from energy, several economic by-products such as reclaimed water, nitrogen, phosphorus, and organic fertilizer can also be recovered. Additionally, a significant portion of the energy consumed in a WWS can be reduced through optimized design and maintenance, increasing sewage load and renovating old systems.

Theoretically, wastewater contains almost ten times the energy that is required for its treatment, which establishes the possibility to develop energy-neutral or even energy-positive WWS. However, several other factors may become crucial in recovering energy from real WWS. Technological barriers, high investment cost, and longer payback period are the major constraints for the implementation of these techniques. Moreover, quantification and evaluation of energy in different stages of the WWS and comparison among different systems have become imperative for the continual energy management of the systems. In this regard, EEIs developed specifically for WWS can bring in the ease and practical applicability to the process of energy evaluation and process benchmarking. Such an index-based energy analysis of WWS helps is fruitful in identifying the stages and factors leading to higher energy efficiency and developing best management practices.

References

Belloir C, Stanford C, Soares A (2015) Energy benchmarking in wastewater treatment plants: the importance of site operation and layout. Environ Technol 36(2):260–269

Borzooei S, Campo G, Cerutti A, Meucci L, Panepinto D, Ravina M, Riggio V, Ruffino B, Scibilia G, Zanetti M (2019) Optimization of the wastewater treatment plant: from energy saving to environmental impact mitigation. Sci Total Environ 691:1182–1189

Bougnom BP, Piddock LJ (2017) Wastewater for urban agriculture: a significant factor in dissemination of antibiotic resistance. Environ Technol 51:5863–5864

Chae KJ, Kang J (2013) Estimating the energy independence of a municipal wastewater treatment plant incorporating green energy resources. Energy Convers Manag 75:664–672

Engelken M, Römer B, Drescher M, Welpe I (2016) Transforming the energy system: why municipalities strive for energy self-sufficiency. Energy Policy 98:365–377

Filipe J, Bessa RJ, Reis M, Alves R, Póvoa P (2019) Data-driven predictive energy optimization in a wastewater pumping station. Appl Energy 252:

Foladori P, Vaccari M, Vitali F (2015) Energy audit in small wastewater treatment plants: methodology, energy consumption indicators, and lessons learned. Water Sci Technol 72 (6):1007–1015

Frijns J, Hofman J, Nederlof M (2013) The potential of (waste) water as energy carrier. Energy Convers Manag 65:357–363

Georgiou E, Raffin M, Colquhoun K, Borrion A, Campos LC (2019) The significance of measuring embodied carbon dioxide equivalent in water sector infrastructure. J Clean Prod 216:268–276

Gikas P (2017) Towards energy positive wastewater treatment plants. J Environ Manage 203:621–629

Gu Y, Li Y, Li X, Luo P, Wang H, Robinson ZP, Wang X, Wu J, Li F (2017) The feasibility and challenges of energy self-sufficient wastewater treatment plants. Appl Energy 204:1463–1475

Gude VG (2015) Energy and water autarky of wastewater treatment and power generation systems. Renew Sustain Energy Rev 45:52–68

Guven H, Ersahin ME, Dereli RK, Ozgun H, Isik I, Ozturk I (2019) Energy recovery potential of anaerobic digestion of excess sludge from high-rate activated sludge systems co-treating municipal wastewater and food waste. Energy 172:1027–1036

Hao X, Li J, van Loosdrecht MC, Han J, Liu R (2019) Energy recovery from wastewater: heat over organics. Water Res 161:74–77

Henriques J, Catarino J (2017) Sustainable value – An energy efficiency indicator in wastewater treatment plants. J Clean Prod 142:323–330

Howells M, Hermann S, Welsch M, Bazilian M, Segerström R, Alfstad T, Gielen D, Rogner H, Fischer G, Van Velthuizen H, Wiberg D (2013) Integrated analysis of climate change, land-use, energy and water strategies. Nat Clim Change 3(7):621–626

Khalkhali M, Mo W (2020) The energy implication of climate change on urban wastewater systems. J Cleaner Prod:121905

Lackner S, Gilbert EM, Vlaeminck SE, Joss A, Horn H, van Loosdrecht MC (2014) Full-scale partial nitritation/anammox experiences—an application survey. Water Res 55:292–303

Liu C, Kroeze C, Hoekstra AY, Gerbens-Leenes W (2012) Past and future trends in grey water footprints of anthropogenic nitrogen and phosphorus inputs to major world rivers. Ecol Ind 18:42–49

Liu YJ, Gu J, Liu Y (2018) Energy self-sufficient biological municipal wastewater reclamation: present status, challenges and solutions forward. Biores Technol 269:513–519

Longo S, d'Antoni BM, Bongards M, Chaparro A, Cronrath A, Fatone F, Lema JM, Mauricio-Iglesias M, Soares A, Hospido A (2016) Monitoring and diagnosis of energy consumption in wastewater treatment plants. A state of the art and proposals for improvement. Appl Energy 179:1251–1268

Longo S, Mauricio-Iglesias M, Soares A, Campo P, Fatone F, Eusebi AL, Akkersdijk E, Stefani L, Hospido A (2019) ENERWATER – A standard method for assessing and improving the energy efficiency of wastewater treatment plants. Appl Energy 242:897–910

Lopes AC, Valente A, Iribarren D, González-Fernández C (2018) Energy balance and life cycle assessment of a microalgae-based wastewater treatment plant: a focus on alternative biogas uses. Biores Technol 270:138–146

Lorenzo-Toja Y, Vázquez-Rowe I, Amores MJ, Termes-Rifé M, Marín-Navarro D, Moreira MT, Feijoo G (2016) Benchmarking wastewater treatment plants under an eco-efficiency perspective. Sci Total Environ 566:468–479

Luo L, Dzakpasu M, Yang B, Zhang W, Yang Y, Wang XC (2019) A novel index of total oxygen demand for the comprehensive evaluation of energy consumption for urban wastewater treatment. Appl Energy 236:253–261

Macintosh C, Astals S, Sembera C, Ertl A, Drewes JE, Jensen PD, Koch K (2019) Successful strategies for increasing energy self-sufficiency at Grüneck wastewater treatment plant in Germany by food waste co-digestion and improved aeration. Appl Energy 242:797–808

Maktabifard M, Zaborowska E, Makinia J (2020) Energy neutrality versus carbon footprint minimization in municipal wastewater treatment plants. Biores Technol 300:

McCarty PL, Bae J, Kim J (2011) Domestic wastewater treatment as a net energy producer—can this be achieved? Environ Sci Technol 45(17):7100–7106

Mo W, Zhang Q (2013) Energy–nutrients–water nexus: integrated resource recovery in municipal wastewater treatment plants. J Environ Manage 127:255–267

Müller MO, Stämpfli A, Dold U, Hammer T (2011) Energy autarky: a conceptual framework for sustainable regional development. Energy policy 39(10):5800–5810

Nagarajan D, Chang JS, Lee DJ (2020) Pretreatment of microalgal biomass for efficient biohydrogen production–Recent insights and future perspectives. Biores Technol 302:122871

Naik KS, Stenstrom MK (2016) A feasibility analysis methodology for decentralized wastewater systems—energy-efficiency and cost. Water Environ Res 88(3):201–209

Niu K, Wu J, Qi L, Niu Q (2019) Energy intensity of wastewater treatment plants and influencing factors in China. Sci Total Environ 670:961–970

Panepinto D, Fiore S, Zappone M, Genon G, Meucci L (2016) Evaluation of the energy efficiency of a large wastewater treatment plant in Italy. Appl Energy 161:404–411

Rojas J, Zhelev T (2012) Energy efficiency optimisation of wastewater treatment: study of ATAD. Comput Chem Eng 38:52–63

Sabia G, Luigi P, Avolio F, Caporossi E (2020) Energy saving in wastewater treatment plants: a methodology based on common key performance indicators for the evaluation of plant energy performance, classification and benchmarking. Energy Convers Manag 220:

Shehabi A, Stokes JR, Horvath A (2012) Energy and air emission implications of a decentralized wastewater system. Environ Res Lett 7(2):024007

Shilpi S, Lamb D, Bolan N, Seshadri B, Choppala G, Naidu R (2019) Waste to watt: anaerobic digestion of wastewater irrigated biomass for energy and fertiliser production. J Environ Manage 239:73–83

Silva C, Rosa MJ (2015) Energy performance indicators of wastewater treatment: a field study with 17 Portuguese plants. Water Sci Technol 72(4):510–519

Simon-Várhelyi M, Cristea VM, Luca AV (2020) Reducing energy costs of the wastewater treatment plant by improved scheduling of the periodic influent load. J Environ Manage 262:110294

Smith K, Guo S, Zhu Q, Dong X, Liu S (2019) An evaluation of the environmental benefit and energy footprint of China's stricter wastewater standards: can benefit be increased? J Clean Prod 219:723–733

Tang J, Zhang C, Shi X, Sun J, Cunningham JA (2019) Municipal wastewater treatment plants coupled with electrochemical, biological and bio-electrochemical technologies: opportunities and challenge toward energy self-sufficiency. J Environ Manage 234:396–403

Vaccari M, Foladori P, Nembrini S, Vitali F (2018) Benchmarking of energy consumption in municipal wastewater treatment plants—a survey of over 200 plants in Italy. Water Sci Technol 77(9):2242–2252

Winfrey BK, Tilley DR (2016) An emergy-based treatment sustainability index for evaluating waste treatment systems. J Clean Prod 112:4485–4496

Winkler MS, Jackson D, Sutherland D, Lim JMU, Srikantaiah V, Fuhrimann S, Medlicott K (2017) Sanitation safety planning as a tool for achieving safely managed sanitation systems and safe use of wastewater. WHO South-East Asia J Public Health 6(2):34–40

Yang X, Wei J, Ye G, Zhao Y, Li Z, Qiu G, Li F, Wei C (2020) The correlations among wastewater internal energy, energy consumption and energy recovery/production potentials in wastewater treatment plant: an assessment of the energy balance. Sci Total Environ 714:136655

Zaman D, Tiwari MK, Mishra S (2020) Low-cost adsorptive removal techniques for pharmaceuticals and personal care products. In: Measurement, analysis and remediation of environmental pollutants. Springer, Singapore, pp 397–421

Chapter 11
Prediction of Hospital Visits for Respiratory Morbidity Due to Air Pollutants in Lucknow

Shubhanshu Tripathi, Himanshu Sharma, and Tarun Gupta

Abstract Various epidemiological and toxicological studies have shown an association between air pollutants and their risk of respiratory morbidity and mortality. Only a few studies have been conducted in India, which evaluates the impact of seasonal air pollution on data on respiratory morbidity. Machine learning models like random forest regression are employed in the present context to predict the change in number of hospital visits for respiratory morbidity associated with the change in concentration of various air pollutants in the atmosphere and to study the effect of potential confounders like temperature and humidity, and also seasonal effect in Lucknow, India, for period of 2017–18. The results of the model revealed that a decrease of 16 patients daily is predicted if there is a reduction in the ambient concentration of $PM_{2.5}$ to National Ambient Air Quality Standards (NAAQS) in the city of Lucknow in one government hospital. SO_2 increases the number of respiratory patients as its ambient air concentration increases. It is observed that with 2 $\mu g/m^3$ increase from 18 to 20 $\mu g/m^3$ increased to nearly four patients. The synergistic effect of $PM_{2.5}$ and NO_2 is the most harmful for the citizens of Lucknow City. This study provides evidence that respiratory morbidity increases with an increase in the concentration of air pollutants in Lucknow. The post-monsoon season is considered as the most polluted period of the year with a higher number of hospital visits.

Keywords Morbidity · Mortality · Lucknow · $PM_{2.5}$ · Particulate matter · Confounders · Predictor · Exposure · NAAQS · CAAQMS

S. Tripathi · H. Sharma · T. Gupta (✉)
Department of Civil Engineering, Indian Institute of Technology Kanpur, Kanpur, Uttar Pradesh 208016, India
e-mail: tarun@iitk.ac.in

S. Singh et al. (eds.), *Pollution Control Technologies*, Energy, Environment, and Sustainability, https://doi.org/10.1007/978-981-16-0858-2_11

11.1 Introduction

There has been a large number of epidemiological studies conducted that assessed short-term health effects of fine particulate matter pollution, with emphasis on mortality and morbidity. Although the adverse health effects vary across the country predominantly due to geographical location, pollution levels, sources of emission, lifestyle, population distribution, etc, these epidemiological studies have focused mainly on particulate matter pollution (Health Effects Institute 2018; Anderson et al. 2001; Garrett and Casimiro 2011). According to the Health Effects Institute (2018), particulate matter (PM) air pollution from several significant sources was responsible for approximately 1.1 million deaths, or 10.6% of the total number of deaths in India in 2015. Combustion burning sources like crop residue burning, residual biomass burning, coal combustion, and transport, distributed diesel are the main contributors to $PM_{2.5}$ (Health Effects Institute 2018) attributable to disease burden. Earlier it was debated that low PM levels are not adversely affecting human health (Holland 1979) while, later, it was seen that PM is harmful even at low levels (Ware et al. 1981). There were several studies relating to ambient air pollution with daily admissions in hospitals (Schwartz 1998; Wang et al. 2013), but only a handful was performed in India. It says that cases of respiratory problems such as asthma in Delhi and Mumbai were predominant in regions that had a higher level of ambient air pollution. (Health Effects Institute 2010). Studies like (Awasthi et al. 2010; Gupta et al. 2016) were focused on episodic extreme air pollution levels attributed to crop burning in the post-monsoon period in India where the higher incidence of acute respiratory ailments was reported. One such study stated that ambient concentration of $PM_{2.5}$, SO_2, O_3, and NO_2 was firmly related to personal $PM_{2.5}$ exposure through which it can be said that ambient level can act as a surrogate for personal exposure (Sarnat et al. 2001).

Many studies incorporated time series to study the effects of air pollution on mortality and morbidity using classical regression models (Katsouyanni et al. 1997; Sarnat et al. 2001; Anderson et al. 2001; Garrett and Casimiro 2011). Time series analysis between air pollution and hospital visits for respiratory problems has shown a significant association between each other (Cropper et al. 1997). Some of these were conducted using machine learning algorithms to estimate the health outcomes from multiple sources and the possible co-relation and interactions between these variables (Patel 2017). A study was conducted to assess the relationship between emergency visits for pediatric asthma from pollutants like $PM_{2.5}$, CO, NO_2, and O_3 using classification and regression trees (CART) algorithm (Gass et al. 2014). Random forest, which work on the similar principle uses several such decision trees, was used in recent epidemiological studies like the one which (Kane et al. 2014) compares the random forest time series analysis model with standard time series method autoregressive integrated moving average (ARIMA) to predict outbreaks of avian influenza H5N1. The random forest approach performed better than the ARIMA model to predict outbreaks of avian influenza. Another study in the UK (Weng et al. 2019), which predicted premature all-cause mortality using random forest and standard epidemiological approaches, found random forest to perform better than existing conventional methods.

Epidemiological and biomedical research study encounters many confounders, and to understand them, the concept of directed acyclic graphs (DAGs) is used in such studies (Shrier and Platt 2008). Judea Pearl (Greenland et al. 1999) introduced DAGs in epidemiology, and since then, it has been used in many such studies to identify confounder as it has proven to reduce bias (Shrier and Platt 2008). A confounder has a causal relationship with both predictor variables and response variable, and it can be explained by a DAG as shown in Fig. 11.1. Ambient temperature has been studied as a cause of morbidity and mortality for hundreds of years (Basu and Samet 2002). In a review study, which identified 38 studies on morbidity due to temperature concluded that there is 2.0% increase in respiratory illnesses with an increase of 10 °C in temperature (Turner et al. 2012). A study in Texas reported that there was a mortality risk between 0.1 and 5% for a 1 °C decrease in temperature below the cold threshold (Chen et al. 2017). Temperature also affects the concentration of pollutants in the ambient atmosphere. During inversion, mixing is constricted during stable atmospheric condition as it decreases vertical airflow and traps the pollutants below inversion layer, which in turn increase the concentration of pollutants at ground level (Oke 2002). In Delhi, it has been found that the concentration of ground-level ozone increases in summers and decreases in winters and monsoon seasons (Jain et al. 2005). Relative humidity also enhances the respiratory problems in people suffering from chronic obstructive pulmonary disorder (COPD) at low temperatures (Mu et al. 2017). In a study done in Cairo, Egypt provides evidence that the relative humidity changes the tropospheric air pollutant concentration in an urban city, the highest average concentrations for sulfur dioxide and particulate matter occurred when relative humidity was above 80% (Elminir 2005).

In this study, we predict the number of patients who are registering at respiratory outpatient department (OPD) by developing a random forest regression model which takes concentration of pollutants as predictor variables and the number of patient visits as the response variable. Temperature and relative humidity were identified as confounders and were taken into account for prediction. Individual and synergistic effects of pollutants were determined by varying the concentration of different air pollutants such as $PM_{2.5}$, SO_2, O_3, and NO_2 in the model and analyzing

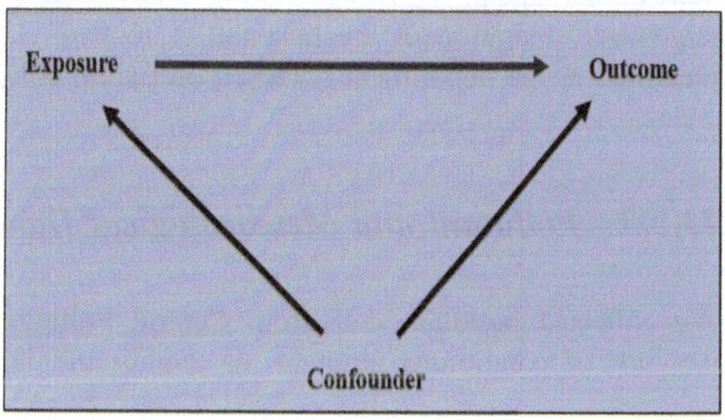

Fig. 11.1 DAG representing a confounder

the trends of patient visits. We also determined at what time of the year the highest and the lowest number of patient visits has been recorded and the corresponding scenario of concentration of air pollutants.

11.2 Materials and Methods

11.2.1 Study Area

This study has been carried out in the city of Lucknow (26° 51′ N, 80° 55′ E), the capital city of Uttar Pradesh, which is the most populated state of India. It is the second-largest city in the northern and central regions of India (Lawrence and Fatima 2014), with an area of 310.1 km^2. It lies in the central part of Indo-Gangetic Plains (IGP), which is a highly polluted region in terms of air quality. Lucknow is one of the most polluted cities in Southeast Asia, the most polluted region in the world. Due to the lack of adequate public transport, most people use their own vehicles. There are around 13 lakh vehicles in Lucknow in 2011–12, with an increase of 8.68% (Kisku et al. 2013). The pollution sources in the city include vehicular emissions, construction activities, industrial emissions, and biomass burning. Lucknow lies in the downwind direction, which brings crop burning residue and other pollutants from the upwind direction (Lawrence and Fatima 2014). The city was chosen for the study because of its variation in pollution levels throughout the year and its administrative and industrial importance.

11.2.2 Hospital Data

Health data was obtained from a government hospital named Shyama Prasad Mukharji (Civil) Hospital (260 50′ 40.83″ N, 800 56′ 56.50″ E), which is located in Hazratganj, Lucknow. We collected data from the registration office of the respiratory outpatient department (OPD) for the daily number of patients who are going to the respiratory department from July 1, 2017, to July 31, 2018. The total number of patients registered during this period was 36,019, which shows the large-scale operation of the hospital. A total of 323 days of observations were identified as the department is closed on public holidays.

11.2.3 Pollutant and Meteorological Data

We collected pollutant data from Central Pollution Control Board's (CPCB's) network of continuous ambient air quality monitoring stations (CAAQMS) in

Lucknow. The stations were located in Lalbagh, Nishatganj, Central School, and Talkatora industries. The location of these four monitoring stations and the hospital is given in Fig. 11.2. The data was retrieved from the CPCBs Web site (*CCR*, n.d.), which generates real-time data from CAAQMS. An average of 96 observations was taken from 00.00 to 23.59 h to calculate the daily average concentration of a pollutant in micrograms per cubic meter (μg/m^3). In this study, daily average concentrations of $PM_{2.5}$, SO_2, NO_2, and O_3 were considered as they are attributed to higher incidences of respiratory illnesses. For calculating the daily average concentration of pollutants in the city of Lucknow, we take the average of four observations and we obtained from four monitoring stations. By doing this, the bias is reduced, which may be increased if we take data from one station.

We take two meteorological parameters temperature and relative humidity (%) as our potential confounders in this study. For relative humidity, we calculate the average of the daily average humidity obtained from the four monitoring stations for the period between July 2017 and July 2018. We take the daily average temperature in degree Celsius from the weather underground Web site (*Local Weather Forecast, News and Conditions | Weather Underground*, n.d.), which provides real-time and historical weather information all around the world.

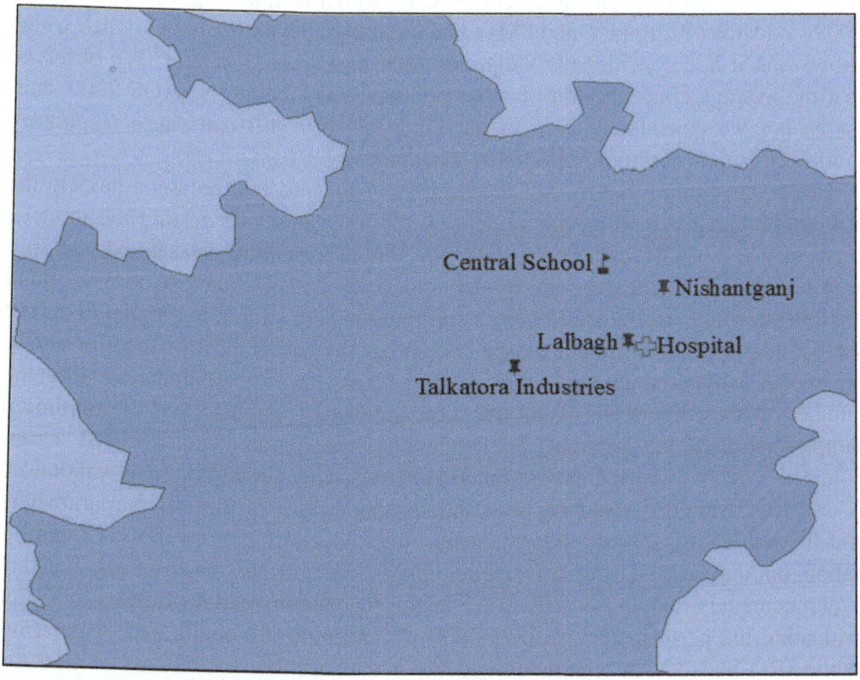

Fig. 11.2 Location of automatic monitoring stations and civil hospital in Lucknow

11.2.4 Data Formatting

We arrange the data in datewise format where the number of respiratory patients, the concentration of pollutants, temperature, and relative humidity were taken in separate columns. The concentration of pollutants and meteorological parameters was taken as predictor variables. Thus, there were in total six predictor variables, namely $PM_{2.5}$, SO_2, NO_2, O_3, relative humidity (%), and temperature (°C). For the time series analysis, we lagged all the six predictor variables by 24 and 48-h such that each predictor had three variables, i.e., for the same day, yesterday, and the day before yesterday.

11.2.5 Data Analysis

We analyze the daily variation of pollutants and meteorological parameters from July 2017 to July 2018 in Lucknow and plotted them, as shown in Fig. 11.3. We observed that the pollution levels are higher during the post-monsoon season from October 2017 to February 2018 and lower during monsoon season (June–July). We then plot a graph shown in Fig. 11.4, the average and standard deviation of predictor variables for 323 data points. It was observed that the average concentration of $PM_{2.5}$ was 114.17 μg/m^3 in Lucknow City, which is much higher than NAAQS 2009 standards (40 μg/m^3), and $PM_{2.5}$ had a standard deviation of 81 μg/m^3, which shows that it had considerable variation much throughout the year. We observed that the average concentration of other pollutants was below NAAQS 2009 standards, but we considered them in our study as they still can cause respiratory morbidity at concentration reported in Lucknow.

Before developing the model, the data was split into two datasets, namely the training dataset for training the model and the test dataset for testing the model's performance. For this, 80% of the data was used as a training dataset and rest 20% as a test dataset to measure the model's accuracy and standard error. We added random state and seed function before splitting the data so that the model gives the same random split next time it is used to compare the model performance for tuning the model's hyperparameters. The random forest has hyperparameters like the number of trees, the depth of the trees, the number of features, and the minimum samples required for the split. For evaluating hyperparameters, we used a randomized search cross-validation function along with threefold cross-validation. A grid was made using starting, end and spacing values of these hyperparameters and the model was cross-validated using threefold combination on every combination on that grid. Using this methodology, we got the best values for that hyperparameters and those were used in the regression model (Table 11.1). For evaluating the performance of the model, we calculate the coefficient of determination (R^2) between the predicted and the actual number of patients to find the

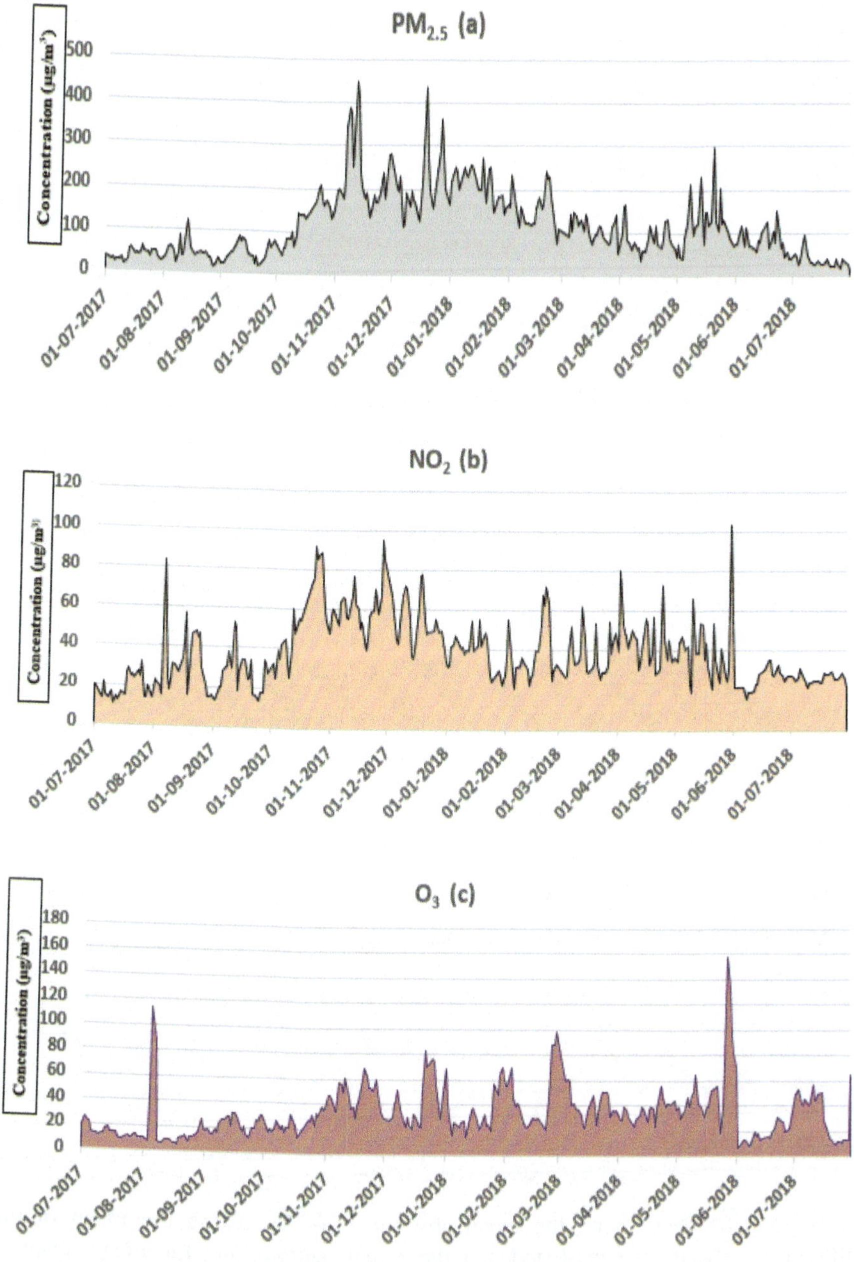

Fig. 11.3 Yearlong variation of predictor variables **a** $PM_{2.5}$. **b** NO_2 (Nitrous oxide). **c** O_3 (Ozone). **d** SO_2 (Sulfur dioxide). **e** Relative humidity. **f** Temperature

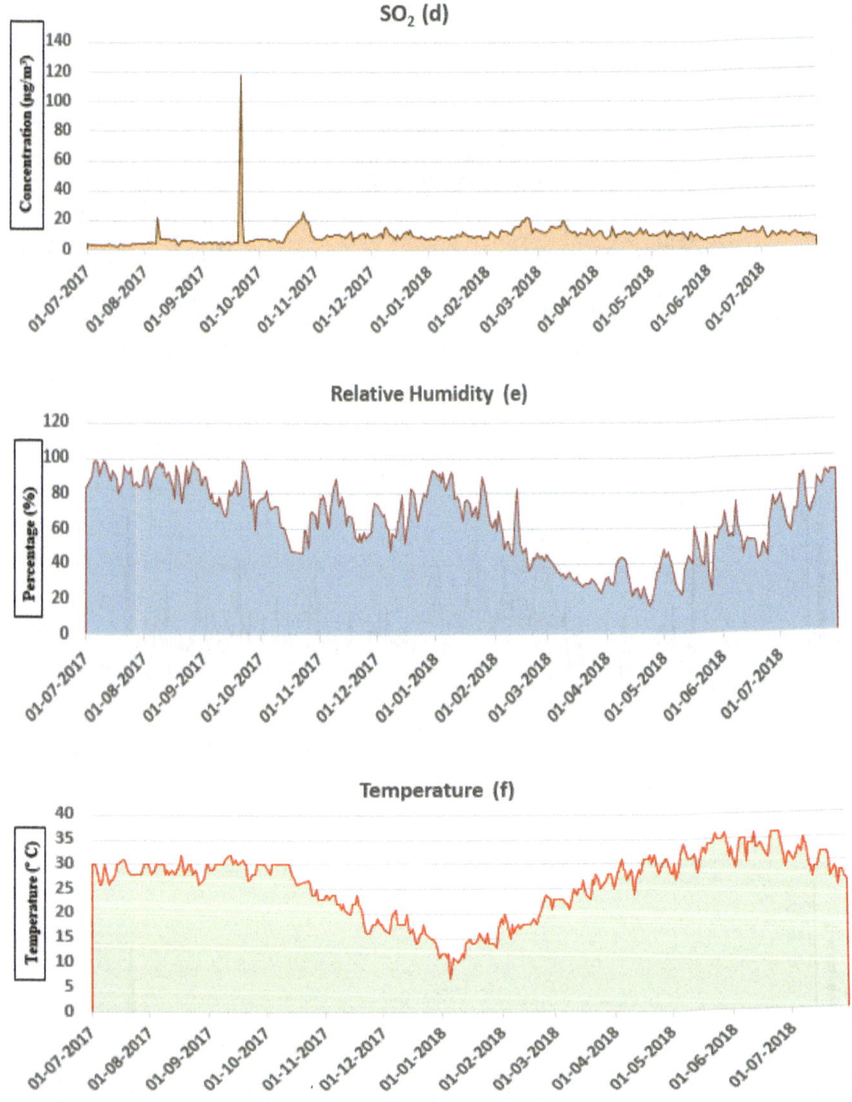

Fig. 11.3 (continued)

association between them, the mean absolute error by taking the mean of the difference between the predicted and the actual patients and the mean absolute percentage accuracy (which is equal to (100—mean absolute percentage error)).

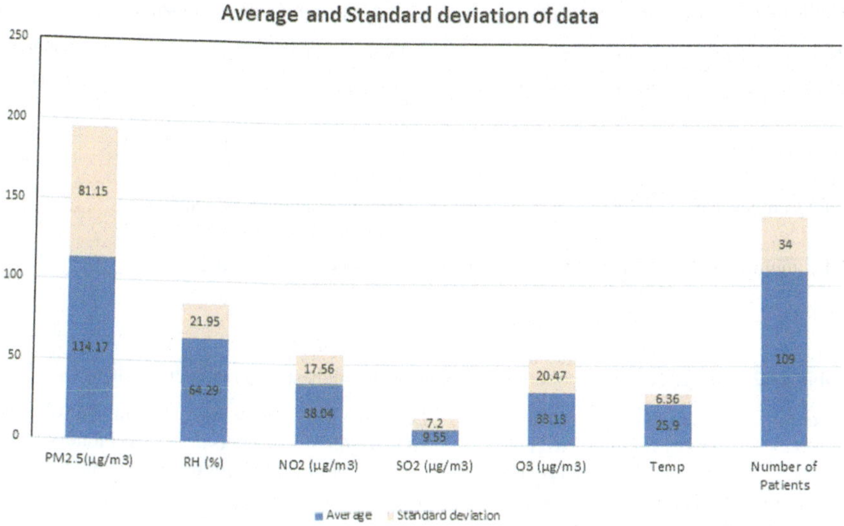

Fig. 11.4 Average and standard deviation of the yearlong observation

11.3 Results and Discussions

11.3.1 Causation Analysis

Before making a regression model, it is necessary to analyze the causation between the predictor and outcome variables. To establish it, we found the concentration of pollutants on the day of the top 10 highest patient visits for 323 observations recorded between July 2017 and July 2018, as shown in Table 11.2. And, in order to compare this, the observations on the days of the top 10 lowest visits as shown in Table 11.3 were sorted. It is seen from Table 11.2 that the highest number of patient visits is observed in most polluted days. We observed that seven out of these ten observations from Table 11.2 are in the month of October, which coincide with large-scale crop residue burning activity rampant throughout the IGP, very prominent in the Punjab and Haryana states upstream of Lucknow. The day when highest patients came to the hospital, the $PM_{2.5}$ concentration was measured to be 199.02 µg/m^3, which is more than three times of NAAQS of 60 µg/m^3, and it was also more than sum (195.32 µg/m^3) of annual average (114.17 µg/m^3) and one standard deviation (81.15 µg/m^3) of the 323 observations in the data given in Fig. 11.4. We then plotted a graph, as shown in Fig. 11.5, in order to compare the average concentration of pollutants on the days of highest and lowest patient visits at the hospital. We had seen from Table 11.3 that five out of the ten days in the lowest patient visits days were from the month of July, which is one of the cleanest months in terms of air quality. Monsoon rains are attributed to settling down of the dust and small particles, which decreases the pollutant

Table 11.1 Best hyperparameters values obtained after threefold cross-validation

Hyperparameters	Grid values	Description	Values
N estimators	Start = 50 Stop = 2000 Spacing = 30	Number of trees	1394
Min samples split	2,5,10	Minimum number of samples to split a node	2
Min samples leaf	1,2,4	Minimum number of samples required at a leaf node	1
Maximum depth	Start = 10 Stop = 200 Spacing = 16	Maximum depth of trees	149

Table 11.2 Concentration of pollutants on days of the highest respiratory morbidity

Date	Daily average $PM_{2.5}$ (μg/m^3)	Daily average O_3 (μg/m^3)	Daily average NO_2 (μg/m^3)	Daily average SO_2 (μg/m^3)	Number of respiratory patients
24-10-2017	199.02	26.64	75.93	21.08	206
16-10-2017	140.13	22.25	55.24	9.52	199
25-10-2017	208.16	31.81	91.43	26.10	196
21-08-2017	47.68	12.45	46.55	6.59	180
05-10-2017	59.55	17.27	39.01	7.40	179
30-10-2017	157.63	37.45	59.05	10.15	178
03-08-2017	42.76	8.39	21.62	4.9	173
06-11-2017	178.65	33.76	51.58	8.59	172
26-10-2017	176.24	21.8	84.93	22.06	171
12-10-2017	91.94	21.00	40.14	6.05	170

concentration. Most of the days shown in Table 11.3 have concentration well below the NAAQS, by which we can say the respiratory morbidity increases with an increase in the concentration of pollutants and vice versa.

11.3.2 Random Forest Regression Model

We made our predictions on the test dataset, which had 66 observations, and the predicted values were rounded off to the nearest natural number as the number of patients cannot be in decimal values. The model gave the predictions of the number of respiratory patients on the corresponding values of predictor variables (without time-lagged variables) given in the test dataset. Figure 11.6 gives a comparison of the predicted number of patients and actual patients that visited the hospital on the randomly selected 66 observations (20% of 323 observations). From Fig. 11.6, it can be observed that the model is able to detect the peaks and lows in the number of patients, but it is not able to reach the highest peaks and lowest lows. For this, the

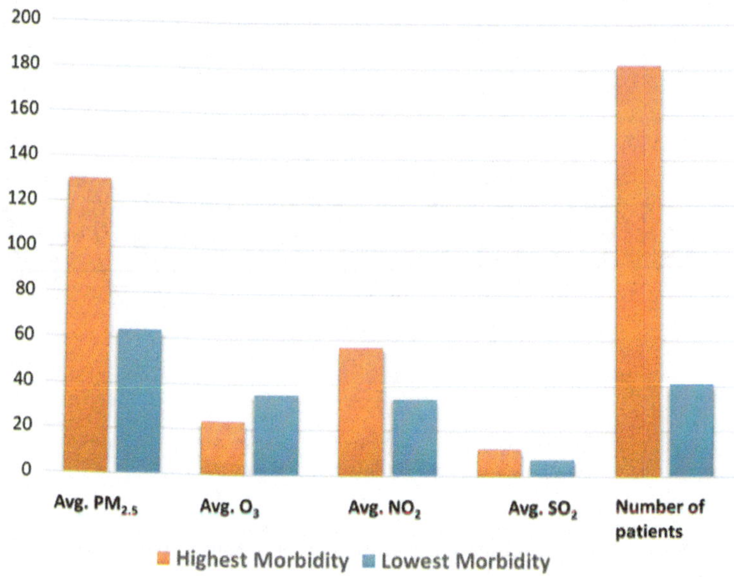

Fig. 11.5 Comparison of the average concentration of pollutants on days of highest and lowest morbidity

Table 11.3 Concentration of pollutants on days of lowest respiratory morbidity

Date	Daily average $PM_{2.5}$ (μg/m³)	Daily average O_3 (μg/m³)	Daily average NO_2 (μg/m³)	Daily average SO_2 (μg/m³)	Number of respiratory patients
02-09-2017	22.39	18.12	13.39	4.69	33
07-08-2017	48.19	86.51	83.15	3.79	38
14-04-2018	59.15	38.52	39.45	10.31	38
14-07-2018	33.66	49.34	24.33	9.56	39
26-07-2018	17.62	11.04	25.61	7.59	40
12-05-2018	121.92	41.45	38.95	10.87	43
13-01-2018	255.70	19.63	39.94	9.39	43
30-07-2018	27.72	12.79	25.79	6.01	45
31-07-2018	13.35	64.61	20.97	5.81	47
28-07-2018	33.69	13.60	29.10	6.85	48

model's performance parameters obtained are shown in Table 11.4. In order to improve the model's performance, we introduced the time-lagged concentration of pollutants and the confounders as additional predictor variables of one day and two days before so that the number of variables is now increased from six to eighteen. The time-lagged concentrations of pollutants were used as additional predictor variables as the pollutant concentration of one day and two days before might affect

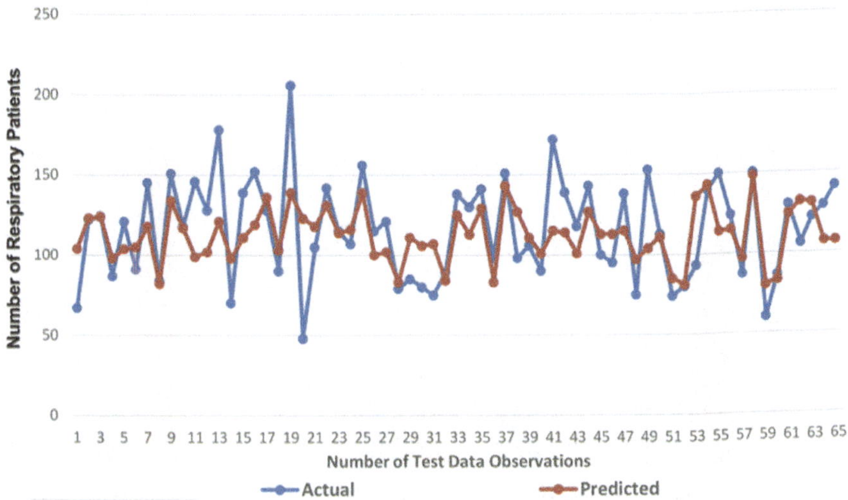

Fig. 11.6 Comparison of the predicted and actual number of patients on test dataset for random forest regression

Table 11.4 Model's performance parameters for random forest regression

Model's performance parameters	Values
Coefficient of determination (R^2)	0.35
Mean absolute error	20 patients
Mean absolute percentage accuracy	81.4%

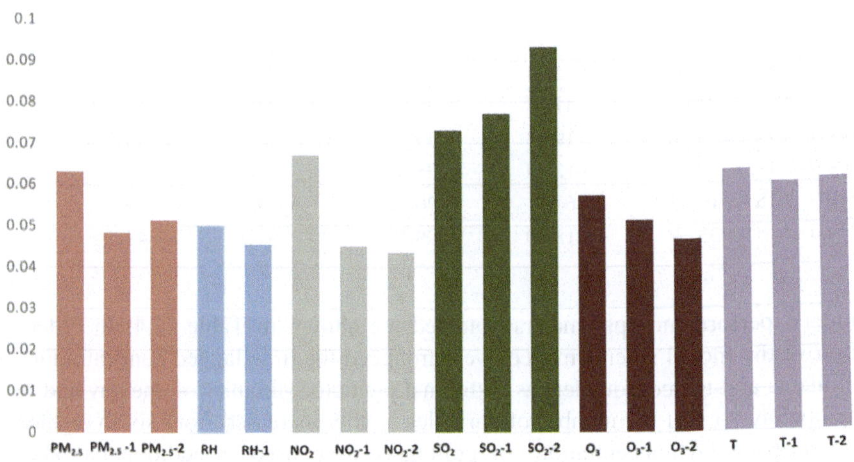

Fig. 11.7 Feature importance of predictor variables in the time-lagged random forest model

the number of respiratory patients coming on a particular day. Feature importance function was used from Scikit-learn library to obtain the importance of each predictor variable. The graph of feature importance is shown in Fig. 11.7. It gives importance to a feature based on how the accuracy of the model decreases if a particular variable is removed. Based on the magnitude of the change in accuracy, it gives the value of the importance of a predictor variable. We used seventeen variables; 48-h lagged relative humidity was the least important variable in predicting, so it was dropped. Hyperparameter tuning using random search threefold cross-validation for the time series lagged model gave the best hyperparameters values as given in Table 11.5. The model with lagged predictor variables was tested on the test dataset, and random state and seed function were applied so that the splitting of data is similar, so that model's performance can be judged with the earlier model. Figure 11.8 gives the comparison obtained between the predicted number of patients and the actual number of patients coming to the hospital from this model. It was observed that this model is giving predictions that are much closer to the actual number of patients, but still, the highest peaks are not reached in the prediction from the model. Seasonal analysis might give better results in giving the highs and lows of the number of patients. The performance parameters are given in Table 11.6, which suggests that after adding the time-lagged predictor variables, the model's performance improved as R^2 score and mean absolute percentage accuracy were higher.

11.3.3 Individual Pollutant and Confounders Effect on Respiratory Morbidity

For interpretation, the random forest is not as good as other models like logistic regression, linear regression, since this model does not gives an equation describing the relationship between predictor variables (pollutants and confounders) and the response variable (daily number of patients visits). This model is a black box (Breiman 2001); there was no equation that could report this. Individual pollutants are taken, and their concentration is then increased by fixed interval (keeping other parameters at the average level obtained for 323 observations), and the output (patients) was recorded to assess its association with respiratory morbidity. Starting with $PM_{2.5}$, it was varied from 30 to 180 μg/m^3 with an interval of 10 μg/m^3 (keeping the same increment in 24-h and 48-h lagged concentration of $PM_{2.5}$) as shown in Fig. 11.9. The increase in the number of patient visits saturated after 160 μg/m^3. According to the model, if the concentration of $PM_{2.5}$ is restricted to NAAQS (60 μg/m^3), a decrease of 16 patients daily at respiratory OPD is observed and with comparing from WHO guidelines (25 μg/m^3), a decrement of 32 patients is observed. The results are from one hospital of Lucknow, and there are several other hospitals. If on an average, the amount spent for treatment of one patient is rupees 100, this will save approximately 5 lakh rupees annually (according to

Table 11.5 Best hyperparameters for time series model obtained after threefold cross-validation

Hyperparameters	Grid values	Description	Values
N estimators	Start = 50 Stop = 2000 Spacing = 30	Number of trees	219
Min samples split	2,5,10	Minimum number of samples to split a node	5
Min samples leaf	1,2,4	Minimum samples required at a leaf node	1
Maximum depth	Start = 10 Stop = 400 Spacing = 50	Maximum depth of trees	352

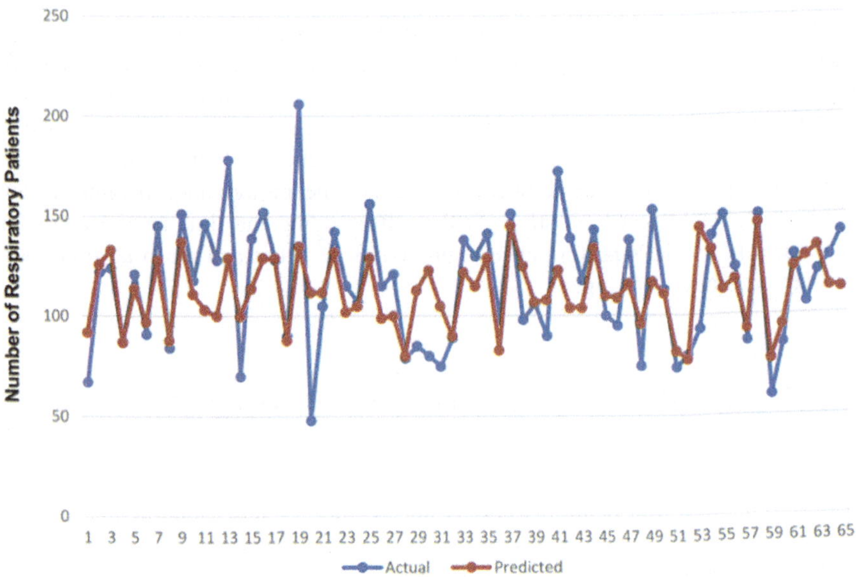

Fig. 11.8 Comparison of the predicted and actual number of patients on test dataset for time-lagged random forest regression

Table 11.6 Model's performance parameters for time-lagged random forest regression

Model's performance parameters	Values
Coefficient of determination (R^2)	0.43
Mean absolute error	18 patients
Mean absolute percentage accuracy	82.9%

NAAQS decrement of $PM_{2.5}$). For assessing the effect of NO_2, $PM_{2.5}$ was kept at 24-hour average standard given in NAAQS for 24-hour average (while others were kept at average obtained from 323 observations), i.e., 60 μg/m^3 because the annual average concentration of $PM_{2.5}$ is very high in Lucknow. NO_2 has an increase in the

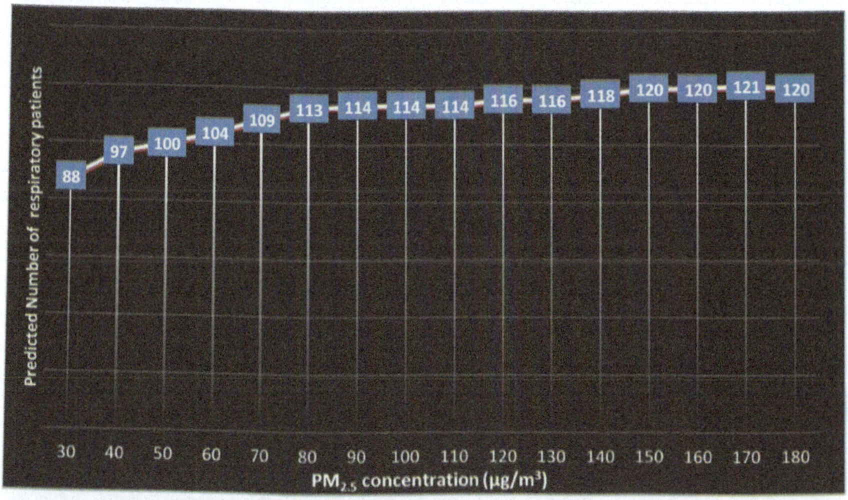

Fig. 11.9 Change in respiratory patients with increment in $PM_{2.5}$ concentration

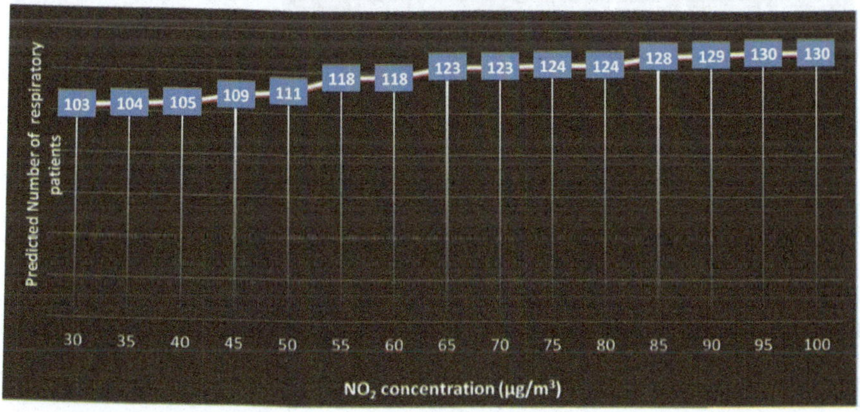

Fig. 11.10 Change in respiratory patients with increment in NO_2 concentration

number of patients as its concentration increases, and it gets saturated after an increase to 90 µg/m^3 to 130 patients, as shown in Fig. 11.10. It was observed from Fig. 11.10 that at NAAQS for NO_2 (80 µg/m^3 for 24-h mean average), the number of patients is 124. This high number of patients suggests that there is a need to lower down the NO_2 levels to even below the NAAQS. At 38 µg/m^3, the increment of 5 µg/m^3 will lead to an increase of 4 patients daily. SO_2 had the highest feature importance among other pollutants. With 2 µg/m^3 increase from 18 to 20 µg/m^3, it was observed that there is an increase of nearly 4 patients, as shown in Fig. 11.11. This was a significant amount of increase in the number of patients with a small increase in SO_2 concentration in the ambient atmosphere. As shown in Fig. 11.12,

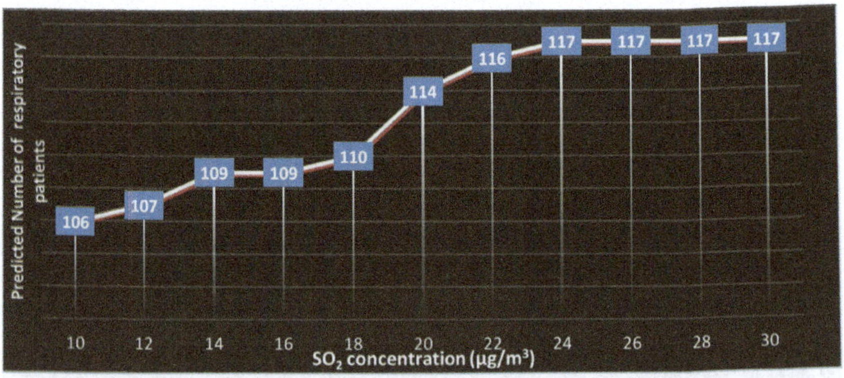

Fig. 11.11 Change in respiratory patients with increment in SO_2 concentration

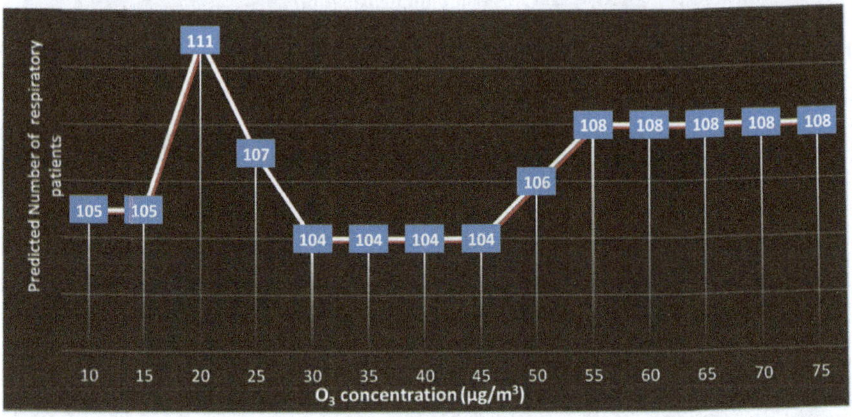

Fig. 11.12 Change in respiratory patients with increment in O_3 concentration

for ground-level ozone, the model suggested that it does not have a direct relationship with the number of patient visits. Some studies have shown the relation of ground-level ozone with respiratory morbidity (Bernstein et al. 2004), which led to hypothesizing that ozone might become harmful in combination with other pollutants. From Fig. 11.13, it can be seen that in the winter season, the number of respiratory patients increased until temperature reached 26 °C. Beyond 26 °C, in summers, it is observed in Fig. 11.13 that the number of respiratory patients is lesser compared to winters. Relative humidity (RH) has also been attributed to an increase in COPD patients and hence increased the number of respiratory patients (Mu et al. 2017). Relative humidity varies from 25 to 99% in the Lucknow City; its effect on the number of patient visits is shown in Fig. 11.14. There was no direct relationship obtained between them. At very high humidity, above 90%, RH increment was accompanied by an increase in the number of respiratory patients. And between 70 and 80%, there was a decrease obtained in the number of patients.

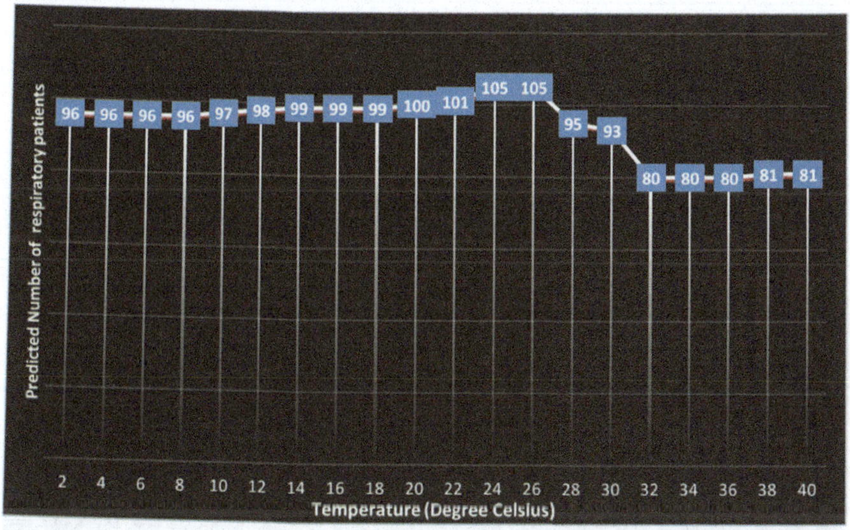

Fig. 11.13 Change in respiratory patients with increment in temperature

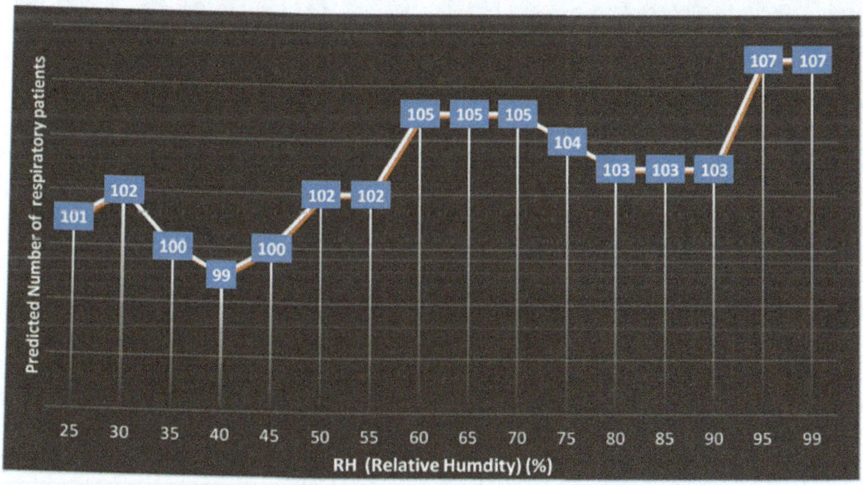

Fig. 11.14 Change in respiratory patients with increment in relative humidity

11.3.4 Synergistic Effect of Pollutants

Health studies have shown short-term exposure to $PM_{2.5}$ increases the risk for hospital admission for cardiovascular and respiratory diseases (Dominici et al. 2006). $PM_{2.5}$ has a very high annual average concentration of 134 $\mu g/m^3$ in Lucknow. This explains that $PM_{2.5}$ is the key causal factor for respiratory morbidity

Table 11.7 Number of patients above daily annual average concentration of pollutants

Pollution levels	Daily average $PM_{2.5}$ ($\mu g/m^3$)	Daily average O_3 ($\mu g/m^3$)	Daily average NO_2 ($\mu g/m^3$)	Daily average SO_2 ($\mu g/m^3$)	Number of respiratory patients
μ	114.17	33.13	38.04	9.55	118
$\mu + \sigma$	195.32	53.6	55.6	16.75	127
$\mu + 2\sigma$	276.47	74.07	73.16	23.95	135

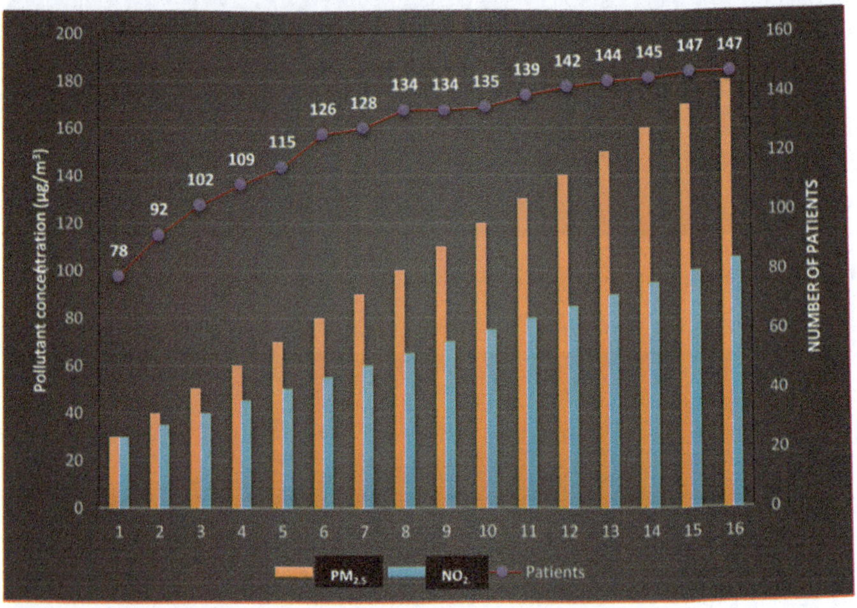

Fig. 11.15 Combine effect of $PM_{2.5}$ and NO_2 on respiratory morbidity

compared to any other pollutants. A combination of $PM_{2.5}$ and other pollutants as given in Table 11.7 were analyzed to evaluate their effect on daily respiratory patient visits in a hospital. The effect of $PM_{2.5}$ and NO_2 is shown in Fig. 11.15. Both $PM_{2.5}$ and NO_2 were varied in the same manner as they were varied during assessing the individual effect. From Fig. 11.15, it was observed that the patients are increasing at a higher slope in the initial phase, where the concentration is low and it reached saturation at 147 patient visits. This suggested that the increase in the number of patients is higher, with a small increase in concentration at lower levels. The combination of $PM_{2.5}$ and SO_2 can also increase respiratory morbidity, as shown in Fig. 11.16. $PM_{2.5}$ was varied in the same manner, and SO_2 increased from 8 to 38 $\mu g/m^3$ with an interval of 2 $\mu g/m^3$. The slope is observed to be higher at low concentrations and reaches almost zero as the concentration increases. The increment saturated at 132 number of patients. If ozone combines with $PM_{2.5}$, then

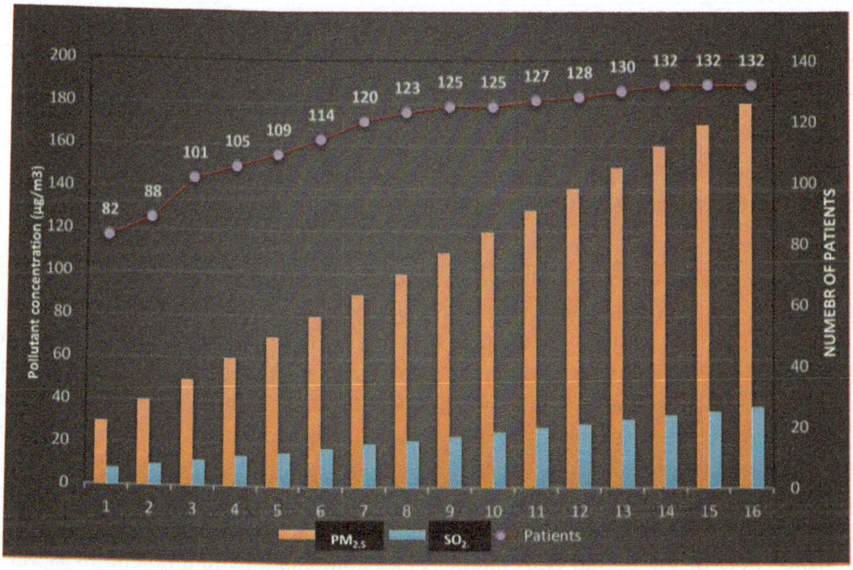

Fig. 11.16 Combine effect of $PM_{2.5}$ and SO_2 on respiratory morbidity

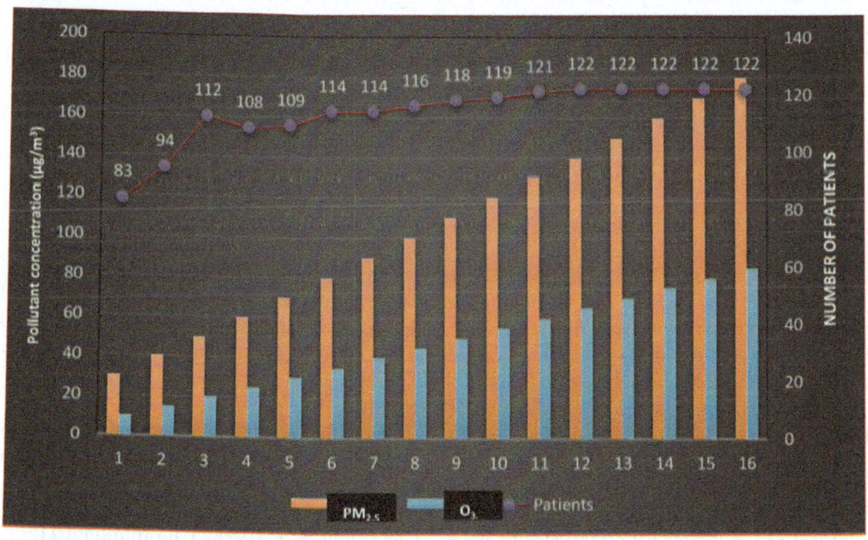

Fig. 11.17 Combine effect of $PM_{2.5}$ and O_3 on respiratory morbidity

there was an increase in respiratory morbidity, as shown in Fig. 11.17. $PM_{2.5}$ was varied in a similar manner, and ozone increased between 10 and 85 µg/m^3 with an interval of 5 µg/m^3. The results are the same; at lower levels, the increment in the patient's visits is observed to be very high and later it saturated at higher levels of concentration of pollutants. $PM_{2.5}$ and NO_2 was considered as more harmful than others as they saturated at 147 patients, which is highest among all other combinations.

11.4 Conclusions

This study demonstrated the change in trends of the number of patients visits for respiratory morbidity in association with the change in concentration of various air pollutants in Lucknow City for the period of 2017–2018. The post-monsoon season which is the most polluted season in the year, recorded the highest daily registration of patients with respiratory ailments. The model results show that $PM_{2.5}$, SO_2, and NO_2 increased the number of patients when their ambient air concentration increased while ground-level ozone did not give a direct relationship with the number of patients. The synergistic effect of NO_2, SO_2 and O_3 with $PM_{2.5}$ shows an increasing trend of the number of patients visits. This study shows that the synergistic effect of $PM_{2.5}$ and NO_2 among other combinations in ambient air is very harmful for the citizens of Lucknow City as the number of patients was saturated at 147 in the regression model, which is the highest. It is not feasible to suggest that the concentration of pollutants can be brought below NAAQS. However, appropriate mitigation measures should be taken to improve air quality which might look like an economic hindrance, but in the long run, it will improve the health of the citizens and the economy of our country.

References

Anderson HR, Bremner SA, Atkinson RW, Harrison RM, Walters S (2001) Particulate matter and daily mortality and hospital admissions in the west midlands conurbation of the United Kingdom: associations with fine and coarse particles, black smoke and sulphate, pp 504–510

Awasthi A, Singh N, Mittal S, Gupta PK, Agarwal R (2010) Effects of agriculture crop residue burning on children and young on PFTs in North West India. Sci Total Environ 408(20):4440–4445. https://doi.org/10.1016/j.scitotenv.2010.06.040

Basu R, Samet JM (2002) Relation between elevated ambient temperature and mortality: a review of the epidemiologic evidence. Epidemiol Rev 24(2):190–202. https://doi.org/10.1093/epirev/mxf007

Bernstein JA, Alexis N, Barnes C, Bernstein IL, Bernstein JA, Nel A, Peden D, Diaz-Sanchez D, Tarlo SM, Williams PB (2004) Health effects of air pollution. J Allergy Clin Immunol 114(5):1116–1123. https://doi.org/10.1016/j.jaci.2004.08.030

Breiman L (2001) Statistical modeling: the two cultures. Stat Sci 16(3):199–215. https://doi.org/10.1214/ss/1009213726

CCR (n.d.) Retrieved 2 Feb 2020, from https://app.cpcbccr.com/ccr/#/caaqm-dashboard-all/caaqm-landing

Chen TH, Li X, Zhao J, Zhang K (2017) Impacts of cold weather on all-cause and cause-specific mortality in Texas, 1990–2011. Environ Pollut 225:244–251. https://doi.org/10.1016/j.envpol.2017.03.022

Cropper ML, Simon NB, Alberini A, Arora S, Sharma PK (1997) Agricultural and applied economics association the health benefits of air pollution control in Delhi the health benefits of air pollution control in Delhi. Am J Agr Econ 79(5):1625–1629

Dominici F, Peng RD, Bell ML, Pham L, McDermott A, Zeger SL, Samet JM (2006) Fine particulate air pollution and hospital admission for cardiovascular and respiratory diseases. J Am Med Assoc 295(10):1127–1134. https://doi.org/10.1001/jama.295.10.1127

Elminir HK (2005) Dependence of urban air pollutants on meteorology. Sci Total Environ 350(1–3): 225–237. https://doi.org/10.1016/j.scitotenv.2005.01.043

Garrett P, Casimiro E (2011) Short-term effect of fine particulate matter ($PM_{2.5}$) and ozone on daily mortality in Lisbon, Portugal. Environ Sci Pollut Res 18(9):1585–1592. https://doi.org/10.1007/s11356-011-0519-z

Gass K, Klein M, Chang HH, Dana Flanders W, Strickland MJ (2014) Classification and regression trees for epidemiologic research: an air pollution example. Environ Health: Global Access Sci Source 13(1):1–10. https://doi.org/10.1186/1476-069X-13-17

Greenland S, Pearl J, Robins JM (1999) Causal diagrams for epidemiologic research. Epidemiology 10(1):37–48. https://doi.org/10.1097/00001648-199901000-00008

Gupta S, Agarwal R, Mittal SK (2016) Respiratory health concerns in children at some strategic locations from high PM levels during crop residue burning episodes. Atmos Environ 137:127–134. https://doi.org/10.1016/j.atmosenv.2016.04.030

Health Effects Institute (2010) Outdoor air pollution and health in the developing countries of Asia: a comprehensive review. Special Report 18, Boston, MA, Nov, p 284

Health Effects Institute (2018) Burden of disease attributable to major air pollution sources in India. Special Report 21. Health Effects Institute, Boston, MA, Jan. https://doi.org/10.1186/cc7871

Holland. (1979). Health effects of participate pollu-tion: reappraising the evidence. *American Journal of Epidemiology*, *63*, 328–331. https://www.cabdirect.org/cabdirect/abstract/19501100562

Jain SL, Arya BC, Kumar A, Ghude SD, Kulkarni PS (2005) Observational study of surface ozone at New Delhi, India. Int J Remote Sens 26(16):3515–3524. https://doi.org/10.1080/01431160500076616

Kane MJ, Price N, Scotch M, Rabinowitz P (2014) Comparison of ARIMA and Random Forest time series models for prediction of avian influenza H5N1 outbreaks. BMC Bioinf 15(1). https://doi.org/10.1186/1471-2105-15-276

Katsouyanni K, Touloumi G, Spix C, Schwarte J, Balducci F, Medina S, Rossi G, Wojtyniak B, Sunyer J, Bacharova L, Schouten JP, Ponka A, Anderson HR (1997) Short term effects of ambient sulphur dioxide and particulate matter on mortality in 12 European cities: results from time series data from the APHEA project. BMJ 314(7095):1658–1663. https://doi.org/10.1136/bmj.314.7095.1658

Kisku GC, Pradhan S, Khan AH, Bhargava SK (2013) Pollution in Lucknow City and its health implication on exposed vendors, drivers and traffic policemen. Air Qual Atmos Health 6(2): 509–515. https://doi.org/10.1007/s11869-012-0190-7

Lawrence A, Fatima N (2014) Urban air pollution and its assessment in Lucknow City—the second largest city of North India. Sci Total Environ 488–489(1):447–455. https://doi.org/10.1016/j.scitotenv.2013.10.106

Local Weather Forecast, News and Conditions| Weather Underground (n.d.). Retrieved 2 Feb 2020, from https://www.wunderground.com/

Mu Z, Chen PL, Geng FH, Ren L, Gu WC, Ma JY, Peng L, Li QY (2017) Synergistic effects of temperature and humidity on the symptoms of COPD patients. Int J Biometeorol 61(11):1919–1925. https://doi.org/10.1007/s00484-017-1379-0

Oke TR (2002) Boundary layer climates. Geog Rev 69(4). https://doi.org/10.2307/214824

Patel CJ (2017) Analytic complexity and challenges in identifying mixtures of exposures associated with phenotypes in the exposome era. Curr Epidemiol Rep 4(1):22–30. https://doi.org/10.1007/s40471-017-0100-5

Sarnat JA, Schwartz J, Catalano PJ, Suh HH (2001) Gaseous pollutants in particulate matter epidemiology: confounders or surrogates? Environ Health Perspect 109(10):1053–1061. https://doi.org/10.1289/ehp.011091053

Schwartz J (1998) Air pollution and hospital admissions for cardiovascular disease. Epidemiology 9(2):220–221. https://doi.org/10.1097/00001648-199803000-00028

Shrier I, Platt RW (2008) Reducing bias through directed acyclic graphs. BMC Med Res Methodol 8:1–15. https://doi.org/10.1186/1471-2288-8-70

Turner LR, Barnett AG, Connell D, Tong S (2012) Ambient temperature and cardiorespiratory morbidity: a systematic review and meta-analysis. Epidemiology 23(4):594–606. https://doi.org/10.1097/EDE.0b013e3182572795

Wang X, Chen R, Meng X, Geng F, Wang C, Kan H (2013) Associations between fine particle, coarse particle, black carbon and hospital visits in a Chinese city. Sci Total Environ 458–460:1–6. https://doi.org/10.1016/j.scitotenv.2013.04.008

Ware JH, Thibodeau LA, Speizer E, Colome S, Benjamin G (1981) Assessment of the health effects of atmospheric sulfur oxides and particulate matter: evidence from observational studies. Environ Health Perspect 41(Oct):255–276

Weng SF, Vaz L, Qureshi N, Kai J (2019) Prediction of premature all-cause mortality: a prospective general population cohort study comparing machine-learning and standard epidemiological approaches. PLoS One 14(3):1–22. https://doi.org/10.1371/journal.pone.0214365

Chapter 12
Technologies for Controlling Particulate Matter Emissions from Industries

Manoranjan Sahu, Vasudev Malyan, and Y. S. Mayya

Abstract In recent years, deterioration of air quality has become a matter of serious concern due to its adverse impacts on human health, local and global climates. These concerns will continue to remain a challenge owing to the inevitable industrial growth to achieve economic progress of the population coupled with inadequate implementation of air pollution control measures. Reducing the emission of pollutants to the atmosphere is indispensable to protect the environment and improve the quality of life. Effective control strategies must be developed to cope up with the existing pollution level and reduce the pollutants emission from various sources. Apart from control strategies such as limiting the emissions at the source and tightening regulations on air pollution, development and deployment of air pollution control devices are necessary to reduce the pollutants before emitting to atmosphere. The aim of this chapter is to review the current available devices used to control particulates in industries and recent development of new emerging technologies to control particulate matter emission.

Keywords Air pollution · Control strategies · Particulate matter · Emerging technologies

12.1 Introduction

Air pollution is one of the leading causes for multitude of adverse impact on human health, ecosystem, and global climate. The World Health Organization (WHO) declared air pollution as the largest environmental health risk causing nearly 7 million deaths in 2012 (WHO 2014). It is one of the leading risk factors to

M. Sahu (✉) · V. Malyan
Environmental Science and Engineering Department, Indian Institute of Technology, Bombay 400076, India
e-mail: mrsahu@iitb.ac.in

Y. S. Mayya
Department of Chemical Engineering, Indian Institute of Technology, Bombay 400076, India

S. Singh et al. (eds.), *Pollution Control Technologies*, Energy, Environment, and Sustainability, https://doi.org/10.1007/978-981-16-0858-2_12

public health responsible for nearly 5 million deaths per year around the world (Ritchie and Roser 2020). From an Indian perspective, out of twenty most polluted cities around the world, thirteen are in India (World Economic Forum 2015), attributable to over 1.2 million early deaths in the country in 2017 (Balakrishnan et al. 2019). PM is emitted from a wide range of natural and anthropogenic sources. PM emissions from natural sources include sea salt, airborne soil or dust, pollen and natural gaseous precursors. However, from a control perspective, anthropogenic sources such as fossil fuel combustion, chemical production, and other industrial sources need attention. India's PM emissions arise mainly from three anthropogenic sources: (1) residential biomass fuel burning (for cooking and heating), (2) burning of coal in power plants and other industries, and (3) open burning in the fields (Venkataraman et al. 2015). These sources also emit precursor gases in large quantities such as SO_2, NO_x and non-methane volatile organic compounds (NMVOCs) that form PM in the atmosphere. Other anthropogenic sources of PM are vehicles and agricultural activities. However, reliable and accurately quantified emission inventories are currently not available for many countries including India.

Air pollutants can be classified as either primary (directly emitted from sources) or secondary (primary pollutants react and transformed in the atmosphere). Out of all the toxic pollutants in the atmosphere, particulate matter (PM) is the principal pollutant in outdoor as well as indoor air pollution (Jang 2012). PM is defined as 'a complex, multipollutant mixture of solid and liquid particles suspended in gas' (Ristovski et al. 2012). They can be emitted directly into the atmosphere (primary PM), or they can be formed from gaseous precursors (secondary PM). PM is classified into three categories: coarse particles (>2.5 μm), fine particles (0.1–2.5 μm), and ultrafine particles (<0.1 μm). PM is further categorized as PM_{10} and $PM_{2.5}$ based on the aerodynamic diameter of particle (EPA 2016). The aerodynamic diameter of a particle is equivalent to the size of a sphere of unit-density and same settling velocity as the particle (Suh 2009). Essentially, particles below 2.5 μm are capable of deeper penetration into the pulmonary region and hence are a cause for greater concern as compared to coarser particles, which essentially deposit in the upper respiratory tract and are generally cleared by mucociliary action (Yang et al. 2019). Apart from size, the properties of the PM such as chemical composition, morphology, surface area, and other physicochemical properties determine the impacts on health and environment and these properties vary depending on the sources of pollutant. The inhalable particles are a major health concern, especially $PM_{2.5}$ as they can penetrate and settle deep in human bronchi due to their very small size and cause chronic respiratory diseases including bronchitis (Yang et al. 2019; Harrison and Yin 2000). Additionally, particulate matter significantly impacts climate change as they act as seed for clouds and scatter or absorb sunlight, cooling and warming up the atmosphere, depending on their physicochemical properties (Lohmann and Feichter 2005). Also, aerosols emissions from wildfires and industrial pollution are likely increasing the melting of ice (Ramanathan and Carmichael 2008; Ménégoz et al. 2014; Warren and Wiscombe 1980). Despite considerable advances in technology and science, estimating direct impacts of aerosols on climate has been difficult.

In order to improve air quality, it is important to identify the contribution of various sources to global PM emissions. Industrial emissions constitute a significant proportion of these emissions in all countries; this is especially so for rapidly developing economies such as India and China (Reddington et al. 2019). To achieve air quality standard and improve the deteriorating air quality, stringent air quality regulations and limit on pollutant emission from industries are necessary. In the past, many suggestions have been given by researchers to switch to natural gas, add-on stack controls, and process fugitive control measures in order to reduce these emissions (EPA 2013). However, changing to different fuels and other process modifications to reduce emissions require significant economic investment. To meet the emission standards, many large industries have also installed control devices such as cyclone separator, scrubbers, electrostatic precipitator (ESP), and fabric filters to treat effluent gas which goes directly into the atmosphere. These methods have been in practice since the nineteenth century and despite these efforts many countries are still not able to meet the given standards. Most often, it is difficult to decide the use of a control device or combination of devices in an industry. There is lack of a detailed study and database, which contains complete information about appropriate control methodology and various pollutants including particle size, composition, and other physicochemical properties of pollutants emitted by different industries. Such detailed information of PM and other pollutants will be helpful in deciding the best suitable technology to capture them in a cost-effective manner.

12.2 Overview of Air Pollution Control Strategies

Air pollution can be controlled by a portfolio of measures. The indoor pollution can, in principle, be regulated by switching over to improved fuel quality and cooking devices. However, in practice, the issues of affordability and economic backwardness of the vast population throw newer challenges to technologies and implementation strategies. The outdoor pollution, on the other hand, originates from multifarious activities, and to tackle them, comprehensive air pollution control and air quality management system are necessary. Three important strategies that are essential for controls are: (1) source control, (2) air quality regulation, (3) use of air pollution control equipment (EPA 2013).

12.2.1 Source Control

Development of air pollution control strategies requires knowledge of the physical and chemical properties of effluents, types of pollutants, quantity, concentration, and sources of origin. Source control (minimizing the emission at the source) is the most preferred approach as a preventive step for minimizing the pollution. Some of the important consideration in this approach are changing of fuels, raw materials, and modification or elimination process steps that generate pollutants (EPA 2013). Additionally, other types of process-oriented changes such as equipment modification or replacement of old process equipment and modification of some specific process steps that generate large quantity of pollutants are essential for source control. For example, coal-fired power plant is one of the main polluters as it emits large quantity of pollutants such as PM, NO_x and SO_x. Changing the fuel composition such as shifting from high sulfur coal to low sulfur coal and changing air combustion to oxy combustion can significantly reduce emissions of pollutants (Meij et al. 2002). Secondly, the renovation of old coal-fired boiler can reduce the emission apart from retrofitting end-of-pipe equipment in the old system (Li 2009).

12.2.2 Air Quality Regulation

Apart from reducing the pollution emission from its source, secondary preventive steps that can be taken through local and central legal regulations and administrative controls. The local and central government can play an important role in reducing the environmental pollution by enforcement of laws and regulations. They can put limitations on stationary sources, mobile sources and on process wide control for concentration limit and total emission amount of various pollutants to limit their discharge and promote pollution control technologies. It should be noted that ambient air quality standards differ from source specific standards. Ambient air quality standards are fixed; however, source specific standards (such as industrial emission standards) vary based on the time at which the unit of a particular industry was installed. For instance, PM emission standard for thermal power plant units which were installed before 2003, between 2003 and 2016 and after 2016 are 100 mg/Nm^3, 50 mg/Nm^3 and 30 mg/Nm^3, respectively (CPCB 2018). On contrary, annual ambient air quality standards for $PM_{2.5}$ and PM_{10} emissions are 40 μg/m^3 and 60 μg/m^3 for industrial areas, respectively (CPCB 2009). This fact is often reflected in regulatory framework adopted by different units in controlling these emissions based on the emission standards imposed on them (Chatto 2017). Regulatory agency can adopt real-time source apportionment to identify the polluting industries and also can deploy high-density sensor-based PM monitoring network for real-time monitoring throughout cities to help identify high emission industries in different area (EPA 2013).

12.2.3 Air Pollution Control Equipment

The secondary approach of pollution control is by application of end-of-pipe treatment pollution control devices (or equipment) that will reduce the pollutants before they are released into the atmosphere. Different types of devices are used for reducing particulates and gaseous pollutants (Boamah et al. 2012). Source apportionment studies have shown that industries are the second largest contributor to direct and indirect PM emissions (Venkataraman et al. 2015). Industries that contribute most to these emissions are coal-fired power plants, iron and steel production units, cement industry pharmaceutical industry and brick production plants (Venkataraman et al. 2015; Meij et al. 2002; Larsson 2014; Lei et al. 2011; Wang 2019; Skinder 2014). Control equipment is used in addition to other control strategies to reduce the emission from these industries. However, addition of control devices is often costly and their selection depends on the process effluent conditions such as exhaust gas temperature, pressure, humidity, and flow rate, and physicochemical properties of pollutants types, sizes, composition, shapes and concentrations of PM (EPA 1998; Mycock et al. 1995; Vallero 2019). The other important considerations are physical and chemical properties of effluents, concentration, and quantity pollutants in air streams and cost for installing the including costs of maintenance.

To achieve required air quality standards across the country, an integrated approach is necessary that combine all engineering approaches and different methods of control in addition to legislative action. A brief review of various air pollution control equipment used in industry is discussed followed by equipment used in various process steps of few specific industry is highlighted. Finally, some of the current emerging technologies for air pollution control are discussed.

12.3 Existing Control Technologies for Particulate Matter

Particulate control became evident with stringent standards coming into picture with growing industrialization. Though it helped to make our lives easier, the threat of health hazards due to air pollution became an emerging problem. Particulates have proved responsible for chronic respiratory diseases in urban areas (Jang 2012; Chen et al. 2019), and many control technologies have been developed to capture particulates emitted from industries. These technologies were based on different collection mechanisms, and they are used to capture particles of different sizes and composition. In this chapter, we will discuss various control devices that have been developed to capture PM emitted from industries.

12.3.1 Gravity Settlers

Gravity settlers are simple control devices for particulates collected under force of gravity. Velocity of particles entering in the chamber reduces, and they settle down under gravitational force. Efficiency is high when velocity of particles is low or the regime of flow is laminar. However, the flow becomes turbulent as velocity increases which decreases the efficiency.

Collection efficiency of particle size (d_p) is expressed as (Buonicore 1976):

$$E = \frac{v_t L}{v_x H} \tag{12.1}$$

where E = fractional efficiency of particle size $d_{p;}$ v_t = vertical settling velocity (m/s); v_x = horizontal gas velocity (m/s); L = chamber length (m); H = chamber height (m) (Fig. 12.1).

An important parameter to characterize performance of a gravity settler is the settling velocity of the particles. It can be calculated using Stokes' law (Kumar 2018):

$$v_t = \frac{g d_p^2 (\rho_p - \rho_g)}{18\mu} \tag{12.2}$$

where v_t = settling velocity in Stokes' law range (m/s); g = acceleration due to gravity (9.81 m^2/s); dp = diameter of a particle (m); ρ_p = particle density (kg/m^3); ρ_g = gas density, usually air (kg/m^3); μ = gas viscosity (kg/ms).

Gravity settlers are not very efficient to control fine particles. Their efficiency is less than 10% for particles of size less than 10 μm (PM_{10}). However, their fractional efficiency approaches 100% for particle size more than or equal to 100 μm (EPA 1982). Efficiency of particle collection can be increased by decreasing gas velocity (increasing residence time of gas in the chamber). However, this superfacial velocity is kept between 0.3 and 3 m/s for best performance of the settler (Vallero 2019; EPA 1982). Due to their simple design, it is possible to construct settling chambers of any desired size. However, to manage flow rate of the gas, their size is restricted to 4.25 m^2 square shipping size which allows them to handle pollutant loading between 20 and 4500 g per standard cubic meter (g/sm^3) (EPA 1998). Gravity settling chambers have been used in the metal refining industries and

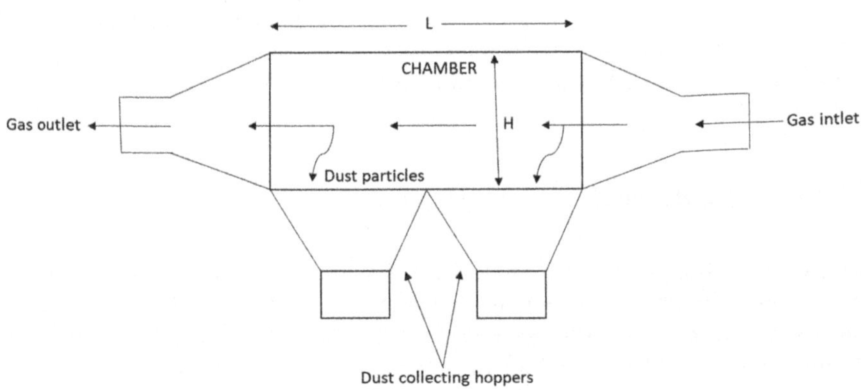

Fig. 12.1 Plan view of a gravity settling chamber. Modified from EPA (1998)

power plants to control large particles. They are most commonly used in industries to clean waste gas before treatment with fabric filters (Mycock et al. 1995). They are also used to treat waste gas in which spark or heated material is present to prevent fabric filter from catching fire (EPA 1998). The efficiency of gravity settling chambers in controlling particles of size less than 10 μm is typically less than 10% (EPA 1998). However, in most industries, they have been in use as pre-collection devices to control larger particles.

12.3.2 Cyclones

Cyclone is a low-cost control technology primarily used in industries for coarse particulate control. Stream of particle-laden gas enters through the inlet of a cyclone and moves in a circular path forming a vortex (Boamah et al. 2012). The centrifugal force on the particles move them away from the center of motion toward the wall. Simultaneously, resistance due to air in the cyclone offers drag force. Particles move and collide to the wall when centrifugal force exceeds the drag force. Collected particles at the dust hopper are then discharged. The typical features of a cyclone are shown in Fig. 12.2 (modified from Mycock et al. 1995), which consists of a rectangular section followed by a part of conical section. They can be constructed using any material (metal or ceramic) which can withstand high temperatures and corrosive nature of the particles. Interior surface of a cyclone is made smooth for particles to easily slide down the dust hopper (Pynadathu et al. 2014;

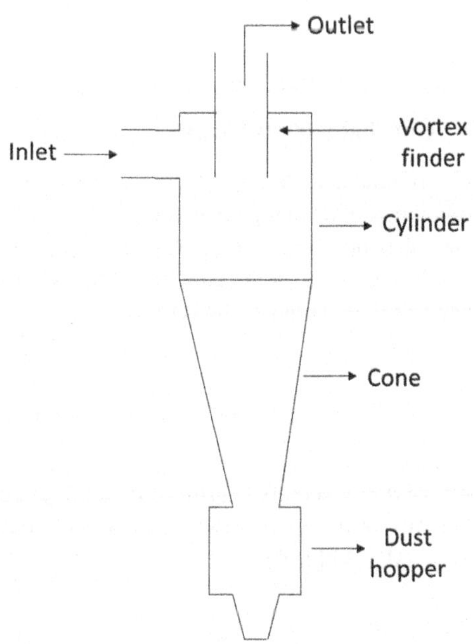

Fig. 12.2 Parts of a cyclone separator. Modified from Mycock et al. (1995)

EPA 1999). Since inertial force varies directly with mass of a particle, therefore larger particles are collected more easily (Mycock et al. 1995). The centrifugal force acting on the particle is represented by,

$$F = \frac{\rho_p d_p^3 v_p^2}{r} \tag{12.3}$$

where F = centrifugal force acting on the particle (kg-m/s); ρ_p = particle density (kg/m^3); d_p = particle diameter (m); v_p = tangential velocity of the particle (m/s); r = radius of the circular path (m).

12.3.2.1 Parameters Affecting Performance of a Cyclone

The efficiency of a cyclone depends on number of turns, particle size distribution at the inlet and inlet gas velocity (Avci et al. 2013). There are three important parameters that are used to characterize cyclone performance:

- **Collection efficiency (E_i):**

Fractional efficiency is defined as the fraction of particles collected of a given size as compared to the size of the particles entering through the inlet. It can be calculated by the expression (Theodore and DePaola 1980),

$$E = \frac{1}{\left[1 + \left(\frac{[d_p]_{cut}}{d_p}\right)^2\right]} \tag{12.4}$$

where E = fractional efficiency; $[d_p]_{\mathrm{cut}}$ = cut diameter (m); d_p = particle diameter (m).

- **Cut diameter ($[d_p]_{\mathrm{cut}}$):**

Cut diameter is referred to the diameter of particles which are collected with 50% efficiency. Cut diameter is a characteristic of the cyclone; thus, it is independent of size distribution of inlet particles. $[d_p]_{\mathrm{cut}}$ depends on gas viscosity, inlet width, inlet gas velocity, particle and gas density and effective number of turns. It can be expressed as (Lapple 1951):

$$[d_p]_{\mathrm{cut}} = \sqrt{\frac{9\mu B_c}{2\pi n_t v_i (\rho_p - \rho_g)}} \tag{12.5}$$

where μ = viscosity (kg/m-s); n_t = effective number of turns (range: 5–10); v_i = gas velocity at the inlet (m/s); ρ_p = particle density (kg/m^3); ρ_g = gas density (kg/m^3); B_c = inlet width (m).

The effective number of turns (n_t) is estimated by calculating $[d_p]_{\text{cut}}$ experimentally and then solving Eq. (12.5) for n_t (Dirgo and Leith 1985). It is evident from Eq. (12.1) that centrifugal force varies inversely with radius of the circular path, so shorter the path (or more the number of turns), more the number of particles collected. As the velocity in the inlet increases and gas viscosity decreases, the cut diameter will decrease for a fixed inlet width.

- **Pressure drop (ΔP):**

Pressure drop largely influences the performance of a cyclone and energy cost for operation. In order to move the influent gas in the cyclone, inlet velocity must be increased sufficiently to oppose the gas already present inside the cyclone. Efficiency of a cyclone increases with increase in pressure drop. As pressure drop increases, the cost of energy required to push the gas also increases. These two parameters, pressure drop and cost, are taken into consideration for optimal performance of a cyclone. The first equation developed to calculate pressure was (Shepherd et al. 1939):

$$\Delta P = \frac{v_i^2 \rho_g}{2g\rho_m} K_c \frac{H_c B_c}{D_e^2} \tag{12.6}$$

where ρ_g = gas density (kg/m^3); ρ_m = density of measuring manometer fluid (kg/m^3); g = gravitational constant (m/s^2); B_c = inlet width (m); H_c = inlet height (m); D_e = exit tube diameter (m) and, K_c = empirical constant = 16 (for standard tangential inlet cyclone) = 7.5 (cyclone with an inlet vane).

There are three types of cyclone separators based on their performance, namely high efficiency, conventional, and high throughput (EPA 1982). Pressure drop of a cyclone depends mainly on its efficiency and size. Low-efficiency, medium-efficiency, and high-efficiency cyclones have pressure drops ranging between 2–4, 4–6, and 8–10 in. H_2O, respectively (Mycock et al. 1995). They are used in different industries based on the operating conditions, target efficiency, and particle size distribution. High-efficiency cyclones can achieve efficiency up to 99% for particles of size more than 40 μm as compared to 95% efficiency of conventional cyclones (Vallero 2019). As particle size becomes less, efficiency of cyclones decreases. Also, single cyclone setup can handle typical industrial throughput ranging from 0.5 to 12 standard cubic meters per second (sm^3/sec). Higher flow rates can cause abrasion of inner layer of the cyclone unit (EPA 1999). To overcome these drawback, a cascade of cyclones (multicyclones) with reduced diameters is used either series or parallel (EPA 1982). Multicyclones can achieve efficiency as high as 90% for particles of size greater than or equal to 10 μm (Buonicore 1976).

Cyclones and gravity settlers are used as pre-cleaning devices. They are cost-effective devices for particulate removal but can be only used to control coarse particulates. Consequently, more sophisticated devices were developed for control of fine particulate matter.

12.3.3 Electrostatic Precipitator (ESP)

Electrostatic precipitators are devices used to control fine particulates and are widely adopted in industries. Particle collection in ESP is the result of five processes mainly, electric field generation, corona generation, ionization of gas stream, charging of particles, and migration of charged particles to the collection electrode (EPA 1982).

Discharge electrode (DE) generates corona by applying high voltage to it. Generation of corona depends on magnitude of electric field. It imparts electrical charge to neutral gas molecules by stripping electrons from them. Particles suspended in the gas stream are charged by bombarding them with gaseous ions (Lawless 1999). Charged particles then migrate from the gas stream and accumulate on the collection electrode (EPA 1982; Dong et al. 2018; Parker et al. 1997). Figure 12.3 describes the working of a typical dry electrostatic precipitator (modified from Ahmad and Smail 2018).

Overall efficiency of modern wire-plate ESPs for submicron particles exceeds 99%. However, conventional ESPs can collect submicron particles with 70–80% efficiency, which is due to low charge carried by small particles (Zhuang et al. 2000). Theoretically, collection efficiency can be calculated by using Deutsch-Anderson equation (Reynolds et al. 1975):

$$E = 1 - \exp\left(-\frac{wA}{Q}\right) \tag{12.7}$$

where E = collection efficiency of ESP; A = effective collecting plate area of the precipitator (m^2); Q = gas flow rate (acms); w = migration velocity (m/s).

One of the major drawbacks of the Deutsch-Anderson equation is that it assumes the particles to be monodispersed and uniform electrical conditions inside the

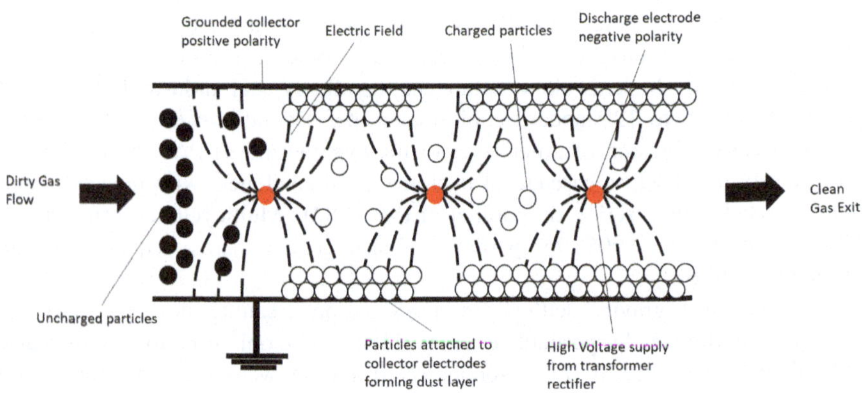

Fig. 12.3 A typical dry electrostatic precipitator. Modified from Ahmad and Smail (2018)

ESP. Deutsch-Anderson equation is modified to overcome these drawbacks and is widely known as the Matts–Ohnfeldt equation (Neundorfer 1996):

$$E = 1 - \exp\left(-\left(\frac{w_k A}{Q}\right)^k\right) \tag{12.8}$$

where all symbols have usual meanings; w_k = modified migration velocity of the particles (m/s) and k = dimensionless parameter related to particle size (range: 0.4–0.6).

12.3.3.1 Parameters Affecting Efficiency of an ESP

Three factors which affect the collection efficiency of an ESP are: migration velocity, effective collecting plate area, and aspect ratio of ESP.

- **Migration velocity (*w*)**:

It is defined as the velocity with which a charged particle moves three the collection electrode. It accounts for the collectability of a particle within the range of a specific collector inside the ESP (Neundorfer 1996; EC/R Incorporated 1998; Leonard et al. 1980). It is represented by the equation:

$$w = \frac{qE_pC}{6\pi\mu r} \tag{12.9}$$

where q = particle charge (C); E_p = strength of field in which particles are collected (V/m); μ = viscosity of gas (Pa s); r = radius of particle (μm); C = Cunningham slip factor.

Migration velocity of a particle varies inversely with its radius therefore larger particles can be collected easily.

- **Specific collection area (SCA)**:

It is the ratio of total collecting surface area and flow rate of the gas into the collector. It is important to describe the A/Q relationship in collection efficiency equations.

$$\text{SCA} = \frac{\text{Total collection surface (m}^2\text{)}}{\text{Flow rate of gas (1000 acms)}} \tag{12.10}$$

It is evident from the Deutsch-Anderson equation that if A/Q increases, collection efficiency also increases. SCA is an important aspect to be kept in mind while designing an ESP.

- **Aspect ratio (AR)**:

The ratio of total length to the height of a collector surface is called the aspect ratio. Length of the collection surface is kept large as compared to the height to achieve better efficiency but increases the cost. So, AR and cost are coupled parameters which are taken care of while designing an ESP. AR ranges from 0.5 to 2.0; however, for practical applications, it is kept greater than 1.0 to achieve an efficiency more than 99.5% (EPA 1982).

Other parameters that affect the efficiency of ESP are particle and gas properties. Particle properties such as resistivity and size distribution of the particles influence the design and operation of ESP. In dry ESPs, high resistivity limits the current and voltage, thus affecting ESP efficiency. When dust of high resistivity is deposited on the electrodes, electric field increases at the dust surface, thereby reducing the voltage at which sparking occurs. At very low values of current and voltage, positive ions are generated that cause a rapid increase in slope of the voltage–current characteristic curve, increasing it to a point where it becomes negative (Lawless and Sparks 1980). This causes a back corona discharge, i.e., positive ions flow in the gas stream attracted by the existing negatively charged ions which decreases efficiency of ESP (EPA 1982). Such cases usually arise when sulfur content of coal is very high. To decrease resistivity of the particles, flue gas desulfurization is done before cleaning flue gas using an ESP (Mycock et al. 1995). Also, wet ESPs have been developed to overcome various disadvantages of dry ESPs. Wet ESP is discussed in Sect. 12.5.3.

ESPs are used in several industries including cement production plants, iron and steel industry, copper production and coal-fired boilers. ESPs can operate at high temperatures (as high as 700 °C). However, high temperature can cause higher resistivity of dust particles which decreases its efficiency. So, operating gas temperature and dust resistivity are also key factors in determining efficiency of a dry ESP. Efficiency of dry ESPs has been found to be as high as 99.2% in controlling PM_{10} as well as $PM_{2.5}$ particles (EPA 2003).

There are five types of ESPs: dry ESP, wet ESP, plate-wire ESP, tubular ESP and flat-plate ESP, and all are used based on different requirements (Woodard and Curran 2006). Earlier, dry ESPs were most widely used in industries due to their advantageous low-pressure drops but as the working time of a dry ESP increases, its efficiency decreases and cost of operation increases. Therefore, ESP alone cannot achieve complete control of particulates emitted from an industry. For this reason, hybrid systems in which ESP is coupled with other technologies like baghouse filters and scrubbers are being developed.

12.3.4 Baghouse Filters

Baghouse filters are long cylindrical bags made of different fabrics. They are a type of fabric filters which divide the flow of gas stream into smaller parts and act as a

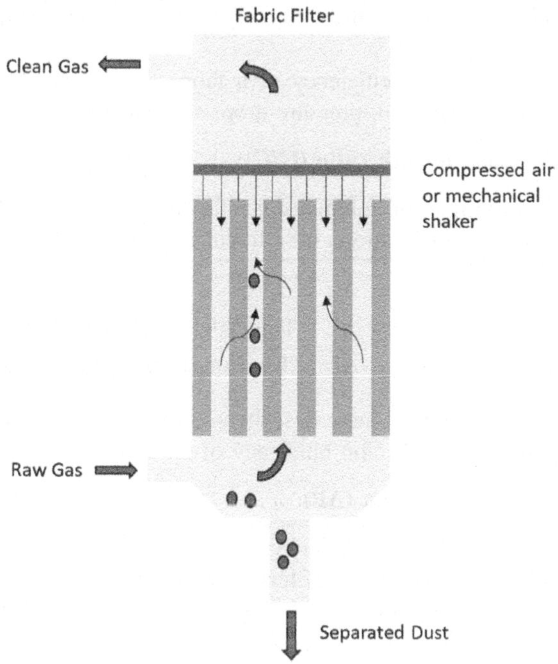

Fig. 12.4 Basic filtration mechanism in a baghouse/ fabric filter. Modified from emis (2019)

medium to filter particles. Collected particles are removed periodically by shaking the filters (Peirce et al. 2006). They are widely used because of their high efficiency and applicability in many industries (EPA 1982). The components of a typical bag house filter are shown in Fig. 12.4 (modified from emis 2019).

There are three underlying mechanisms by which particle collection occurs in a fabric filter: inertial impaction, diffusion due to Brownian motion of particles, and interception (EPA 1982). Impaction is due to the inertia of particles in motion. Heavy particles of size larger than 50 μm resist changes in their path due to flow of the gas molecules and hence move toward the obstacle. Very small particles cannot be separated by inertial impaction under normal conditions. Particle can also be separated from a gas stream by an obstacle if the size of particle larger than its streamline displacement. Interception can be considered as a form of impaction. Also, particles which are as large as the gas molecules (10^{-3} μm) undergo random movement due to collisions along the gas stream called the Brownian motion or diffusion (EPA 1982). The trajectory of the particle depends on these three processes. However, high diffusion rates are observed for very fine particles of size 0.001 to 0.1 μm and it is practically negligible for large particles. Filters are useful in filtration to some extent; however, the dust which is accumulated on the filters contributes to the filtration capacity of a baghouse filter. This layer of dust is called the filter cake (Roth et al. 2002). Once the filters are loaded and significant pressured drop is observed, back pulse high pressure air jet is used to clean up the bag house filter (Peirce et al. 2006).

12.3.4.1 Parameters Affecting Efficiency of a Baghouse Filter

The collection efficiency of a fabric filter depends on three parameters, namely gas-to-cloth ratio, pressure drop, and material of the filter.

- **Gas-to-Cloth ratio (G/C):**

It is the ratio of volume of gas flowing in a filter to the total filter cloth area in the collector. G/C ratio is used to calculate required size of a baghouse filter (Mycock et al. 1995).

$$\frac{G}{C} = \frac{\text{Total inlet volume} + \text{cleaning volume}}{\text{Total area of filter cloth in the collector}} \tag{12.11}$$

Also, as G/C increases, the pressure drop in a baghouse increases too. Thus, it influences collection efficiency of the filter.

- **Pressure drop (ΔP):**

To determine the efficiency of a fabric filter, usually single fiber efficiency model is applied (Chang 2019; Chen et al. 2014; Bahk et al. 2013; Sen and Otani 2013). Pressure drop for an operating fabric filter with accumulated dust cake is predicted by using basic Darcy equation (Mycock et al. 1995):

$$\Delta P = S_E V + K_2 C_i V^2 \tag{12.12}$$

where ΔP = pressure drop (in H_2O); S_E = effective residual drag (in H_2O); V = velocity (fpm); K_2 = specific cake coefficient; t = time period.

The first part of the pressure drop equation (Eq. 12.12) is due to the velocity of the particles through the filter, and the latter is the contribution of filter cake deposited on filter surface. Coefficient is calculated by using the equation (Cheng and Tsai 1998):

$$K_2 = \frac{18\mu}{\rho_P D_{VS}^2 C} \frac{2k(1-\varepsilon)}{\varepsilon^3} \tag{12.13}$$

where k is an empirical constant = 4.8 (for spherical particles) = 5.0 (for irregular particles) and, D_{VS} = volume–surface mean diameter or Sauter diameter, ε = specific dust cake resistance coefficient, μ = gas viscosity, ρ_P = particle density, C = Cunningham slip correction factor.

This equation describes the pressure drop after some period through a definite area of fabric. Area can be distributed across many compartments, and these compartments are arranged in a parallel fashion. This type of filter is called multicompartment baghouse.

- **Material of the filter:**

Selection of filter medium is another important parameter which decides the efficiency of a baghouse filter. The choice of filter is further dependent on various parameters such as operating temperature, tensile strength of the fabric, resistance to certain chemicals and cost. All these parameters are considered while designing a baghouse filter.

Other important parameters that affect the removal efficiency of a baghouse filter are cleaning frequency and intensity (EPA 1999). If the cleaning is too frequent or too intense, it will decrease removal efficiency due to cleaning of the dust cake. On the other hand, pressure drop of the baghouse will be very high if the cleaning is too infrequent, which will increase operational cost. Cleaning of baghouse filters can be done by mechanical shaking, reverse air cleaning, sonic horns, and reverse-jet cleaning methods (Mycock et al. 1995; EPA 1999). Mechanical shaking is the oldest method used to clean filters by subjecting filters to high mechanical load. But cleaning efficiency of this method is not at par, and it increases operational cost. For this reason, other cleaning methods were developed such as reverse air cleaning and compressed air cleaning or reverse-jet cleaning methods in which rinsing air is allowed to flow in the opposite direction inside the filter. The mechanical load applied on the filter is less in these methods due to low air pressure applied in reverse blowing, hence making them more efficient (EPA 1998; Mycock et al. 1995; Emis 2019).

Baghouse filters have few key advantages over other control devices. They have high efficiency, high flexibility for different particle sizes and can operate at very low-pressure drops. But they are not very efficient in filtering dry gas streams and maintenance of fabric is costly. However, optimum choice of fabric for a certain requirement can minimize cost. Recent developments in filtration by using electrically augmented filters have shown promising results in controlling particle pollutants from industrial stacks. Vanosdell et al. (1985) and Ouari et al. (2019) have investigated electrically augmented filters proving that corona charging improves particle collection efficiency of the filter medium (Vanosdell 1985; Ouari et al. 2019). More data and investigation are required to effectively employ such filters for industrial use.

12.3.5 Wet Scrubbers

Wet scrubbers are control devices which collect particles by spraying them with water or any other liquid (Boamah et al. 2012; Woodard and Curran 2006). Their basic principle of collection is inertial impaction and Brownian motion (Vallero 2019). Particle-laden gas stream moving in a scrubber is sprayed with a liquid, wetting the particles which are removed by a collecting surface which can be either a bed or a wetter surface (Vallero 2019; EPA 1982). Wet scrubbers are of various types such as venturi scrubbers, orifice scrubbers, and spray-tower scrubbers.

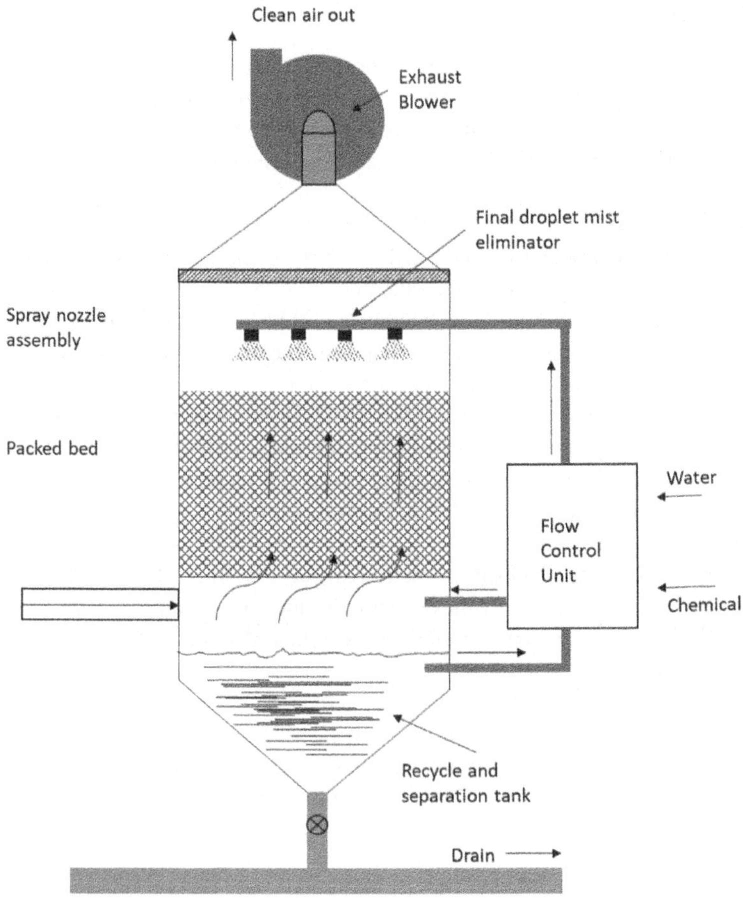

Fig. 12.5 Components of a wet scrubber. Modified from Rogoff and Screve (2011)

Venturi scrubbers are the most efficient among the wet scrubbers with an efficiency of 98% for particles of size 0.5 μm (Boamah et al. 2012). Venturi scrubbers can be used for simultaneous removal of gaseous as well as particle pollutants. They work extensively on inertial impaction mechanism for PM removal and sorption for gas removal (Mycock et al. 1995). Figure 12.5 shows components and basic principle of working of a wet scrubber (modified from Rogoff and Screve 2011).

Collection efficiency for a venturi scrubber can be calculated as a function of penetration of particles of a given size. It assumes that inertial impaction is the dominant mechanism of the removal of particles in a venturi scrubber (Calvert et al. 1972). Overall penetration can be calculated as (Peirce et al. 2006):

$$P_0 = \int_0^L \frac{P_d}{L} dL \tag{12.14}$$

where P_0 = Overall penetration; P_d = particle penetration of size d_p; L = Total dust loading.

Another important parameter which is considered while designing a venturi scrubber is the pressure drop. Pressure drop for a large venturi scrubber used for industry is given by (Peirce et al. 2006; Calvert et al. 1972; Mohebbi et al. 2003):

$$\Delta P = 2\rho_L V_G^2 \left(\frac{Q_L}{Q_G}\right)\left(1 - X^2 + \sqrt{X^4 - X^2}\right) \tag{12.15}$$

where ΔP = pressure loss (N/m^2); V_G = gas velocity (m/s); ρ_L = density of the liquid (kg/m^3); X = throat length (dimensionless).

and,

$$X = \frac{3I_t C_D \rho_G}{16 d_d \rho_L} + 1 \tag{12.16}$$

where I_t = venturi throat length (m); C_D = drag coefficient.

Wet scrubbers have advantages for industries which handle flammable particulates. Collection of dust can be achieved in a single unit; thus, space requirement for scrubbers is less as compared to other technologies. But rusting of scrubbers due to corrosive particles increases their maintenance cost (EPA 1982; Peirce et al. 2006). They are used in various industries such as coal power plants and cement industries.

The above discussion on existing air pollution control technologies implies that selection of air pollution control devices must be made case by case basis as the particulate characteristics will vary significantly in air stream coming from different applications. Particle characteristics that will influence the selection are corrosivity, reactivity, shape, density, and size and size distribution. The air stream characteristics that should be considered are pressure, temperature, viscosity, flow rate, and removal efficiency requirements. Based on the technology employed, efficient PM control technologies require higher investment and higher operational costs. However, stringent regulations and improving emission standards will be the driving force implementing these technologies within the industry. With increasing urbanization and transport sector, air pollution caused by multiple pollutants has become a leading concern around the globe. Serious actions have to be taken by the authorities to improve air quality by decreasing pollution emission load below a permissible range.

12.4 Particulate Matter Emissions from Industries

For effective air quality management, it is important to understand the pollutant emission and quantify the emission from various sectors contributing to pollution and projected emissions under current scenarios. A detailed emission inventory studies estimated the emissions from various sectors and projected the emissions quantitatively up to 2050 (Venkataraman et al. 2015; Sadavarte and Venkataraman 2014). The major sources contributing to PM come from mainly three sources: residential biomass fuel use, coal burning in power generation and heavy industry, and (iii) open burning of agricultural residue. Many coal burning power plants and industry emit significant amount of particulate matter. For understanding of the particulate emission and control devices used in industry, three industries have been chosen for discussion mainly coal-fired plants, cement industry, and iron and steel industry.

12.4.1 Cement Industry

Cement industry has undergone an immense development since nineteenth century majorly in countries like China and India (Schneider et al. 2011). It is a direct indication of increased emission of pollutants from cement production which include particulate matter (PM), SO_x, NO_x, CO_2, and others. The process flow diagram in Fig. 12.6 shows the steps at which PM is generated in a cement industry (modified from Huntzinger and Eatmon 2009).

The largest emission source of PM within cement plants is the pyro-processing system which is due to the kiln and clinker cooler. Particulate control devices are

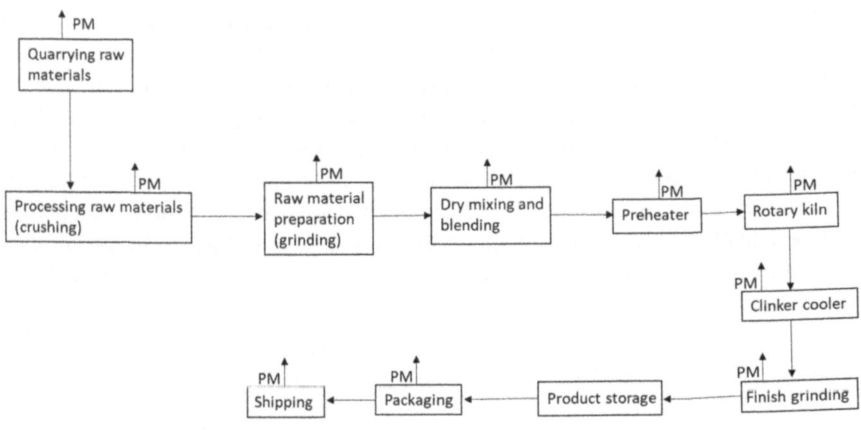

Fig. 12.6 Process flowchart: cement manufacturing process. Modified from Huntzinger and Eatmon (2009)

installed within the preheater and kiln systems to collect unburned and partially burned particles (EPA 1994). These collected particulates are called cement kiln dust (CKD), and it imparts damage to central nervous system, respiratory tract, cardiovascular tissues, skin, and digestive system (Raffetti et al. 2019). Decline in lung functioning has also been observed in workers upon long-term exposure to dust in cement production plants (Nordby et al. 2016).

As the process of cement production has evolved, particulate production has also increased due to processes like crushing and grinding of raw materials under dry conditions. Thermal processing which includes preheating and sintering in a rotary kiln also plays a major role in emission of particulates. Particulates are emitted in large amount every year from cement production units. Total amount of PM production from cement plants in China was reported as 1.95 million tons in 2014 (Tang et al. 2018). The generated particles have different size based on the process involved. For instance, processes like dry kiln, raw mill, and kiln by-pass produce large particles of size greater than 50 μm. On the other hand, clinker cooler produces particles in the submicron range (less than 5 μm) in the form of CKD (Bapat 2001). Apart from their size, the chemical composition of these particles is also different based on factors like temperature, humidity, and the process in which they are formed. In general, PM generated from kiln stacks has iron and calcium in higher quantities followed by nitrate and sulfate. Other metals found in clinker dust that are a concern to environment are arsenic, cadmium, chromium, and nickel (Gupta et al. 2012). Their varying size and composition make it difficult to control them at the point of emission (Table 12.1).

These are unabated emission factors so it can be inferred that the emission range of these particulates from a standard cement production plant is close to the calculated EF range (Lei et al. 2011). With increasing concern toward the environment and human health, incorporating control technologies to meet the effluent standards is necessary (Bapat 2001).

Control technologies including cyclones, wet scrubbers, ESPs, and baghouses are used to control PM emissions in cement industry. Each of these technologies is used at different point of emission based on the size and chemical nature of particles emitted. Table 12.2 shows control technologies used to control particulates emitted from different processes in the production of cement (EPA 1983).

Use of hybrid technologies has been observed and proved to be useful in controlling PM emissions in cement production units. Electrostatic precipitators and

Table 12.1 PM emission factors (EF) for cement production in China (unit: g kg^{-1} cement) (Lei et al. 2011)

Emission sources	Total PM	$PM_{2.5}$	$PM_{2.5-10}$	$PM_{>10}$	EF range
Kilns-Pre-calciner kilns	105	18.9	25.2	60.9	58.2–317.9
Other rotary kilns	98	14.0	21.0	63.0	24–330
Shaft kilns	30	3.3	6.0	20.7	13.4–91.2
Other sources as a whole	140	9.5	23.8	107.7	63–235

Table 12.2 Use of control technologies at different point of emissions in a cement production plant (EPA 1983)

Process	Control technology used
Crushing and grinding	Baghouse filters
Raw material storage	Baghouse filters
Integral preheater and kiln	Mechanical collectors (to meet opacity requirements) Baghouse filters/ESPs
Kiln	Mechanical collectors (as pre-cleaners) Baghouse filters/ESPs
Clinker cooler	Mechanical collectors (as pre-cleaners) Baghouse filters/ESPs
Finish grinding	Baghouse filters or ESPs (for large mills)
Finished material storage	Baghouse filters
Packaging and shipping	Baghouse filters

fabric filter (ESP-FF) combination is reported in a typical Indian cement plant, and it is being used for processes like raw grinding and blending, coal grinding and cement grinding (Bapat 2001). This is because, most of the dust particles are collected in first few stages of the ESP; however, it is inefficient to control very fine particles. Reportedly, first two ESP fields collect more than 98% of the dust particles. On the other hand, fabric filters (FF) are used to collect remaining fine particles since their collection efficiency depends on the fiber element. Their combination has proved to be useful in large cement plants (having production of >10^6 tonnes/year) and has been able to achieve an efficiency of 100%, accounted by the size of ESP to treat the large gas volumes (Bapat 2001).

12.4.2 *Iron and Steel Industry*

Iron and steel industry is one of major contributors to the development of human civilization. Iron and steel industry consumes large quantities of fossil fuels and other raw materials. Due to rapid economic expansion of this industry, it has become one of the highest pollutant emission industries which includes PM, SO_2, NO_X, TSP, and VOCs (Zhao et al. 2011). According to the International Iron and Steel Institute (2011), India is the largest producer of sponge iron and second largest producer of crude iron in the world (Fletcher 2001). Figure 12.7 shows process flow diagram of iron and steel production and sources of PM emissions in the process (modified from Wang et al. 2019).

It is important to identify emission sources so that suitable control devices can be used. In typical iron and steel production plant, technologies like cyclones, ESPs, wet ESPs, and baghouse filters are most frequently used. It is evident from Fig. 12.7 that there are five major processes which cause PM emissions: sintering, pelletizing, blast furnace (BF), basic oxygen furnace (BOF), and electric arc furnace

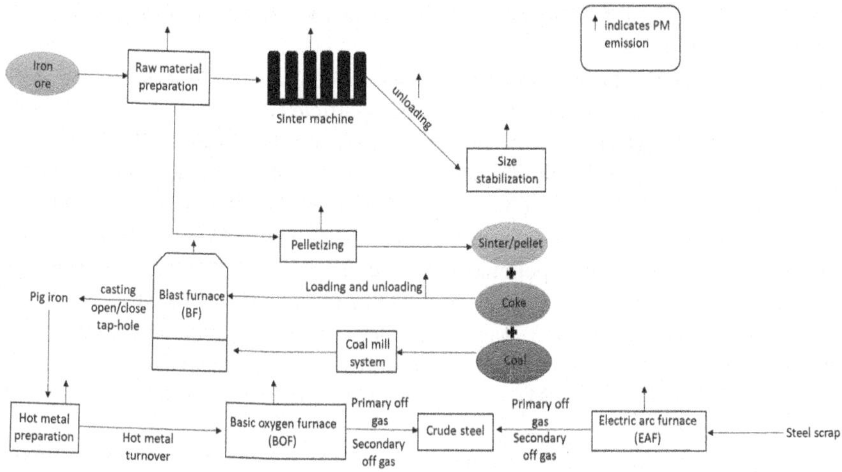

Fig. 12.7 Schematic diagram of iron and steel production and sources of PM emissions of PM, SO_2 and NO_X. Modified from Wang et al. (2019)

Table 12.3 Unabated emission factors of $PM_{2.5}$, PM_{10} and $PM_{>10}$ for iron and steel sector (unit: gm kg^{-1} product) (Wang et al. 2010)

Process	Emission source	$PM_{2.5}$	PM_{10}	$PM_{>10}$
Sintering	Total	1.76	2.99	26.98
	Raw material preparation	0.26	0.35	3.40
	Sinter machine head (wind-box exhaust)	0.63	1.24	10.66
	Sinter machine tail (unloading off gas)	0.39	0.76	6.52
	Size stabilization system	0.49	0.64	6.41
Pelletizing	Total	0.48	0.94	8.08
BF	Total	4.69	2.88	14.27
	Coal mill system	1.29	0.82	0.05
	Load and unload material system	0.51	0.32	0.02
	Casting	2.35	1.41	11.53
	Tap-hole opening/closing	0.54	0.33	2.66
BOF	Total	16.89	6.77	10.55
	Hot metal pre-treatment	0.06	0.04	0.45
	Hot metal turnover	0.10	0.04	0.45
	Primary BOF off gas	11.35	4.54	6.81
	Secondary BOF off gas	5.38	2.15	3.23
EAF	Total	8.92	3.11	8.70
	Primary EAF off gas	6.35	2.21	6.19
	Secondary EAF off gas	2.57	0.90	2.51

(EAF). Table 12.3 shows unabated emission factors of $PM_{2.5}$, PM_{10} and $PM_{>10}$ (Wang et al. 2010).

Table 12.3 shows that coarse particle mass concentration ($PM_{>10}$), and fine is the highest in the sintering process; however, fine particle (PM_{10}) and very fine particle mass concentration ($PM_{2.5}$) are highest in the basic oxygen furnace (BOF) (Jia et al. 2017). Particulate control technologies are used in the plant based on particle size to be collected in a process and working conditions. In a Chinese iron and steel production unit, efficiency of control devices to capture PM was calculated. For processes like sintering and pelletizing, particulate emissions were best controlled by ESPs with a capturing efficiency of 90% for $PM_{2.5}$, 94% for PM_{10}, and 98.5% for $PM_{>10}$. On the other hand, baghouse filters were most efficient in controlling PM emitted from other sources with a capturing efficiency range between 85–99% for $PM_{2.5}$, 87.5–99.5% for PM_{10} and 97.3–99.9% for $PM_{>10}$ particles (Wang et al. 2010) (Table 12.4).

Thus, it can be concluded that different control technologies are used at different steps in an iron and steel production unit. This is because of the particle characteristics which are inherent due to the process involved. These characteristics which include temperature, viscosity, chemical nature, size and charge play a key role in deciding the use of these devices in an industry. However, it has been observed that ESP and fabric filters in particular are used in almost all industries to achieve maximum removal of generated particles at different steps of the process.

Table 12.4 Control technologies used to control particulates emitted from different processes in the production of steel (EPA 1983)

Process	Control technology used
Blast Furnace • Blast furnace gas • Cast house emissions	Cyclone followed by Wet scrubber Cyclone followed by ESP Baghouse filters at the exhaust
Coke oven • Combustion stack • Charging • Pushing	ESPs or Baghouse filters Emissions control is not applied Scrubber/ESP/Baghouse filter
Basic oxygen furnace (BOF) • Oxygen blow period • Tapping or charging • Reladling stations	Scrubber or ESP Scrubber/ESP/Baghouse filter Baghouse filters
Electric furnace • Melt period • Charging and tapping	Scrubber/ESP/Baghouse filter Baghouse filters
Sinter plant • Wind-box • Coolers	Baghouse filters ESPs Wet Scrubbers Gravel-bed filter Scrubber Baghouse filters

12.4.3 Coal-Fired Power Plants

Coal-fired power plants are one of the major sources of fine particulate matter with size less than 3 μm which cause several respiratory problems (Wang et al. 2016; Liu et al. 1990. The flue gas generated from coal-fired power plants contains a mixture of particles such as $PM_{2.5}$, PM_{10}, condensable PM (CPM), and filterable PM (FPM) (Lu et al. 2010). Additionally, combustion of coal release pollutants such as VOCs, CO, NO_x, and SO_x, mercury and other heavy metals (Guttikunda and Jawahar 2014). According to some reports, coal combustion in power plants is the major source of airborne mercury emissions in developed countries such as USA and China (Zhang et al. 2012; Srivastava et al. 2006; Wang et al. 2012). Ash generated by combustion of coal contains 10–20% bottom ash and nearly 90% fly ash (Meij et al. 2011). Coal fly ash is incombustible and may cause asthma and chronic bronchitis (Grahame and Schlesinger 2007). These particles are generated at the flue gas desulfurization (FGD) unit and stack of the power plant. In India, coal is the primary fuel choice in thermal power plants and accounts for 50–55% of power generation (Mittal 2010; Chikkatur et al. 2011). Thermal power plants account for 15% of the total $PM_{2.5}$, 30% of NO_x and 50% of SO_2 emissions annually in India (Guttikunda and Jawahar 2014). For a typical coal-fired power plant, it was found that filterable PMs are the highest emissions at FGD inlet and stack (Lu et al. 2010). Figure 12.8 shows process flow diagram of a DB Riley design low-emission boiler system used for commercial purpose (modified from Spath et al. 1999).

It is also evident from the literature that control technologies such as ESPs, fabric filters, and scrubbers are most widely used in coal-fired power plants for particulate

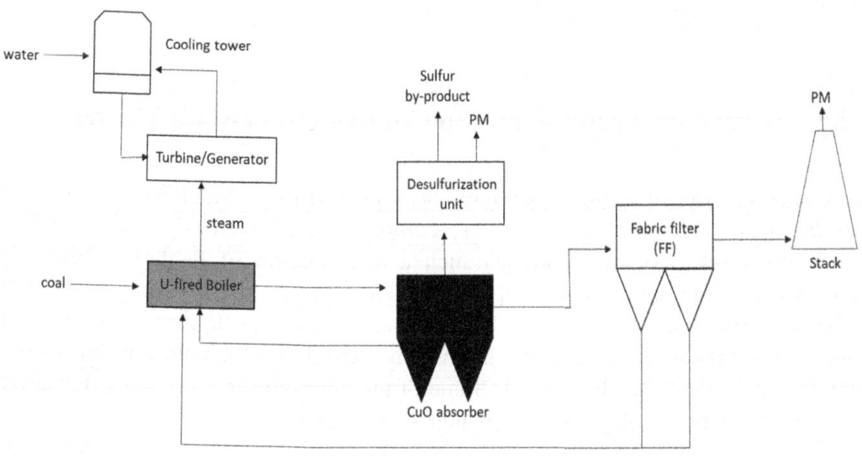

Fig. 12.8 Process flow diagram of a DB Riley design low-emission boiler system used for commercial purpose. Modified from Spath et al. (1999)

Table 12.5 Uncontrolled emission factors for PM from different type of furnaces used for coal combustion (Li 2009)

Furnace			Uncontrolled emission, g/kg coal (wt%)
Boiler category	Heat input 10^6 kcal/hr	Type	Particulate matter
Utility or large Industrial	>25	Pulverized Cyclone	8 1
Commercial or industrial	2.5–25	Underfeed and cross-feed Overfeed	2.5 6.5
Small commercial or domestic	<2.5	Overfeed Hand fired	1 20

control (Lu et al. 2010; Baig et al. 2017; Bandyopadhyay and Mandal 2015) (Table 12.5).

It has been reported that often cyclones are used as pre-cleaners in these power plants before subjecting flue gas to baghouse filters or ESPs (Bandyopadhyay and Mandal 2015). Existing technologies are efficient in controlling particles generated from power plants. ESPs have efficiency greater than 99.5% for fine particles but fabric filters are found better in controlling very fine particles of fly ash (Lu et al. 2010).

An overview of particulate control technologies in mentioned industries indicates that conventional control devices are most frequently used in industrial sector to meet emission standards. Devices like mechanical separators are used to control coarse particles, while sophisticated devices such as ESPs, baghouse filters, and scrubbers are used to control fine particles. Use of these devices is based on a number of parameters, most important of which are particle size distribution, chemical nature of particles and cost involved.

12.5 Emerging Technologies to Control Particulate Matter

Conventional control technologies have been proved to be inefficient for fine and ultrafine particles. In many instances, combination of technologies is used to effectively capture smaller particles such a combination of baghouse filter and ESP. Most of the cases, separate process/devices are used for capturing gaseous pollutants. Simultaneous removal of gaseous and particle pollutants is not possible using conventional devices. Additionally, any hybrid system with combination of various mechanisms has been used. Some of the emerging technologies developed for effective removal of particulate matter is discussed.

12.5.1 Self-priming Venturi Scrubbers

Gaseous and particle pollutants are usually emitted from industries in association with one-another. Their individual effects are severe as pollutants but in the presence of particulate pollutants, gaseous pollutants are synergistically modified, and their combined effects are much severe (Meikap and Biswas 2004). So, simultaneous removal of particulate and gaseous pollutants in industries is very imperative. Self-priming venturi scrubbers are devices which can be used for this purpose. The self-priming method is based on the pressure difference between hydrostatic pressure applied by liquid in the tank and static pressure of gas in the scrubber (Ali et al. 2013). The liquid in the venturi scrubber captures the particles and absorbs gaseous pollutants. Figure 12.9 shows self-priming working method of a venturi scrubber (modified from Ahmed et al. 2018). Earlier, multistage bubble column scrubbers which remove particulate and gaseous pollutants sequentially were also in practice (Meikap and Biswas 2004). However, this technique is not useful for industrial use as sequential removal technique cannot be used to treat particle-laden gas stream (Bal et al. 2019).

A self-priming venturi scrubber was developed for simultaneous removal of HCl gas and gas-borne particulates from the flue gas. They have reported maximum fly-ash removal to be 99.91% at 60 m/s of gas velocity at the throat, 0.77 m of water height, 500 ppm of HCL and 0.3×10^{-3} of fly-ash concentration. Also, maximum HCL removal at same working condition was found to be 98.3%. It was also reported that efficiency fly-ash removal increases due to scrubbing of HCl gas. This is because when HCl gas is scrubbed with a scrubbing liquid, it adsorbs on the surface of fly-ash particles (Bal et al. 2019).

Another study shows variation of aerosol removal efficiency in a self-priming venturi scrubber under different operating conditions (Ali et al. 2020). Removal efficiency of venturi scrubber is directly proportional to inlet gas velocity and available head of liquid. Scrubbing liquid is injected in the venturi throat as tiny droplets and encapsulate particulates. Gaseous pollutants like I_2, SO_2, Ch_3I and HCl

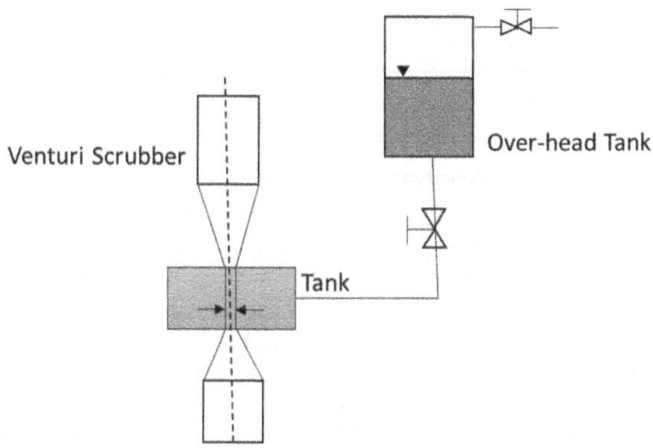

Fig. 12.9 Self-priming working method of a venturi scrubber. Modified from Ahmed et al. (2018)

are adsorbed by scrubbing liquid. Aerosols in the size range 0.1–100 μm can be removed using venturi scrubbers (Rust and Tfinnler 1993). Many investigations have also been done for simultaneous removal of SO_2 gas and particulates from a particle-laden gas stream using wet scrubbers. It was also reported that nearly 100% removal efficiency of SO_2 from a particle-laden flue gas was achieved in a modified venturi scrubber under certain operating conditions (Bandyopadhyay 2009). These types of scrubbers are not commonly used in industries, and they are effective devices for controlling gaseous and particle pollutants (Bal et al. 2019).

12.5.2 Wet Electrostatic Scrubbers

Conventional ESPs and other technologies which remove particles by precipitation are not able to remove very fine particles with high efficiencies (Jaworek et al. 2007). Hybrid technologies can be adopted to control very fine particles with high efficiencies. Wet electrostatic scrubber is a hybrid device made by combination of electrostatic precipitator and wet scrubber. It is one of the most effective hybrid technologies to control fine particulate (Carotenuto et al. 2010). Figure 12.10 shows an experimental setup of a WES (modified from Jaworek et al. 2006). The basic

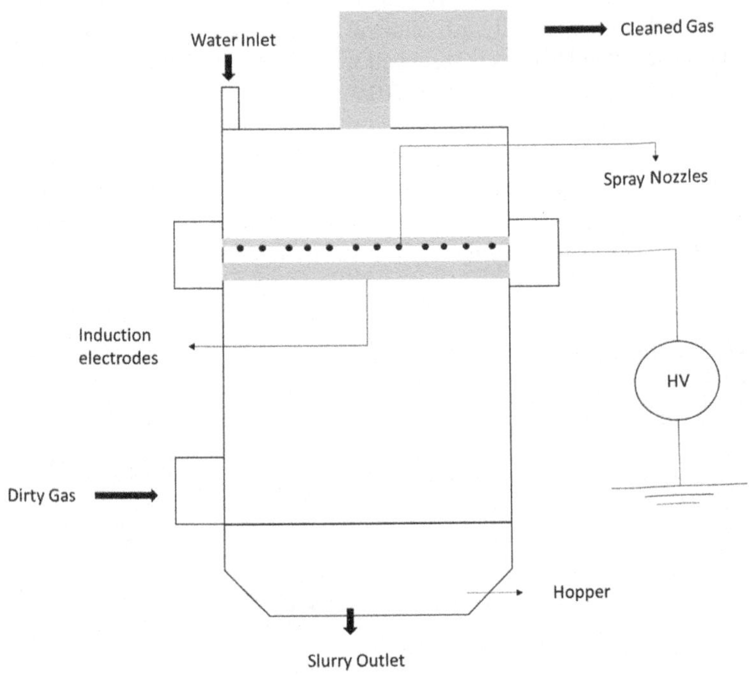

Fig. 12.10 Parts of a wet electroscrubber. Modified from Jaworek et al. (2006)

principle of WES is spraying a charged liquid in the flue gas stream to attract particles of opposite charge (Carotenuto et al. 2010).

WES can remove combustion particles finer than 500 nm. Under certain operating conditions, their removal efficiency was observed to be 91% for particles smaller than 300 nm and as high as 95% for particles coarser than 300 nm (Di Natale et al. 2015). Use of WES has also been extended to control bioaerosols including potentially harmful bacteria and virus (Di Natale et al. 2017). Besides, they are also useful to remove gaseous pollutants like SO_2 (Di Natale et al. 2016). Earlier experiments show removal efficiency of WES in capturing a bacterium in 500-1500 nm range to be as high as 99% under certain operating conditions (Di Natale et al. 2017).

12.5.3 Wet Electrostatic Precipitators (WESP)

Electrostatic precipitator is a control device which utilizes the principle of electrical charging to capture particles. Dry ESPs are not able to capture very fine with high efficiency (typically less than 90%). Since they operate at high temperature, it is not possible to capture condensable PM in a dry ESP. Therefore, their application is limited to capture filterable particulate matter (PM_{10}) like fly ash (Seetharama et al. 2013). Back corona and particle re-entrainment are other factors that decrease their efficiency (Sobczyk et al. 2017). Additionally, in ESP when resistive power is high, particle capture is reduced due to re-entrainment of particles at the time of particle removal from collection plates. In such cases, wet ESPs can be used to overcome these drawbacks. In WESP, the removal of particles from collecting plates is done by a flushing liquid which avoids re-entrainment of particles due to high degree of adhesion between water and particles. Washing the collected particles prevents back corona since the resistive particles are constantly being washed off (Bayless et al. 2012). Particle agglomeration and condensation play an important role in improving efficiency of a WESP. These two processes result in enlargement of particle size and thus increase the particle capturing efficiency (Najafabadi et al. 2014; Yang et al. 2018). Another advantage of using a WESP is that the particle capturing efficiency does not vary much over a long period of operation. On the other hand, collection efficiency dry ESP decreases gradually when operated even for short period of time (Darcovich et al. 1997). Figure 12.11 shows plan view of a WESP schematic (modified from Chen et al. 2014).

WESP has higher particle collection efficiency as compared to dry ESP. In a typical case, the efficiency of a dry ESP in collecting fly-ash particles emitted from a coal-fired power plant was found to be 93.75%. However, under same working conditions, the efficiency of a WESP in controlling the same particles was 97.86%, which means that only by changing the removal mechanism of particles from the collection plate, the particle capturing efficiency varied by 4.11% (Yang et al. 2017). Also, WESP has proved to be useful in capturing filterable ($PM_{2.5}$) and condensable particles. They have demonstrated efficiency greater than 99% in

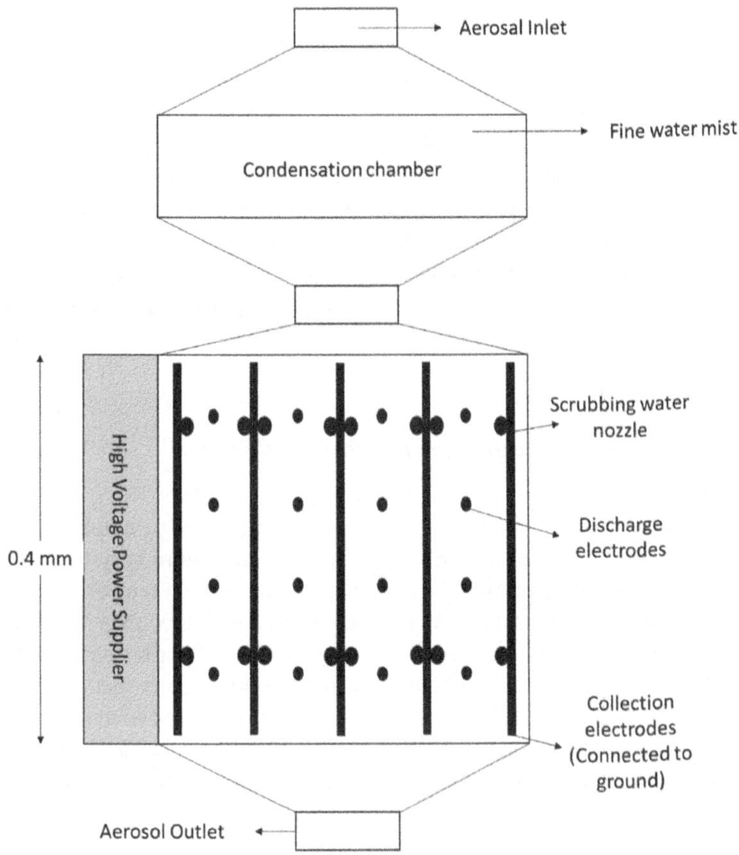

Fig. 12.11 Plan view of a wet electrostatic precipitator. Modified from Chen et al. (2014)

cleaning flue gas containing $PM_{2.5}$ particles, droplets of H_2SO_4, and mercury (Seetharama et al. 2013).

12.5.4 High-Efficiency Particulate Air (HEPA) Filters

As the health threats posed by fine aerosols are increasing, new filter materials are also being developed. Direct filtration of fine aerosol particles ($PM_{2.5}$) is one of the most effective ways to control them. Filtration mechanism is being used in many places including nuclear power plants, pharmaceutical processing units, aircraft cabins and indoor environment (Lee et al. 2020; Barn et al. 2018; Lencioni et al. 1997; Rosenberger et al. 2018. The performance of a filter depends on two factors, namely filtration efficiency and pressure drop. High filtration efficiency and low-pressure drop are the desired combination of an ideal filter. But usually, these two parameters are not found in a filter because in order to achieve high filtration efficiency, high pressure drop is also required (Chen 1955).

According to standards set by US Department of Energy (DOE), a high-efficiency particulate air (HEPA) filter must have minimum efficiency of 99.97% for capturing particles of size 0.3 μm (USDOE 2015). They are particularly used in places where capturing of fine and ultrafine aerosols is required. Over the past few years, many new materials have been used to make HEPA filters including nanofibers and carbon nanotubes (CNTs). Small fibers (in submicron range) are known to have higher filtration efficiency as compared to large fibers at the same pressure drop. But in case of nanofibers, there is another factor that comes into picture which is the effect of slip at fiber surface. They have a larger surface area available as compared to microfibers due to which the velocity becomes practically zero at the fiber surface due to slip (Shou et al. 2014). Despite these advantages, nanotubes-based HEPA filters are considered better since air filtration can occur in the free-molecular flow regime; i.e., the disturbance of CNTs is negligible on flow pattern (Shou et al. 2014; Zhang et al. 2017).

HEPA filters are used in many industries to control fine and ultrafine particles. They are used in nuclear power plants to capture radioactive particles with efficiency as high as 99.85% for high particle-laden gas (Lee et al. 2020). HEPA filters can also be used in industries which treat flue gas at high temperatures. In pharmaceutical industry, HEPA filters have been proved to capture fine particles (0.2–1.0 μm) with efficiency greater than 99% at 300 °C (Lencioni et al. 1997). A case study was conducted using charcoal equipped HEPA filters in aircrafts to improve air quality. They have been proved to improve air quality and efficiently capture pollutants such as aldehydes, volatile organic compounds (VOCs), and oligomeric proanthocyanidin complexes (OPCs) (Rosenberger 2018).

12.5.5 Electrostatically Assisted Fabric Filtration

Electrically stimulated fabric filtration is another hybrid system that combines the efficiency of filtration and low-pressure drop concept of ESP. In this method, a high voltage electrode is co-axially placed inside the filtration system to establish electric filed just between centra electrode and the outer bag surface of filter (Thorpe and Brown 2003; Tomitatsu et al. 2014; Dutta 2019). It combines the efficiency of fabric filters and low-pressure drop of an ESP. By using this approach, the deposition pattern of particulate is altered compared to normal filtration process, results in lower pressure drop compared to conventional filtration process. Because of this, the bags can be operated at higher air flow rate. Similar modification applying the electric filed on the fabric surface can help accumulate the particles on the surface. Numerous attempts have been made to use electrostatic charge effects to improve the filtration, and results show varying degree of performance of filters (Vanosdell et al. 1985). Vanosdell et al. (1985) employed this strategy in pilot plant and found that baghouse could be operated at twice face velocity compared to conventional baghouse. Several other studies also reported that electrically charge-assisted deposition is beneficial in achieving lower pressure drop due to a tendency forming dendritic structures as the particles collect and result in lower power consumption (Thorpe and Brown 2003; Tomitatsu et al. 2014; Dutta 2019). It also helps in providing better cake formation on the surface that can be easily be dislodged during cleaning, resulting lesser load on filter media and improve the filter life (Tomitatsu et al. 2014; Dutta 2019). These results look very promising and need further research to translate to use this in industrial scale.

Stringent regulations and emission standards imposed by the government; it will become the driving force for implementing these technologies within industries. The case studies conducted using these emerging technologies have shown promising annd result in capturing fine particles and gaseous pollutants. In future, ambient air quality and stringent standards of emission can only be met by using these technologies.

12.6 Current Emerging Air Pollution Challenges

Apart from various existing issues and challenges associated with air pollution management, several new air pollution problems are emerging. Currently, few major additional challenges are (a) Indoor air pollution due to inappropriate cooking technologies. (b) Outdoor air pollution, in many cities and urban areas are deteriorating (c) Bioaerosol pandemic, like COVID-19. New strategies and technologies and mixed bag of intervention strategies may be effective to reduce the pollution. The technologies discussed in this chapter are used traditionally in various industries to control the emission of pollutants. Some of these technologies after appropriate modification or in combination can be used in current situation to address these new challenges which need further research and development.

12.6.1 Indoor Air Pollution

Several studies have reported that the major emissions of PM originate from indoor are residential biomass fuel use (cooking and heating) (Venkataraman et al. 2015; Sadavarte and Venkataraman 2014) from inappropriate stove technologies used. Sahu et al. (2011) investigated the emissions from various types of stoves in rural India and compared the emissions and found that the emission of pollutants depends on the stage of combustion and mixing in the combustion chamber and ventilation. Also reports indicate that in many scenarios although the $PM_{2.5}$ emission is significantly less, ultrafine particle emitted are in large numbers which accounts less to $PM_{2.5}$ concentration. Particle emitted from various combustion scenarios will have different physicochemical properties

Several approaches haven been adopted such as improved stove technologies, shifting to high-quality fuel such as LPG, biogas. However, implementation and sustainability of new technologies have been difficult. In many instances, many indoor air cleaners are used to reduce the exposure to particulate in indoor environment, which are mostly based on filtration-based and ionization-based technology with the concept similar to technologies used for industrial pollution control. However, there I considerable scope for further improvement and development of new technologies to address indoor air pollution. Overall, there is a need to develop technologies to reduce and control indoor particulate exposure.

12.6.2 Outdoor Air Pollution

Outdoor air pollution is becoming a big concern in many countries and also in India. Currently, 129 non-attainment cities are declared by Central Pollution Control Board (CPCB) India. Again, this does not include cities having smaller population less than 10 lakh people. It means there will be many more cities around India which need significant attention to control air pollution. Most of the non-attainment declaration are also based on data from only few sampling locations. Many of cities in India do not have continuous sampling stations to understand the existing level of pollution. There are several places across the world experience smog events every year. For example, Delhi experience significant smog hanging over the cities in winter during Oct–Dec period and poor air flow and dispersion allows to build up the pollutants. Exposure to outdoor air pollution is associated with various diseases discussed earlier in the chapters. During these episodic events, many susceptible individuals undergo increased medical attention and admission to hospital.

Various control strategies required to address for mitigating these concerns. Development of technologies for outdoor pollution is researched recently. Only in China, a large cleaning tower is demonstrated to be used to control air pollutants around the surrounding (Cao et al. 2018a, b) in addition to control measures from sources and policy changes. More research and technology development required in this area as outdoor air pollution may continue to rise as economic growth increases.

12.6.3 Biological Aerosols

Of greater concern recently has been modern type of pollution in today's environments, which are indiscernible such as viruses, bacteria, and other toxins in the air but manifests in chronic health effects. In recent years highly, transmissible biological aerosols emerged that has caused significant epidemics leading to morality and disease globally. For example, recent COVID-19 pandemic, aerosol transmission is considered to be one pathway for spreading. Airborne transmission is mainly occurred by respiratory activity or medical procedures and once released may travel longer distances and remain airborne longer. The current strategy has been using of fitted personal protective mask (PPE) to reduce the exposure. Additionally, the virus is predominantly spread by droplet and contact routes. Since the airborne transmission for particles (<5 μm), it remains in the air for prolonged periods and is most likely to penetrate deeply into the lung and cause health hazard, and it creates significant risk to healthcare workers.

Currently, there is no control strategy available to control aerosol transmission apart from using face masks. These type of bioaerosol pollution will be significant in hospitals. The transition dynamics of these types of aerosols are not clearly know yet. A study by Santarpia et al. (2020) by taking air samples at different places in hospitals, where COVID-19 patients are treated indicates that SARS-CoV-2 can shed to the environment as expired particles. They also mentioned that stated that disease can spread by both direct contact (person to person and droplet) and indirect contact though contaminated objects and airborne transmission. This rapid emergence poses a threat to global health and environment. It is very critical to understand transmission dynamics of bioaerosols such as SARS-CoV-2 and implementing effective control strategies to protect healthcare workers. Control technology in terms of spraying disinfectant and wide deployment of facemask and their effectiveness to all these viruses are still under research. Some of the previous control technologies which has been described earlier can be redesigned and customized to address such bioaerosol capture and control in future.

12.7 Conclusion

This book chapter compressively reviewed various types of air pollution control devices used in industries. Various control technologies and approaches used in few selected industries that contributes significantly to PM emissions is discussed. Additionally, the chapter discusses various types of advanced technologies and also highlights the current emerging air pollution problems such as indoor pollution, outdoor pollution, and biological aerosols, for which control method needed to be developed and research should be pursued. The selection of control devices and approaches needed for PM control are dependent on the types of particles, size, shape, and physicochemical composition. In many cases, pretreatments are necessary to remove the larger particle followed by high-efficiency ESP and filtration to control the PM emissions. Particle emission from different industries varies greatly in their size, composition, shape that dictates type of control devices and control strategies requirement. The review clearly indicates that in order to reduce the pollutant emissions, a combination of control strategies needs to be implemented and more research and technology development are essential to address the current emerging issues of air pollution.

References

Ahmad S, Smail J (2018) The optimum operation conditions for electrostatic precipitator (ESP) in particulate emitter industries. Icpas:145–153. https://doi.org/10.14500/icpas2018.nmt106

Ahmed S, Mohsin H, Qureshi K et al (2018) Investigation of dust particle removal efficiency of self-priming venturi scrubber using computational fluid dynamics. Nucl Eng Technol. 50 (5):665–672. https://doi.org/10.1016/j.net.2018.01.016

Ali M, Yan C, Sun Z, Gu H, Wang J, Mehboob K (2013) Iodine removal efficiency in non-submerged and submerged self-priming venturi scrubber. Nucl Eng Technol. 45(2):203–210. https://doi.org/10.5516/NET.03.2012.047

Ali S, Waheed K, Qureshi K et al (2020) Experimental investigation of aerosols removal efficiency through self-priming venturi scrubber. Nucl Eng Technol. https://doi.org/10.1016/j.net.2020.03.019

Avci A, Karagoz I, Surmen A, Camuz I (2013) Experimental investigation of the natural vortex length in tangential inlet cyclones. Sep Sci Technol 48(1):122–126. https://doi.org/10.1080/01496395.2012.675003

Bahk YK, Buha J, Wang J (2013) Determination of geometrical length of airborne carbon nanotubes by electron microscopy, model calculation, and filtration method. Aerosol Sci Technol 47(7):776–784. https://doi.org/10.1080/02786826.2013.791745

Baig KS, Yousaf M (2017) Coal fired power plants: emission problems and controlling techniques. J Earth Sci Clim Change 08(07). https://doi.org/10.4172/2157-7617.1000404

Bal M, Siddiqi H, Mukherjee S, Meikap BC (2019) Design of self priming venturi scrubber for the simultaneous abatement of HCl gas and particulate matter from the flue gas. Chem Eng Res Des 150:311–319. https://doi.org/10.1016/j.cherd.2019.08.005

Balakrishnan K, Dey S, Gupta T et al (2019) The impact of air pollution on deaths, disease burden, and life expectancy across the states of India: the global burden of disease study 2017. Lancet Planet Heal 3(1):e26–e39. https://doi.org/10.1016/S2542-5196(18)30261-4

Bandyopadhyay A (2009) Abatement of particulate-laden SO_2 in tapered bubble column with internals, 147–155. https://doi.org/10.1007/s11869-009-0041-3

Bandyopadhyay A, Mandal PK (2015) Investigation on multiple benefits of wet ESP for emission control in an Indian coal fired thermal power plant. Clean Technol Environ Policy 17(2):421–431. https://doi.org/10.1007/s10098-014-0795-7

Bapat JD (2001) Application of ESP for gas cleaning in cement industry—with reference to India. J Hazard Mater 81(3):285–308. https://doi.org/10.1016/S0304-3894(00)00352-6

Barn P, Gombojav E, Ochir C et al (2018) The effect of portable HEPA filter air cleaners on indoor $PM_{2.5}$ concentrations and second hand tobacco smoke exposure among pregnant women in Ulaanbaatar, Mongolia: The UGAAR randomized controlled trial. Sci Total Environ 615:1379–1389. https://doi.org/10.1016/j.scitotenv.2017.09.291

Bayless DJ, Shi L, Kremer G et al (2012) Membrane-based wet electrostatic precipitation membrane-based wet electrostatic precipitation:2247. https://doi.org/10.1080/10473289.2005.10464658

Boamah PO, Onumah J, Takase M, Bonsu PO, Salifu T (2012) Air pollution control techniques. Integr Air Qual Manag 1(2):251–252. https://doi.org/10.1201/b12235-11

Calvert S, Lundgren D, Mehta DS (1972) Venturi scrubber performance. J Air Pollut Control Assoc 22(7):529–532. https://doi.org/10.1080/00022470.1972.10469674

Cao Q, Huang M, Kuehn TH et al (2018a) Urban-scale SALSCS, part II: a parametric study of system performance. Aerosol Air Qual Res 18(11):2879–2894. https://doi.org/10.4209/aaqr.2018.06.0239

Cao Q, Kuehn TH, Shen L et al (2018b) Urban-scale SALSCS, part I: experimental evaluation and numerical modeling of a demonstration unit. Aerosol Air Qual Res 18(11):2865–2878. https://doi.org/10.4209/aaqr.2018.06.0238

Carotenuto C, Di Natale F, Lancia A (2010) Wet electrostatic scrubbers for the abatement of submicronic particulate. Chem Eng J 165(1):35–45. https://doi.org/10.1016/j.cej.2010.08.049

Chang D-Q, Tien C-Y, Peng C-Y, Tang M, Chen S-C (2019) Development of composite filters with high efficiency, low pressure drop, and high holding capacity $PM_{2.5}$ filtration

Chatto V (2017) Reinventing air quality, May. https://doi.org/10.13140/rg.2.2.36357.55525

Chen CY (1955) Filtration of aerosols by fibrous media. Chem Rev 55(3):595–623. https://doi.org/10.1021/cr50003a004

Chen SC, Wang J, Bahk YK, Fissan H, Pui DYH (2014a) Carbon nanotube penetration through fiberglass and electret respirator filter and nuclepore filter media: experiments and models. Aerosol Sci Technol 48(10):997–1008. https://doi.org/10.1080/02786826.2014.954028

Chen T, Tsai C, Yan S, Li S (2014b) An efficient wet electrostatic precipitator for removing nanoparticles, submicron and micron-sized particles. Sep Purif Technol 136:27–35. https://doi.org/10.1016/j.seppur.2014.08.032

Chen CH, Wu CD, Chiang HC et al (2019) The effects of fine and coarse particulate matter on lung function among the elderly. Sci Rep 9(1):1–8. https://doi.org/10.1038/s41598-019-51307-5

Cheng YH, Tsai CJ (1998) Factors influencing pressure drop through a dust cake during filtration. Aerosol Sci Technol 29(4):315–328. https://doi.org/10.1080/02786829808965572

Chikkatur AP, Chaudhary A, Sagar AD (2011) Coal power impacts, technology, and policy: connecting the dots. Annu Rev Environ Resour 36(1):101–138. https://doi.org/10.1146/annurev.environ.020508.142152

CPCB (2009) National air quality standards. Gov Gaz 534(32816):6–9. 24 Dec 2009

CPCB (2018) TPP EMISSION STANDARDS.pdf

Darcovich K, Jonasson KA, Capes CE (1997) Developments in the control of fine particulate air emissions. Adv Powder Technol 8(3):179–215. https://doi.org/10.1016/S0921-8831(08)60463-X

Di Natale F, Carotenuto C, D'Addio L et al (2015) Capture of fine and ultrafine particles in a wet electrostatic scrubber. J Environ Chem Eng 3(1):349–356. https://doi.org/10.1016/j.jece.2014.11.007

Di Natale F, Carotenuto C, Manna L et al (2016) A publication of IIETA water electrified sprays for emission control in energy production processes 34(2):597–602

Di Natale F, Manna L, La Motta F et al (2017) Capture of bacterial bioaerosol with a wet electrostatic scrubber. J Electrostat 2018(93 October):58–68. https://doi.org/10.1016/j.elstat.2018.04.003

Dirgo J, Leith D (1985) Cyclone collection efficiency: comparison of experimental results with theoretical predictions. Aerosol Sci Technol 4(4):401–415. https://doi.org/10.1080/02786828508959066

Dong M, Zhou F, Zhang Y, Shang Y, Li S (2018) Numerical study on fine-particle charging and transport behaviour in electrostatic precipitators, vol 330. Elsevier. https://doi.org/10.1016/j.powtec.2018.02.038

Dutta S (2018) Hybrid aerosol filtration systems—a review 5(Sep 2018):535–555

EC/R Incorporated (1998) Stationary source control techniques document for fine particulate matter. Organization 1(68):2

Emis (2019) Fabric filter| EMIS

Emis (2019) Fabric filter| EMIS. https://emis.vito.be/en/techniekfiche/fabric-filter

EPA (1982) Control techniques for particulate emissions from stationary sources, vol 2. 1 Sep

EPA (1983) Air pollution control systems for selected industries: self industrial guidebook

EPA (1994) Background documentation, AP-42 Section 11.6, Portland Cement Manufacturing (4601):123. https://www3.epa.gov/ttnchie1/ap42/ch11/bgdocs/b11s06.pdf

EPA (1998) Stationary source control techniques document for fine particulate matter. Organization 1(68):2

EPA (1998) Air pollution control technology fact sheet—settling chambers. http://infohouse.p2ric.org/ref/50/49195.pdf

EPA (1998) Air pollution control technology fact sheet:1–5

EPA (1999) Air pollution control technology fact sheet—cyclones. InfohouseP 2RicOrg. 1990–1994. http://infohouse.p2ric.org/ref/50/49195.pdf

EPA (1999) Pollution control technology. Environ Sci Technol 3(10):879. https://doi.org/10.1021/es60033a602

EPA (2003) Air pollution control technology fact sheet—electrostatic precipitator, dry wire-pipe type:1–6. http://www.epa.gov/ttn/catc/dir1/fdespwpi.pdf

EPA (2013) Regulatory Impact Analysis for the Final Revisions to the National Ambient Air Quality Standards for Particulate Matter On February 28, 2013, the PM RIA Is Being Replaced to Add Appendix 3. A, Which Was Inadvertently Left out And

EPA (2016) Particulate Matter (PM) Basics Particulate Matter (PM) Pollution US EPA. USEPA. https://www.epa.gov/pm-pollution/particulate-matter-pm-basics#PM

Fletcher IF (2001) Overview of iron and steel industry

Grahame TJ, Schlesinger RB (2007) Health effects of airborne particulate matter : do we know enough to consider regulating specific particle types or sources ? 457–481. https://doi.org/10.1080/08958370701382220

Gupta RK, Majumdar D, Trivedi JV, Bhanarkar AD (2012) Particulate matter and elemental emissions from a cement kiln. Fuel Process Technol 104:343–351. https://doi.org/10.1016/j.fuproc.2012.06.007

Guttikunda SK, Jawahar P (2014) Atmospheric emissions and pollution from the coal-fired thermal power plants in India. Atmos Environ 92:449–460. https://doi.org/10.1016/j.atmosenv.2014.04.057

Hao J, Li G (2009) Air pollution caused by industries. Point Sources Pollut Local Eff Control I (I):171–179

Harrison RM, Yin J (2000) Particulate matter in the atmosphere: which particle properties are important for its effects on health? Sci Total Environ 249(1–3):85–101. https://doi.org/10.1016/S0048-9697(99)00513-6

Huntzinger DN, Eatmon TD (2009) A life-cycle assessment of Portland cement manufacturing: comparing the traditional process with alternative technologies. J Clean Prod 17(7):668–675. https://doi.org/10.1016/j.jclepro.2008.04.007

Jang A-S (2012) Particulate air pollutants and respiratory diseases. Intech. i(tourism):13. http://dx.doi.org/10.5772/57353

Jaworek A, Balachandran W, Krupa A, Kulon J, Lackowski M (2006) Wet electroscrubbers for state of the art gas cleaning 40(20):6197–6207

Jaworek A, Krupa A, Czech T (2007) Modern electrostatic devices and methods for exhaust gas cleaning : a brief review. 65:133–155. https://doi.org/10.1016/j.elstat.2006.07.012

Jia J, Cheng S, Yao S et al (2017) Emission characteristics and chemical components of size-segregated particulate matter in iron and steel industry. Atmos Environ 2018(182 Sepr):115–127. https://doi.org/10.1016/j.atmosenv.2018.03.051

Kumar D, Kumar D (2018) Dust control 12.1:265–278. https://doi.org/10.1016/b978-0-12-812632-5.00012-4

Lapple CE (1951) Processes use many collection types. Chem Eng 58:145–151

Larsson DGJ (2014) Pollution from drug manufacturing: review and perspectives. Philos Trans R Soc B Biol Sci 369(1656). https://doi.org/10.1098/rstb.2013.0571

Lawless PA, Sparks LE (1980) A mathematical model for calculating effects of back corona in wire-duct electrostatic precipitators. J Appl Phys 51(1):242–256. https://doi.org/10.1063/1.327416

Lee MH, Yang W, Chae N, Choi S (2020) Performance assessment of HEPA filter against radioactive aerosols from metal cutting during nuclear decommissioning. Nucl Eng Technol 52 (5):1043–1050. https://doi.org/10.1016/j.net.2019.10.017

Lei Y, Zhang Q, Nielsen C, He K (2011) An inventory of primary air pollutants and CO_2 emissions from cement production in China, 1990–2020. Atmos Environ 45(1):147–154. https://doi.org/10.1016/j.atmosenv.2010.09.034

Lencioni E, Travagli V, Pecchioli L, Bechini C (1997) Performance qualification of a HEPA filter at high temperature. Pharm Acta Helv 72(1):31–35. https://doi.org/10.1016/S0031-6865(96)00044-1

Leonard G, Mitchner M, Self SA (1980) Particle transport in electrostatic precipitators. Atmos Environ 14(11):1289–1299. https://doi.org/10.1016/0004-6981(80)90230-9

Liu F, Zhang Q, Tong D et al (2015) High-resolution inventory of technologies, activities, and emissions of coal-fired power plants in China from 1990 to 2010, (x):13299–13317. https://doi.org/10.5194/acp-15-13299-2015

Lohmann U, Feichter J (2005) Global indirect aerosol effects: a review. Atmos Chem Phys 5 (3):715–737. https://doi.org/10.5194/acp-5-715-2005

Lu P, Wu J, Pan WP (2010) Particulate matter emissions from a coal-fired power plant. In: 2010 4th International Conference of Bioinformatics and Biomedical Engineering iCBBE 2010, pp 8–11. https://doi.org/10.1109/icbbe.2010.5517175

Meij R, Vredenbregt LHJ, te Henk W (2002) The fate and behavior of mercury in coal-fired power plants. J Air Waste Manag Assoc 52(8):912–917. https://doi.org/10.1080/10473289.2002.10470833

Meij R, Vredenbregt LHJ, Winkel H, Meij R, Vredenbregt LHJ, Winkel H (2011) The fate and behavior of mercury in coal-fired power plants PAPER the fate and behavior of mercury in coal-fired power plant:2247. https://doi.org/10.1080/10473289.2002.10470833

Meikap BC, Biswas MN (2004) Fly-ash removal efficiency in a modified multi-stage bubble column scrubber. Sep Purif Technol 36(3):177–190. https://doi.org/10.1016/S1383-5866(03)00213-2

Ménégoz M, Krinner G, Balkanski Y et al (2014) Snow cover sensitivity to black carbon deposition in the Himalayas: from atmospheric and ice core measurements to regional climate simulations. Atmos Chem Phys 14(8):4237–4249. https://doi.org/10.5194/acp-14-4237-2014

Mittal ML (2010) Estimates of emissions from coal fired thermal power plants in India, 39 (Nov):1–22

Mohebbi A, Taheri M, Fathikaljahi J, Talaie MR (2003) Simulation of an orifice scrubber performance based on Eulerian/Lagrangian method. J Hazard Mater 100(1–3):13–25. https://doi.org/10.1016/S0304-3894(03)00066-9

Mycock JC, Mckenna JD, Theodore L (1995) Handbook of air pollution control engineering and technology. CRC Press

Najafabadi MM, Tabrizi HB, Aramesh A (2014) Effects of geometric parameters and electric indexes on performance of a vertical wet electrostatic precipitator. J Electrostat 72(5):402–411. https://doi.org/10.1016/j.elstat.2014.06.005

Neundorfer M (1996) Electrostatic precipitator design and operation

Nordby KC, Notø H, Eduard W et al (2016) Thoracic dust exposure is associated with lung function decline in cement production workers. Eur Respir J 48(2):331–339. https://doi.org/10.1183/13993003.02061-2015

Ouari A, Miloua F, Labair H, Tilmatine A (2019) Experimental analysis of a corona-charged baghouse filter. Int J Environ Stud 00(00):1–10. https://doi.org/10.1080/00207233.2019.1696595

Parker KR (1997) Jl, vol 95. https://doi.org/10.1080/13102818.2016.1232606

Peirce JJ, Weiner RF, Vesilind PA (2006) Air pollution control. Arch Environ Health 18(5):834–838. https://doi.org/10.1080/00039896.1969.10665495

Pynadathu N, Thomas P, Arjun P (2014) Engineering design and specifications of cyclone separator. MTech, Energy Environ Eng SMBS, VIT-University, India, June 2014. https://doi.org/10.13140/rg.2.2.30512.40966

Raffetti E, Treccani M, Donato F (2019) Cement plant emissions and health effects in the general population: a systematic review. Chemosphere 218:211–222. https://doi.org/10.1016/j.chemosphere.2018.11.088

Ramanathan V, Carmichael G (2008) Global and regional climate changes due to black carbon. Nat Geosci 1(4):221–227. https://doi.org/10.1038/ngeo156

Reddington CL, Conibear L, Knote C et al (2019) Exploring the impacts of anthropogenic emission sectors on $PM_{2.5}$ and human health in South and East Asia. Atmos Chem Phys 19 (18):11887–11910. https://doi.org/10.5194/acp-19-11887-2019

Reynolds JP, Theodore L, Marino J (1975) Calculating collection efficiencies for electrostatic precipitators. J Air Pollut Control Assoc 25(6):610–616. https://doi.org/10.1080/00022470.1975.10470116

Ristovski ZD, Miljevic B, Surawski NC et al (2012) Respiratory health effects of diesel particulate matter. Respirology 17(2):201–212. https://doi.org/10.1111/j.1440-1843.2011.02109.x

Ritchie H, Roser M (2020) Air pollution—our world in data. OurWorldInData. https://ourworldindata.org/air-pollution

Rogoff MJ, Screve F (2011) Permitting issues. Waste-to-Energy:89–116. https://doi.org/10.1016/b978-1-4377-7871-7.10008-5

Rosenberger W (2018) Effect of charcoal equipped HEPA filters on cabin air quality in aircraft: a case study including smell event related in-flight measurements. Build Environ 143:358–365. https://doi.org/10.1016/j.buildenv.2018.07.031

Roth KW, Dieckmann J, Hamilton SD, Goetzler W (2002) Energy consumption characteristics of commercial building HVAC systems volume III : energy savings potential. Build Technol Progr III(68370)

Rust H, Tfinnler C (1995) Engineering and design pressure release of containments during severe accidents in Switzerland, 157(Dec 1993):337–352

Sadavarte P, Venkataraman C (2014) Trends in multi-pollutant emissions from a technology-linked inventory for India: I. Industry and transport sectors. Atmos Environ 2014(99):353–364. https://doi.org/10.1016/j.atmosenv.2014.09.081

Sahu M, Peipert J, Singhal V, Yadama GN, Biswas P (2011) Evaluation of mass and surface area concentration of particle emissions and development of emissions indices for cookstoves in rural India. Environ Sci Technol 45(6):2428–2434. https://doi.org/10.1021/es1029415

Santarpia JL, Rivera DN, Herrera V et al (2020) Transmission potential of SARS-CoV-2 in viral shedding observed at the University of Nebraska Medical Center. medRxiv. https://doi.org/10.1101/2020.03.23.20039446

Schneider M, Romer M, Tschudin M, Bolio H (2011) Sustainable cement production-present and future. Cem Concr Res 41(7):642–650. https://doi.org/10.1016/j.cemconres.2011.03.019

Seetharama AS, Benedict A, Reynolds JB (2013) Comparison of wet and dry electrostatic precipitator (ESP) technologies, Sep

Shepherd CB, Lapple CE (1939) Flow and Education.pdf

Shou D, Ye L, Fan J (2014) Gas transport properties of electrospun polymer nanofibers. Polymer (Guildf). https://doi.org/10.1016/j.polymer.2014.05.016

Skinder BM (2014) Brick kilns: cause of atmospheric pollution. J Pollut Eff Control 02(02). https://doi.org/10.4172/2375-4397.1000112

Sobczyk AT, Marchewicz A, Krupa A et al (2017) Enhancement of collection efficiency for fly ash particles ($PM_{2.5}$) by unipolar agglomerator in two-stage electrostatic precipitator 187:91–101. https://doi.org/10.1016/j.seppur.2017.06.039

Spath PL, Mann MK, Kerr DR (1999) Life cycle assessment of coal-fired power production

Srivastava RK, Hutson N, Martin B, Princiotta F, Staudt J (2006) Control of mercury emissions from coal-fired electric utility boilers. Environ Sci Technol 40(5):1385–1393. https://doi.org/10.1021/es062639u

Suh HH (2009) Particulate matter. Expo Assess Occup Environ Epidemiol 1:1–40. https://doi.org/10.1093/acprof:oso/9780198528616.003.0014

Tang Q, Chen X, Xia X et al (2018) Scenario study on PM emission reduction in cement industry. IOP Conf Ser Earth Environ Sci 111(1). https://doi.org/10.1088/1755-1315/111/1/012014

Theodore L, Buonicore AJ (1976) Industrial Air Pollution Control Equipment. CRC Press

Theodore L, DePaola V (1980) Selenium: a water pollutant from flue gas desulfurization. J Air Pollut Control Assoc 30(10):1134. https://doi.org/10.1080/00022470.1980.10465160

Thorpe A, Brown RC (2003) Performance of electrically augmented fibrous filters, measured with monodisperse aerosols. Aerosol Sci Technol 37(3):231–245. https://doi.org/10.1080/02786820300944

Tomitatsu K, Kato M, Ueda Y, Nagata C (2014) Development of advanced electrostatic fabric filter. Int J Plasma Environ Sci Technol. 8(1):37–44

Turner JH, Lawless PA (1999) Section 6 particulate matter controls chapter 3 electrostatic precipitators. Part Matter Control:70. EPA/452/B-02-001

USDOE (2015) DOE technical standard specification for HEPA filters used by DOE contractors (Dec 2005)

Vallero DA (2019) Air pollution control technologies. https://doi.org/10.1016/b978-0-12-814934-8.00013-2

Vanosdell DW, Furlong DA, Hovis LS (1985) Pilot plant study of the effect of a surface electric field on fabric filter operation. IEEE Trans Ind Appl IA-21(1):62–68. https://doi.org/10.1109/tia.1985.349644

Venkataraman C, Brauer M, Tibrewal K et al (2018) Source influence on emission pathways and ambient $PM_{2.5}$ pollution over India (2015–2050). Atmos Chem Phys 18(11):8017–8039. https://doi.org/10.5194/acp-18-8017-2018

Wang C Sen, Otani Y. Removal of nanoparticles from gas streams by fibrous filters: A review. *Ind Eng Chem Res*. 2013;52(1):5–17. https://doi.org/10.1021/ie300574m

Wang S, Zhang L, Zhao B, Meng Y, Hao J (2012) Mitigation potential of mercury emissions from coal-fired power plants in China. Energy Fuels 26(8):4635–4642. https://doi.org/10.1021/ef201990x

Wang Y, Jia C, Tao J et al (2016) Science of the total environment chemical characterization and source apportionment of $PM_{2.5}$ in a semi-arid and petrochemical-industrialized city, Northwest China. Sci Total Environ 573:1031–1040. https://doi.org/10.1016/j.scitotenv.2016.08.179

Wang X, Lei Y, Yan L, Liu T, Zhang Q, He K (2019) A unit-based emission inventory of SO_2, NO_x and PM for the Chinese iron and steel industry from 2010 to 2015. Sci Total Environ 676 (x):18–30. https://doi.org/10.1016/j.scitotenv.2019.04.241

Warren S, Wiscombe W (1980) A model for Spectral Albedo of Snow. J Atmos Sci 37:2734–2745

WHO (2014) WHO| 7 million premature deaths annually linked to air pollution. http://www.who.int/mediacentre/news/releases/2014/air-pollution/en/

Woodard and Curran I (2006) Treatment of air discharges from industry. Ind Waste Treat Handb (03):335–361. https://doi.org/10.1016/b978-0-7506-7963-3.50010-2

World Economic Forum (2015) Which is the world's most polluted city?| World Economic Forum

Yang Z, Zheng C, Chang Q, Wan Y, Wang Y (2017) Fine particle migration and collection in a wet electrostatic precipitator. J Air Waste Manage Assoc 67(4):498–506. https://doi.org/10.1080/10962247.2016.1260074

Yang Z, Zheng C, Liu S, Guo Y, Liang C, Wang Y (2018) A combined wet electrostatic precipitator for efficiently eliminating fine particle penetration 180(Aug):122–129. https://doi.org/10.1016/j.fuproc.2018.08.013

Yang Y, Ruan Z, Wang X et al (2019) Short-term and long-term exposures to fine particulate matter constituents and health: a systematic review and meta-analysis. Environ Pollut 247:874–882. https://doi.org/10.1016/j.envpol.2018.12.060

Zhang L, Wang S, Meng Y, Hao J (2012) Influence of mercury and chlorine content of coal on mercury emissions from coal-fired power plants in China. Environ Sci Technol 46(11):6385–6392. https://doi.org/10.1021/es300286n

Zhang R, Liu C, Zhou G et al (2017) Morphology and property investigation of primary particulate matter particles from different sources:1–11. https://doi.org/10.1007/s12274-017-1724-y

Zhao Y, Nielsen CP, Lei Y, McElroy MB, Hao J (2011) Quantifying the uncertainties of a bottom-up emission inventory of anthropogenic atmospheric pollutants in China. Atmos Chem Phys 11(5):2295–2308. https://doi.org/10.5194/acp-11-2295-2011

Zhuang Y, Kim YJ, Lee TG, Biswas P (2000) Experimental and theoretical studies of ultra-fine particle behavior in electrostatic precipitators. J Electrostat 48(3–4):245–260. https://doi.org/10.1016/S0304-3886(99)00072-8

The manufacturer's authorised representative in the EU is Springer Nature Customer Service Centre GmbH, Europaplatz 3, 69115 Heidelberg, Germany. If you have any concerns regarding our products, please contact ProductSafety@springernature.com

Printed and bound by CPI Group (UK) Ltd, Croydon, CR0 4YY

30/09/2026

02225288-0004